RENEWALS 458-4574
DATE DUE

GAYLORD			PRINTED IN U.S.A.

Central Nervous System Diseases and Inflammation

Thomas E. Lane • Monica Carson
Conni Bergmann • Tony Wyss-Coray
Editors

Central Nervous System
Diseases and Inflammation

 Springer

Thomas E. Lane
Center for Immunology
Department of Molecular
  Biology & Biochemistry
University of California
Irvine, CA
USA
tlane@uci.edu

Monica Carson
Department of Biomedical Sciences
University of California
Riverside, CA
USA
monica.carson@ucr.edu

Conni Bergmann
Department of Neurosciences
Lerner Research Institute
Cleveland Clinic Foundation
Cleveland, OH
USA
bergmac@ccf.org

Tony Wyss-Coray
Department of Neurology and Neurological
  Sciences
Stanford University School of Medicine
Stanford, CA
USA
twc@stanford.edu

ISBN-13: 978-0-387-73893-2   e-ISBN-13: 978-0-387-73894-9

Library of Congress Control Number: 2007938685

Preface

Up until approximately 20 years ago, the idea that the central nervous system (CNS) and components of the immune system were dynamically interactive was considered impossible (or at least highly unlikely) as the CNS was judged an immunosuppressive environment based upon experimental evidence highlighting the survival of tissue grafts within the brain. Additional evidence supporting this viewpoint included (i) the presence of the blood–brain barrier (BBB) which provides a physical and physiological obstruction that is difficult for cells and macromolecules to cross, (ii) the relative absence of MHC class I and II expression on CNS cells like astrocytes and neurons, and (iii) lack of abundant antigen presenting cells (APC) which are required for the generation of an adaptive immune response. However, in spite of these obstacles, it is now well-accepted that the CNS is routinely subject to immune surveillance under both normal as well as diseased conditions. Indeed, activated cells of the immune system such as T and B lymphocytes and monocyte/macrophages readily infiltrate and accumulate within the CNS following microbial infection, injury, or upon development of autoimmune responses directed toward resident antigens of the CNS.

The importance of studying events surrounding the initiation and maintenance of neuroinflammation is now recognized by scientists and clinicians alike as critical not only in characterizing the complex mechanisms associated with host defense following infection but also in contributing to neurologic disease. Much of our understanding of neuroinflammation has been derived from numerous animal studies of neurologic disease including autoimmune models of demyelination such as experimental autoimmune encephalomyelitis (EAE) and transgenic mice, microbial e.g. virus and bacteria infections, spinal cord injuries in mice and rats, and mouse models of Alzheimer's to highlight just a few. Moreover, the underlying molecular and cellular mechanisms governing inflammation are just now being understood and the importance of chemokines and chemokine receptors in recruiting targeted populations of leukocytes into the CNS is now appreciated. In addition, resident cells of the CNS e.g. microglia are recognized as important mediators in regulating innate defense mechanisms as well as disease.

This volume highlights important advances in our understanding of different aspects of neuroinflammation with a concentration on specific areas focusing on glial activation, molecular signals regulating inflammation and neurotoxicity, immune responses concentrating within the CNS, and the emergence of transgenic models of neurologic disease. It was the goal of the editors to provide timely and insightful comments on these particular aspects of neuroinflammation and disease while recognizing that it was impossible to adequately address other equally important and relevant aspects of neuroinflammation not covered but certainly deserving of attention. In addition, the editors feel this text will be useful for researchers, clinicians, as well as a valuable resource for students interested in the fascinating arena of neuroinflammation.

Thomas E. Lane
Irvine, CA
May 2006

Contents

Contributors

Katerina Akassoglou
Department of Pharmacology
University of California
San Diego
La Jolla, CA
USA

K.I. Andreasson
Department of Neurology and Neurological Sciences
Stanford University School of Medicine
Stanford, CA
USA

S.A. Austin
Department of Pharmacology, Physiology and Therapeutics School
  of Medicine and Health Sciences
University of North Dakota
Grand Forks, ND
USA

Samantha L. Bailey
Department of Microbiology–Immunology, Feinberg School of Medicine
Northwestern University
Chicago, IL
USA

Ingo Bechmann
Institute of Clinical Neuroanatomy
Frankfurt/Main
Germany

Cornelia Bergmann
Department of Neurosciences
Lerner Research Institute
Cleveland OH
USA

Iain L. Campbell
School of Molecular and Microbial Biosciences
The University of Sydney
Sydney
Australia

Pamela A. Carpentier
Department of Microbiology–Immunology, Feinberg School of Medicine
Northwestern University
Chicago, IL
USA

Monica J. Carson
Division of Biomedical Sciences, Center for Glial-Neuronal Interactions
University of California Riverside
Riverside, CA
USA

Colin K. Combs
Department of Pharmacology, Physiology and Therapeutics School
    of Medicine and Health Sciences
University of North Dakota
Grand Forks, ND
USA

Dimitrios Davalos
Department of Pharmacology
University of California, San Diego
La Jolla, CA
USA

Jonathan M. Doose
Division of Biomedical Sciences, Center for Glial-Neuronal Interactions
University of California Riverside
Riverside, CA
USA

D'Anne S. Duncan
Department of Microbiology–Immunology, Feinberg School of Medicine
Northwestern University,
Chicago, IL
USA

Michaela Fux
Medical Biotechnology Center
University of Southern Denmark
Denmark

Maya Hatch
Department of Anatomy and Neurobiology
University of California
Irvine, CA
USA

Michelle J. Hickey
Department of Molecular Biology and Biochemistry
University of California
Irvine, CA
USA

Hans S. Keirstead
Department of Anatomy and Neurobiology
University of California
Irvine, CA
USA

Martin Krüger
Institute of Clinical Neuroanatomy
Frankfurt/Main
Germany

Thomas E. Lane
Department of Molecular Biology and Biochemistry
University of California
Irvine, CA
USA

Pedro Lowenstein
Cedars-Sinai Medical Center
University of California, Los Angeles
Los Angeles, CA
USA

Stephen D. Miller
Department of Microbiology–Immunology, Feinberg School of Medicine
Northwestern University
Chicago, IL
USA

Jason Millward
Medical Biotechnology Center
University of Southern Denmark
Denmark

Thomas J. Montine
Division of Neuropathology
University of Washington, Oregon Health & Sciences University
Portland, OR
USA

Marcus Müller
School of Molecular and Microbial Biosciences
The University of Sydney
Sydney
Australia

Trevor Owens
Medical Biotechnology Center
University of Southern Denmark
Denmark

Karntipa Pisalyaput
Department of Molecular Biology and Biochemistry
University of California,
Irvine, CA
USA

Shweta S. Puntambekar
Division of Biomedical Sciences, Center for Glial-Neuronal Interactions
University of California Riverside
Riverside, CA
USA

Chris S. Schaumburg
Department of Molecular Biology and Biochemistry
University of California
Irvine, CA
USA

Linda N. Stiles
Department of Molecular Biology and Biochemistry
University of California
Irvine, CA
USA

Malu G. Tansey
Department of Physiology
U. T. Southwestern Medical School
Dallas, TX
USA

Andrea J. Tenner
Department of Molecular Biology and Biochemistry
University of California
Irvine, CA
USA

Tony Wyss-Coray
Department of Neurology and Neurological Sciences
Stanford University School of Medicine
Stanford, CA
USA

1
Microglia: A CNS-Specific Tissue Macrophage

Shweta S. Puntambekar, Jonathan M. Doose, and Monica J. Carson

1 Introduction

The central nervous system (CNS) is a complex integrated organ comprised of:

1. Neurons (10% of total number of CNS cells and 50% of total cell mass).
2. Glia (90% of the total cell number but only 50% of the total mass of the CNS) (Simons and Trajkovic, 2006; Verkhratsky and Toescu, 2006; He and Sun, 2007).

Glia can be further divided into macroglia (oligodendrocytes and astrocytes) which are of neuroectodermal origin and microglia which are of mesenchymal origin. While neurons and macroglia are endogenous cells of the CNS, microglia and/or their progenitors appear to invade CNS tissue very early during embryonic development (Carson and Sutcliffe, 1999; Carson et al., 2004, 2006).

The primary function of all glia is to maintain the optimal operation of the CNS information circuit (the neuronal network). This involves: active regulation of the network's operations, ongoing maintenance to deal with normal wear and tear as well as active defense and repair following injury or pathogen attack (Carson and Sutcliffe, 1999; Carson et al., 2004, 2006). Neurons, macroglia and microglia all coordinately and dynamically participate in these processes. All of these cells also interact with CNS-infiltrating immune cells as part of their regulation of inflammatory responses in the CNS. Until recently, the importance of microglia in homeostatic CNS function was not fully recognized.

In this chapter, we will explore the experimental basis of the many suggested beneficial versus detrimental functions of microglia in both the healthy and injured/diseased CNS. We will also explore to what extent aberrant microglial function is due to primary microglial dysfunction versus primary dysfunctions in neurons and glia.

2 What are Microglia?

Microglia express most common macrophage markers and are often referred to as the tissue macrophage of the brain. In the healthy brain, microglia have a stellate morphology and are found in all areas of the brain and spinal cord (Fig. 1.1). As commonly

T.E. Lane et al. (eds.), *Central Nervous System Diseases and Inflammation.*
© Springer 2008

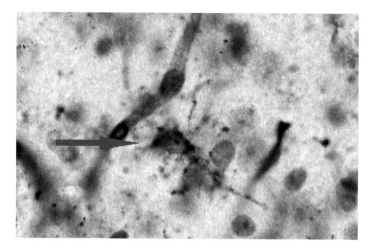

Fig. 1.1 A typical parenchymal microglia extending processes to all elements of its environment. Microglia and blood vessels are visualized in brown using tomato lectin. Nuclei are visualized in blue using hematoxylin (*See also color plates*).

used in the literature, the term microglia has been applied to at least three different types of myeloid cells: parenchymal microglia, perivascular microglia and acutely blood-derived inflammatory macrophages that display a stellate morphology within the CNS. In this chapter, we will operationally define microglia as parenchymal cells that are largely from either a self-renewing population or only rarely replenished from adult bone marrow derived cells. Perivascular cells and myeloid cells which are acutely derived from the blood and which have the demonstrated potential to emigrate from the CNS shortly after entry (period of days to weeks), we will refer to as perivascular macrophages and CNS-infiltrating macrophages.

As yet there are no reagents able to distinguish acutely infiltrating macrophages from CNS-resident microglia in histological preparations. However, parenchymal microglia are unique in the adult in that they continue to express the very low levels of CD45 normally expressed during embryonic development of the hematopoetic system (Sedgwick et al., 1991; Ford et al., 1995; Carson et al., 1998). In the adult, all other nucleated differentiated macrophages and immune cells express high levels of CD45. While this is a useful biomarker to purify and separate CNS-resident microglia from CNS-infiltrating macrophages, it also indicates that these two populations have distinct functions. CD45 is an inhibitory receptor for CD22 (Mott et al., 2004; Han et al., 2005). While this CD45 ligand was long recognized as being expressed by B cells, Tan and colleagues have recently demonstrated than CNS neurons secrete a soluble form of CD22 from axonal terminals (Mott et al., 2004). Functionally, these authors also found that this was the mechanism by which neurons in culture were able to inhibit LPS-induced TNF-alpha production by microglia. Interestingly, the differential expression between microglia and macrophages suggests that CNS neurons may be more effective at inhibiting the functions of macrophages than those of microglia!

To date, most research on microglial function has focused on their roles during injury, pathogen infection or chronic neurodegeneration (Bechmann et al., 2001, 2005). In part, this focus is a consequence of the dramatic and rapid changes in microglial morphology and gene expression observed immediately following CNS injury or infection. Conversely, microglia in the healthy CNS have often been presumed to be quiescent and largely inactive. However, important non-defense oriented functions of microglia are suggested by the human disease referred to as Nasu–Hakola disease (Paloneva et al., 2000, 2001; Cella et al., 2003).

Nasu–Hakola disease is a genetic disorder leading to bone spurs, early onset cognitive dementia in the 1920s and death in the late 1930s. Positional cloning identified mutations in the TREM-2 pathway as the genetic cause of the disease (Paloneva et al., 2000, 2001; Cella et al., 2003). A primary neuronal defect was long speculated to be the cause of this neurological disorder due to the early onset of cognitive symptoms. However, we found that in the murine CNS TREM-2 RNA could only be detected in microglia (Schmid et al., 2002)! These data, in conjunction with similar findings from other groups illustrate that a primary disease of microglia has the potential to lead to a disease with primary psychological manifestations (not just primary inflammatory or autoimmune manifestations!) (Bouchon et al., 2001; Schmid et al., 2002; Daws et al., 2003; Melchior et al., 2006). As yet, it is unclear what functions are dysregulated in microglia with a dysfunctional TREM-2 pathway or how microglial dysfunction would lead to cognitive dementia after two decades of life.

3 What Do Microglia Do in the Normal CNS?

Recently, a variety of relatively non-invasive techniques, have been used to explore microglial function in the healthy CNS and to quantify changes in cellular activity upon acute injury(Davalos et al., 2005; Nimmerjahn et al., 2005). Using two-photon imaging of fluorescently labeled cells, Nimmerhan et al (2005) and Davalos et al. (2005) have monitored the extension and motility of microglia and their processes before and after introduction of a focal injury. In the healthy CNS, microglial cells had small, rod shaped cell bodies with many thin and highly ramified processes symmetrical extending from the cell body. Using time-lapse imaging, they observed that while microglial cell bodies remained relatively fixed, their processes were remarkably motile. The processes underwent continuous cycles of de novo formation and withdrawal, apparently surveying all elements of the CNS every 6h!

As part of this analysis, the authors also noted that not all microglial processes were highly motile(Nimmerjahn et al., 2005). A subset of microglial processes provided a stable scaffold, perhaps anchoring the microglia in place. These data suggest that the branching of microglial processes may not be random and may serve to integrate homeostatic signals throughout the entire CNS. It is tempting to speculate that this may be a mechanism by which microglia help modulate the extensive network of neuronal synapses and functional plasticity of the healthy CNS.

Microglia are the resident immunocompetent cell of the CNS are provide the first line of defense in response to pathogens and neuronal injury. As such they can produce a wide variety of cytokines, proteases, reactive oxygen species as well regulate CNS-infiltrating T cells in an antigen-specific manner. Nimmerhan et al proposed that the constant microglial surveying of the brain is a necessary consequence of their defense functions (Nimmerjahn et al., 2005). Using a targeted disruption of the blood brain barrier (BBB), the authors demonstrated that microglial responses to injury were rapid (within minutes), spatially directed to the focal injury and not dependent on the presence of pathogenic molecules. Astrocytes are also an important neuronal support cell, playing a key role in CNS immune responses as well as in glutamate uptake in response to neuronal activity(Parpura et al., 2004; Volterra and Steinhauser, 2004). Somewhat surprisingly, these authors found that the basal motility of microglial processes was much higher than those of astrocytes (Nimmerjahn et al., 2005). In addition, while microglial processes rapidly extended toward an acute focal injury, astrocytic processes did not. These data provide a dramatic demonstration of the cell-type specific support provided by microglia and astrocytes.

4   What are Microglia Monitoring in the CNS?

Microglia are known to recognize pathogens using evolutionarily conserved pathogen recognition receptors, such as the Toll-like receptors (TLRs) (Lee and Lee, 2002; Schiller et al., 2006). However, the obvious question is how do microglia recognize changes in CNS function and neuronal health? And can they directly detect neuronal activity? Over 10 years ago, Neumann and colleagues demonstrated that blocking neuronal activity in slice cultures with the sodium channel blocker tetrodotoxin resulted in rapid microglial activation (Neumann et al., 1996, 1998). More recently, Nimmerjhan et al reported that application of the GABA receptor blocker bicucullin (resulting in an upregulation the neuronal activity) dramatically increased the region being sampled by microglia in the otherwise uninjured CNS (Nimmerjahn et al., 2005). How directly microglia detect neuronal activity is still a subject of debate. However, microglia do express inward rectifying potassium channels as well as receptors for many of the CNS neurotransmitters (Kettenmann et al., 1990, 1993; Schmidtmayer et al., 1994; Chung et al., 1999; Schilling et al., 2000).

Neuman and his colleagues found in their studies that microglia were responding in part to the neurotrophins being secreted by the neurons in these cultures (Neumann et al., 1996, 1998). Since then several neuronally expressed cues have been identified, including CD200, fractalkine, polyamines, CCL21 and ATP (reviewed in Carson et al., 2006, Melchior et al., 2006). Interestingly, these cues can be divided into those that are expressed by healthy neurons and that suppress pro-inflammatory microglial responses, and those that are expressed by damaged and/or dying neurons. These latter cues by in large augment the pro-inflammatory responses of microglia.

A dramatic of demonstration neuronally directed regulation is provided by the studies of Cardona et al. examining the function of fractalkine (Cardona et al., 2006). Fractalkine is a chemokine expressed as a transmembrane glycoprotein. It can be proteolytically cleaved from the membrane to generate a soluble fragment retaining the ability to bind its receptor (Cook et al., 2001). Within the CNS the expression of the fractalkine receptor (CXCCR1) is restricted to Iba1-positive cells (microglia and perivascular macrophages). Neurons (NeuN + cells), oligodendrocytes (NG2 + cell), and astrocytes (GFAP + cells) do not express CXCCR1 nor is the CXCCR1 promoter active in these cells (Cardona et al., 2006).

By comparing microglial responses in Cx3cr1$^{+/-}$ or Cx3cr1$^{-/-}$ mice, Cardona et al. were able to demonstrate the dual role of fractalkine to limit microglial responses and recruit microglia to the site of injury (Cardona et al., 2006). Systemic injection of LPS in mice lacking the fractalkine receptor resulted in a large increase in hippocampal neuronal cell death (Cardona et al., 2006). To prove that this was a consequence of dysregulated microglial function, microglia were isolated from the CNS of LPS-treated Cx3cr1$^{+/-}$ and Cx3cr1$^{-/-}$ mice. These cells were than transferred into the frontal cortex of wild-type littermates. Transferred wild-type microglia were highly motile and trafficked along white matter tracts throughout the wild-type recipient CNS. Strikingly, microglia from knockout mice failed to migrate and persisted at the injection site for at least 36h following injection! Moreover, apoptotic neurons were observed at the site of injection only in animals injected with activated microglia lacking the fractalkine receptor (Cardona et al., 2006).

Cardona et al. (2006) identified IL-1B as playing a key role in the dysregulated responses in KO microglia. Coadministrating an IL1 receptor antagonist at the same time as the injection of KO microglia into wild-type mice significantly reduced the number of apoptotic neurons. Transfer of these microglia into the CNS of IL-1 receptor knock-out mice partially restored migration in Cx3cr1$^{-/-}$ microglia and completely prevented the previously observed neuronal apoptosis! Fractalkine is known to be a strong microglial chemoattractant. However, the restoration of migration suggests that fractalkine also serves to amplify microglial chemoattractant responses to other injury signals that are otherwise blocked by IL-1.

Cardona et al. (2006) further demonstrated that fractalkine regulates more than microglial responses to bacterial components such as LPS. They also examined microglial responses in the MPTP model of Parkinson's disease and in the transgenic SOD1^{G93A} model of ALS. In both models, neuronal loss was much greater in Cx3cr1$^{-/-}$ mice than in Cx3cr1$^{+/-}$ mice.

5 Is it Important to Distinguish Between CNS-Resident Microglia and Acutely Infiltrating Macrophages?

Cardona et al. (2006) observed heightened cytokine responses to LPS in both peripheral macrophages (peritoneal macrophages) as well as in microglia from Cx3cr1$^{+/-}$ mice. Due to the experimental paradigm, they were able to specifically examine the responses of microglia (the CD45low cells) separate from CNS-infiltrating macrophages (CD45high cells). In many studies, microglia and macrophages are grouped together in one population and their differential functions are not examined. However, over the last 10 years several studies have illustrated that these two populations are both molecularly and functionally distinct.

Initially, CNS-resident microglia have been distinguished from those that acutely infiltrate the CNS by their lower expression levels of many of the molecules required to interact with T cells (reviewed in Carson et al., 2006). Specifically, microglia tend to express lower levels of MHC and co-stimulatory molecules. As early as 1988, Hickey and Kimura (1988) demonstrated that antigen-specific interactions between CNS-microglia and myelin-specific T cell were not required to initiate or sustain destructive autoimmune responses during experimentally induced autoimmune encephalomyelitis (EAE), a rodent model of multiple sclerosis. This and several subsequent studies have definitively demonstrated that perivascular macrophages, CNS-infiltrating macrophages and/or dendritic cells are by themselves sufficient to trigger the onset and progression of EAE (Greter et al., 2005; McMahon et al., 2005). Conversely, antigen-specific interactions between CNS-resident microglia and myelin-specific T cells in the absence of peripheral antigen-presenting cells were by themselves insufficient trigger and sustain EAE. In part, this inability may be due to the failure of microglia to leave the CNS parenchyma at the same rates as perivascular macrophages or other CNS-infiltrating cells (Carson et al., 1999b).

6   Are Microglia Just an Incomplete or Redundant Macrophage Population?

From these types of studies, it may be tempting to refer to microglia as partial macrophages and to presume that they merely play redundant functions in CNS defense. Two studies, one using an EAE model and one using a facial axotomy model suggest otherwise.

In the first, Magnus and colleagues demonstrated that a B7 family member, B7 homologue-1 (also known as PD-L1) is abundantly expressed on the surface of CNS-resident microglia (Magnus et al., 2005). Microglial expression of PD-L1 is dramatically upregulated during the recovery phase of MOG- and PLP-forms of EAE and by direct treatment with IFNg. Several studies have revealed that PD-L1 acts in a negative feedback loop suppressing T-cell activation by decreasing IFNγ and IL-2 production and by down-regulating the expression ICOS, a T-cell activation marker (Magnus et al., 2005). PD-L1 knock-out mice developed more severe inflammation in the MOG-induced EAE (Latchman et al., 2004). Thus, while microglia may have comparatively weak APC function as compared to dendritic cells, their dialogue between peripheral immune cells and microglia through PD-L1 may not serve to amplify pro-inflammatory T cell responses. Rather, microglia may help to modulate local inflammation within the CNS by limiting the severity and spread of pro-inflammatory T cell responses. Consistent with this conclusion is the observation that CNS-resident microglia produce much higher levels of molecules such as prostaglandins and NO that repress antigen-presentation and T cell activation than infiltrating peripheral immune cells (Carson et al., 1998, 1999a).

Using the facial axotomy model, Byram et al. (2004) have demonstrated an essential and non-redundant function of CNS-resident microglia. In this model, the facial motoneuron cell body resides within the CNS brainstem, while its axon transverses the skull to innervate the vibrissae in the face. Slicing the axon at the point it transverses the skull, causes the axon to withdraw and prevents it from subsequently regenerating and finding its natural target. Serpe et al. (1999) had previously shown that in this model, CD4 + T cells limits the rate of facial motoneuron cell death. In the study by Byram et al. (2004), the authors demonstrated that peripheral antigen-presenting cells (presumably macrophages and dendritic cells) were required to initiate a neuroprotective T cell response. While these cells clearly infiltrate the site of the facial motoneuron nucleus (FMN), they could not sustain the protective T cell response. CNS-resident cells (presumably microglia) were absolutely essential to either evoke or sustain the protective lymphocyte response!

7 Are CNS-Infiltrating Macrophages Always Bad for CNS Function?

In both the EAE and facial axotomy models just discussed, microglia express high levels of MCP-1/CCL2 (reviewed in Carson et al., 2006). Indeed, in many models of CNS injury and pathogen exposure, microglia are induced to express both MHC class II (a perquisite to present antigen to CD4 + T cells) and CCL2 (a potent macrophage chemoattractant). Since macrophages are highly effective producers of free radicals *AND* have demonstrated pro-inflammatory roles in EAE, can this be considered a beneficial response of microglia designed to maintain optimal CNS function? Two studies using murine models of amyloid/Alzheimer's disease (AD) pathology suggest that microglial production of CCL2 and thus microglial recruitment of macrophages to the CNS may be an essential mechanism to impede the rate of AD pathogenesis (Simard et al., 2006; Khoury et al., 2007).

Microglia and macrophages surround the amyloid plaques in both human AD tissue and in rodent models of AD. In the first study, Rivest and colleagues sought to identify the relative contribution of CNS-resident versus hematogenously derived macrophages in the cells surrounding the plaque (Simard et al., 2006). To this end, the authors generated bone marrow chimeric mice, in which the hematogenously derived macrophages express green-fluorescent protein (GFP) while CNS-resident microglia did not. Not unexpectedly, the authors found that early in the formation of amyloid plaques, peripheral macrophages were readily recruited into the CNS. Somewhat surprisingly, macrophage recruitment did not continue to increase with age and plaque deposition, rather the reverse. The authors subsequently illustrated that the peripheral macrophage population was more effective at phagocytosis than the resident cells. From these data, the authors conclude that the late stage failure to recruit peripheral macrophages contributes to the progression of AD pathogenesis.

El Khoury et al. (2007) have recently confirmed and extended these studies. In this study, the authors studied amyloid responses in mice lacking the CCL2 receptor (ccr2 KO mice). They found that in ccr2 KO mice, fewer peripheral macrophages were recruited and that as a consequence amyloid pathogenesis developed much more rapidly, deposition within the vascular was much more severe and lethality occurred a much earlier ages than in mice expressing normal levels of ccr2. Interestingly, heterozygotes for the receptor expressed an intermediate phenotype, suggesting that the strength of the recruitment signal is carefully titrated in CNS immune responses!

8 Microglial Activation Gone Awry: Effects of Peripheral Infection and Aging!

In the AD studies just discussed, if may be that microglia are not effective in maintaining a sustained recruitment of macrophages. Recently, Cunningham and colleagues have presented a reciprocal problem in which peripheral inflammation may prime microglia to respond in an overly aggressive fashion to neuronal insults (Perry et al., 2002; Cunningham et al., 2005).

For this study, the authors chose to examine the effects of peripheral inflammation in a mouse model for transmissible spongiform encephalopathy prion disease (ME7) (Perry et al., 2002; Cunningham et al., 2005). In this ME7 model, mice are injected with symptomatic brain homogenates within the hippocampus and develop vacuolation, loss of hippocampal CA1 neurons and extracellular deposition of an insoluble protein (PrP^{Sc}) (Perry et al., 2002; Cunningham et al., 2005). However in contrast to many neurodegenerative models, microglial activation is atypical and mostly characterized by an overexpression of the anti-inflammatory cytokine, TGF-b (Cunningham et al., 2005).

ME7 mice were subsequently challenged with LPS, by either direct injection into the CNS or into the peritoneal cavity. Microglia in control prion brains (ME7 injected with saline) had a very similar appearance to those receiving prions and LPS intracerebrally (Cunningham et al., 2005) but expressed very different patterns of cytokine expression. Those challenged with LPS ICV expressed much higher levels of IL-1b and inducible nitric-oxide synthase (iNOS) (Cunningham et al., 2005). In addition, much higher numbers of neutrophils were found in the CNS of mice receiving LPS icv. Strikingly, the authors found that LPS injection intraperitoneally (ip) exacerbated the levels of IL-1b, COX-2, and TNF. Furthermore, neuronal cell death was doubled in ME7 mice receiving peripheral LPS injection. The precise mechanism underlying the observed exacerbation is as yet not fully defined. It is likely to be due to multiple mechanisms including vagal nerve stimulation from the spleen to the hypothalamus, induced BBB alterations and systemic increases in chemokine and cytokine levels.

9 So What Goes Right and What Goes Wrong with Microglial Activation?

Microglia are found in all mammalian brains and spinal cords. Conversely, there are no spontaneous animal models in which microglia are severely deficient or absent. These two facts suggest an evolutionarily conserved function. However, like all myeloid cells, microglia are highly plastic and are able to summate cues from all aspects of their environment (Fig. 1.2). Thus, at any point in time, the phenotype of an individual microglia is determined as a function of its environmental cues. This observation suggests two possible outcomes:

1. Microglial phenotypes are likely to be unstable and highly heterogeneous throughout the CNS and throughout the lifespan of the individual as a direct consequence of the many different local CNS microenvironments.
2. Dysfunctional microglial responses may be a direct consequence of dysfunctional neurons and macroglia.

Lastly, in contrast to most peripheral macrophage populations, CNS-resident microglia are relatively long-lived. Thus their dysfunction may have more long-lasting consequences than for other peripheral macrophage populations. Recently Sierra et al. have demonstrated that with age, microglial pro-inflammatory responses do become more robust (Sierra et al., 2007). As yet it remains unexplored if this is a primary dysfunction of the aged microglia or a consequence of the aging neuronal and macroglial population providing inappropriate regulatory cures! In the end, it is apparent that for most of us, for most of our lives, microglial activation is either a benign or a beneficial event. However, for therapeutic intervention, it is important to discern whether the dysfunction apparent in many chronic neurodegenerative diseases is due to inherent deficits in microglia or whether inappropriate microglial activation is a consequence of a dysfunctional CNS microenvironment!

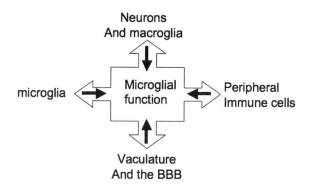

Fig. 1.2 Microglial phenotype and function are not stable and are determined by their interactions with their environment.

References

Bechmann I (2005) Failed central nervous system regeneration: a downside of immune privilege? *Neuromolecular Med* 7:217–228.

Bechmann I, Priller J, Kovac A, Bontert M, Wehner T, Klett FF, Bohsung J, Stuschke M, Dirnagl U, Nitsch R (2001) Immune surveillance of mouse brain perivascular spaces by blood-borne macrophages. *Eur J Neurosci* 14:1651–1658.

Bouchon A, Hernandez-Munain C, Cella M, Colonna M (2001) A dap12-mediated pathway regulates expression of cc chemokine receptor 7 and maturation of human dendritic cells. *J Exp Med* 194:1111–1122.

Byram SC, Carson MJ, Deboy CA, Serpe CJ, Sanders VM, Jones KJ (2004) CD4 + T cell-mediated neuroprotection requires dual compartment antigen presentation. *J Neurosci* 24:4333–4339.

Cardona AE, Pioro EP, Sasse ME, Kostenko V, Cardona SM, Dijkstra IM, Huang D, Kidd G, Dombrowski S, Dutta R, Lee JC, Cook DN, Jung S, Lira SA, Littman DR, Ransohoff RM (2006) Control of microglial neurotoxicity by the fractalkine receptor. *Nat Neurosci* 9:917–924.

Carson MJ, Sutcliffe JG (1999) Balancing function vs. self defense: the CNS as an active regulator of immune responses. *J Neurosci Res* 55:1–8.

Carson MJ, Reilly CR, Sutcliffe JG, Lo D (1998) Mature microglia resemble immature antigen-presenting cells. *Glia* 22:72–85.

Carson MJ, Sutcliffe JG, Campbell IL (1999a) Microglia stimulate naive T-cell differentiation without stimulating T-cell proliferation. *J Neurosci Res* 55:127–134.

Carson MJ, Reilly CR, Sutcliffe JG, Lo D (1999b) Disproportionate recruitment of CD8 + T cells into the central nervous system by professional antigen-presenting cells. *Am J Pathol* 154:481–494.

Carson MJ, Thrash JC, Lo D (2004) Analysis of microglial gene expression: identifying targets for CNS neurodegenerative and autoimmune disease. *Am J Pharmacogenomics* 4:321–330.

Carson MJ, Doose JM, Melchior B, Schmid CD, Ploix CC (2006) CNS immune privilege: hiding in plain sight. *Immunol Rev* 213:48–65.

Cella M, Buonsanti C, Strader C, Kondo T, Salmaggi A, Colonna M (2003) Impaired differentiation of osteoclasts in TREM-2-deficient individuals. *J Exp Med* 198:645–651.

Chung S, Jung W, Lee MY (1999) Inward and outward rectifying potassium currents set membrane potentials in activated rat microglia. *Neurosci Lett* 262:121–124.

Cook DN, Chen SC, Sullivan LM, Manfra DJ, Wiekowski MT, Prosser DM, Vassileva G, Lira SA (2001) Generation and analysis of mice lacking the chemokine fractalkine. *Mol Cell Biol* 21:3159–3165.

Cunningham C, Wilcockson DC, Campion S, Lunnon K, Perry VH (2005) Central and systemic endotoxin challenges exacerbate the local inflammatory response and increase neuronal death during chronic neurodegeneration. *J Neurosci* 25:9275–9284.

Davalos D, Grutzendler J, Yang G, Kim JV, Zuo Y, Jung S, Littman DR, Dustin ML, Gan WB (2005) ATP mediates rapid microglial response to local brain injury in vivo. *Nat Neurosci* 8:752–758.

Daws MR, Sullam PM, Niemi EC, Chen TT, Tchao NK, Seaman WE (2003) Pattern recognition by TREM-2: binding of anionic ligands. *J Immunol* 171:594–599.

Ford AL, Goodsall AL, Hickey WF, Sedgwick JD (1995) Normal adult ramified microglia separated from other central nervous system macrophages by flow cytometric sorting. Phenotypic differences defined and direct ex vivo antigen presentation to myelin basic protein-reactive CD4 + T cells compared. *J Immunol* 154:4309–4321.

Greter M, Heppner FL, Lemos MP, Odermatt BM, Goebels N, Laufer T, Noelle RJ, Becher B (2005) Dendritic cells permit immune invasion of the CNS in an animal model of multiple sclerosis. *Nat Med* 11:328–334.

Han S, Collins BE, Bengtson P, Paulson JC (2005) Homomultimeric complexes of CD22 in B cells revealed by protein-glycan cross-linking. *Nat Chem Biol* 1:93–97.

He F, Sun YE (2007) Glial cells more than support cells? *Int J Biochem Cell Biol* 39:661–665.

Hickey WF, Kimura H (1988) Perivascular microglial cells of the CNS are bone marrow-derived and present antigen in vivo. *Science* 239:290–292.

Kettenmann H, Hoppe D, Gottmann K, Banati R, Kreutzberg G (1990) Cultured microglial cells have a distinct pattern of membrane channels different from peritoneal macrophages. *J Neurosci Res* 26:278–287.

Kettenmann H, Banati R, Walz W (1993) Electrophysiological behavior of microglia. *Glia* 7:93–101.

Khoury JE, Toft M, Hickman SE, Means TK, Terada K, Geula C, Luster AD (2007) Ccr2 deficiency impairs microglial accumulation and accelerates progression of Alzheimer-like disease. *Nat Med* 13:432–438

Latchman YE, Liang SC, Wu Y, Chernova T, Sobel RA, Klemm M, Kuchroo VK, Freeman GJ, Sharpe AH (2004) PD-L1-deficient mice show that PD-L1 on T cells, antigen-presenting cells, and host tissues negatively regulates T cells. *Proc Natl Acad Sci U S A* 101:10691–10696.

Lee SJ, Lee S (2002) Toll-like receptors and inflammation in the CNS. *Curr Drug Targets Inflamm Allergy* 1:181–191.

Magnus T, Schreiner B, Korn T, Jack C, Guo H, Antel J, Ifergan I, Chen L, Bischof F, Bar-Or A, Wiendl H (2005) Microglial expression of the B7 family member B7 homolog 1 confers strong immune inhibition: implications for immune responses and autoimmunity in the CNS. *J Neurosci* 25:2537–2546.

McMahon EJ, Bailey SL, Castenada CV, Waldner H, Miller SD (2005) Epitope spreading initiates in the CNS in two mouse models of multiple sclerosis. *Nat Med* 11:335–339.

Melchior B, Puntambekar SS, Carson MJ (2006) Microglia and the control of autoreactive T cell responses. *Neurochem Int* 49:45–53

Mott RT, Ait-Ghezala G, Town T, Mori T, Vendrame M, Zeng J, Ehrhart J, Mullan M, Tan J (2004) Neuronal expression of CD22: Novel mechanism for inhibiting microglial proinflammatory cytokine production. *Glia* 46:369–379.

Neumann H, Boucraut J, Hahnel C, Misgeld T, Wekerle H (1996) Neuronal control of MHC class II inducibility in rat astrocytes and microglia. *Eur J Neurosci* 8:2582–2590.

Neumann H, Misgeld T, Matsumuro K, Wekerle H (1998) Neurotrophins inhibit major histocompatibility class II inducibility of microglia: involvement of the p75 neurotrophin receptor. *Proc Natl Acad Sci U S A* 95:5779–5784.

Nimmerjahn A, Kirchhoff F, Helmchen F (2005) Resting microglial cells are highly dynamic surveillants of brain parenchyma in vivo. *Science* 308:1314–1318.

Paloneva J, Kestila M, Wu J, Salminen A, Bohling T, Ruotsalainen V, Hakola P, Bakker AB, Phillips JH, Pekkarinen P, Lanier LL, Timonen T, Peltonen L (2000) Loss-of-function mutations in TYROBP (DAP12) result in a presenile dementia with bone cysts. *Nat Genet* 25:357–361.

Paloneva J, Autti T, Raininko R, Partanen J, Salonen O, Puranen M, Hakola P, Haltia M (2001) CNS manifestations of Nasu–Hakola disease: a frontal dementia with bone cysts. *Neurology* 56:1552–1558.

Parpura V, Scemes E, Spray DC (2004) Mechanisms of glutamate release from astrocytes: gap junction "hemichannels", purinergic receptors and exocytotic release. *Neurochem Int* 45:259–264.

Perry VH, Cunningham C, Boche D (2002) Atypical inflammation in the central nervous system in prion disease. *Curr Opin Neurol* 15:349–354.

Schiller M, Metze D, Luger TA, Grabbe S, Gunzer M (2006) Immune response modifiers–mode of action. *Exp Dermatol* 15:331–341.

Schilling T, Quandt FN, Cherny VV, Zhou W, Heinemann U, Decoursey TE, Eder C (2000) Upregulation of Kv1.3 K(+) channels in microglia deactivated by TGF-beta. *Am J Physiol Cell Physiol* 279:C1123–C1134.

Schmid CD, Sautkulis LN, Danielson PE, Cooper J, Hasel KW, Hilbush BS, Sutcliffe JG, Carson MJ (2002) Heterogeneous expression of the triggering receptor expressed on myeloid cells-2 on adult murine microglia. *J Neurochem* 83:1309–1320.

Schmidtmayer J, Jacobsen C, Miksch G, Sievers J (1994) Blood monocytes and spleen macrophages differentiate into microglia-like cells on monolayers of astrocytes: membrane currents. *Glia* 12:259–267.

Sedgwick JD, Schwender S, Imrich H, Dorries R, Butcher GW, ter Meulen V (1991) Isolation and direct characterization of resident microglial cells from the normal and inflamed central nervous system. *Proc Natl Acad Sci U S A* 88:7438–7442.

Serpe CJ, Kohm AP, Huppenbauer CB, Sanders VM, Jones KJ (1999) Exacerbation of facial motoneuron loss after facial nerve transection in severe combined immunodeficient (scid) mice. *J Neurosci* 19: RC7.

Sierra A, Gottfried-Blackmore AC, McEwen BS, Bulloch K (2007) Microglia derived from aging mice exhibit an altered inflammatory profile. *Glia* 55:412–424.

Simard AR, Soulet D, Gowing G, Julien JP, Rivest S (2006) Bone marrow-derived microglia play a critical role in restricting senile plaque formation in Alzheimer's disease. *Neuron* 49:489–502.

Simons M, Trajkovic K (2006) Neuron-glia communication in the control of oligodendrocyte function and myelin biogenesis. *J Cell Sci* 119:4381–4389.

Verkhratsky A, Toescu EC (2006) Neuronal-glial networks as substrate for CNS integration. *J Cell Mol Med* 10:826–836.

Volterra A, Steinhauser C (2004) Glial modulation of synaptic transmission in the hippocampus. *Glia* 47:249–257.

2
Mechanisms of Microglial Activation by Amyloid Precursor Protein and its Proteolytic Fragments

S.A. Austin and C.K. Combs

1 Reactive Microglia are a Characteristic Histopathology of Alzheimer's Disease Brains

Alzheimer's disease (AD) is the most prevalent neurodegenerative disease (Selkoe, 2005). Histologically, it is characterized by the deposition of extracellular senile plaques composed primarily of beta amyloid (Aβ) peptides and intracellular inclusions, termed neurofibrillary tangles, made up of primarily hyperphosphorylated tau protein (Braak and Braak, 1997a, b; Grundke-Iqbal et al., 1986; Selkoe, 2001). In addition, AD brains demonstrate significant neuron loss and abundant gliosis (McGeer et al., 1986). The mechanisms by which these pathology occur, however, is debatable. It has been hypothesized that inflammatory events contribute to both the histological and behavioral progression of disease (Akiyama et al., 2000). The histological data demonstrating gliotic changes in AD brains as compared to age-matched controls certainly supports the notion that microglia, in particular, may mediate the changes that are observed. Reactive microglia with swollen bodies and shortened, thickened processes are histologically identified in close association with the fibrillar or congophilic plaques in the AD brain (Itagaki et al., 1989; Miyazono et al., 1991). Although the percentage of microglia associated with fibrillar plaques is greater, they are also localized, in a more ramified phenotype, with the diffuse plaques (Itagaki et al., 1989; Mattiace et al., 1990; Sasaki et al., 1997). These data suggest that microglia develop a specific reactive phenotype in association with plaques as Aβ undergoes a transition from a nonfibrillar to fibrillar, congophilic conformation (Sheng et al., 1997). In fact, some studies suggest that microglia are involved in the earliest stages of plaque deposition perhaps even dictating where plaques are depositing in the brain (Griffin et al., 1995; Sheng et al., 1995, 1998). Moreover, AD brains have increased protein levels of several proinflammatory mediators commonly associated with reactive microgliosis, including cytokines: interleukin (IL)-1β, IL-6, and tumor necrosis factor (TNF)-α, activated complement components, and cyclooxygenase (COX)-2 when compared to controls (Akiyama et al., 2000; Dickson et al., 1993; Eikelenboom et al., 1989; Luterman et al., 2000; Mrak and Griffin, 2000; O'Banion et al., 1997; Strauss et al., 1992; Xiang et al., 2006;). Strikingly similar observations have been made while examining transgenic mouse models of disease over the last decade. The majority of the mouse models that have been created over-express human mutant forms of the amyloid precursor protein (APP) and/or mutant forms of the proteins responsible for gamma secretase cleavage of APP, presenilin (PS) 1 and PS2. These animal models have consistently demonstrated that reactive microgliosis occurs in association with fibrillar plaque formation as detected histologically with multiple immuno-markers (Morgan et al., 2005). Collectively, a voluminous body of data strengthens the proposition that APP and its proteolytic fragments are involved in not just plaque deposition but also the reactive microgliosis observed in AD brains.

2 Amyloid Precursor Protein and its Relationship to Alzheimer's Disease

APP is a ubiquitously expressed type I transmembrane protein that structurally resembles a cell surface receptor (Kang et al., 1987). The x-ray and crystal structure of the extracellular domain suggests that the protein can homodimerize in a cis (on the same cell surface) or trans (opposing cell surfaces) fashion (Rossjohn et al., 1999; Wang and Ha, 2004). Indeed, multimerization of APP occurs in neuronal cell lines basally (Scheuermann et al., 2001) and following ligand dependent stimulation (Lu et al., 2003). Furthermore, cross-linking APP with antibodies against the extracellular domain stimulates changes in intracellular signaling in vitro reminiscent of ligand dependent-receptor activation (Hashimoto et al., 2003; Okamoto et al., 1995). APP binds to extracellular matrix components including collagen (Beher et al., 1996) and laminin (Kibbey et al., 1993) as well as proteoglycans (Williamson et al., 1995) suggesting a role in mediating cell adhesion. A role as an adhesion receptor is further supported by the fact that APP levels increase on the neurite surface of differentiating neurons (Hung et al., 1992) and localize to points of focal adhesion in cell culture (Sabo et al., 2001). The short cytoplasmic tail contains a Y(682)ENPTY(687) (695 numbering) motif commonly employed by cell surface receptors as a docking site for SH2 and PTB domain containing proteins. Not surprisingly, several adaptor proteins including FE65, X11, JIP-1b and Shc have reported associations with this domain in a variety of different paradigms further supporting its role as a cell surface receptor (Borg et al., 1996; Bressler et al., 1996; Matsuda et al., 2001; Tarr et al., 2002).

As already mentioned, clinical interest in APP derives from the fact that its proteolytic processing leads to generation of the Aβ peptides that accumulate as extracellular plaques in AD brains (Masters et al., 1985). Moreover, a variety of APP missense mutations have been identified which result in a rare, autosomal dominant form of AD (Cai et al., 1993; Chartier-Harlin et al., 1991; Hendriks et al., 1992; Mullan et al., 1992; Murrell et al., 1991). The best characterized consequence of these mutations is an alteration in proteolytic processing of APP leading to elevated secretion of the longer Aβ peptide, amino acids 1–42 (Citron et al., 1992, 1994; Scheuner et al., 1996; Suzuki et al., 1994;). This peptide forms the fibrillar core of the amyloid plaques in AD brains (Jarrett et al., 1993). Additionally, the fibrillar peptide is potently toxic to neurons in a variety of paradigms (Lorenzo and Yankner, 1996; Pike et al., 1993). These collective data led to the formulation of the amyloid cascade hypothesis which proposed that fibrillization of the amyloid peptide is a key event in the pathophysiology of disease and critically important in the death of neurons leading to dementia (Hardy and Higgins, 1992).

3   Fibrillar Aβ is an Activating Stimulus for Microglia

Because Aβ peptide forms the fibrillar core of the senile plaques in both sporadic and autosomal dominant disease, it has been hypothesized that fibrillar plaque deposition represents a mechanistically critical process in disease progression (Hardy and Higgins, 1992; Jarrett et al., 1993). As already mentioned the fibrillar peptides exhibit a direct toxic action on neurons and the biology of this process, although certainly of relevance to AD, is outside of the scope of this discussion. On the other hand, the close association of reactive microglia with amyloid plaques as they transition from diffuse to fibrillar dense core (mature) plaques has suggested that fibrils are direct stimuli for activating microglia. Indeed, a large body of data exists demonstrating that fibrillar peptides stimulate microglia to acquire a reactive, neurotoxic phenotype.

Since peptide stimulation often requires interaction with a cell surface protein, many groups have worked to identify putative Aβ "receptors" on cells within the nervous system. Aβ has been shown capable of interacting with a truly diverse set of cell surface proteins including parent APP (Lorenzo et al., 2000; Van Nostrand et al., 2002; Wagner et al., 2000) the receptor for advanced glycation end products (RAGE) (Yan et al., 1996), scavenger receptor A (El Khoury et al., 1996), CD36 (Coraci et al., 2002; Moore et al., 2002), CD47 (Koenigsknecht and Landreth, 2004), β1 integrins (Koenigsknecht and Landreth, 2004), glypican (Schulz et al., 1998), N-Methyl-D-Aspartate (NMDA) receptors (Bi et al., 2002), α-7 nicotinic acetylcholine receptors (Wang et al., 2000), serpin-enzyme complex receptor (Boland et al., 1995), N-formyl peptide receptor-like (FPRL) 1 (Yazawa et al., 2001), and the insulin receptor (Xie et al., 2002a). Importantly, a number of these studies were performed using microglia and microglial cell lines offering some insight into the mechanism by which the peptide interact with the cell surface of microglia. These studies demonstrate that the Aβ peptide has the potential, particularly in its fibrillar form, to interact with a large array of structurally and functionally distinct proteins suggesting that fibril-cell interactions may be somewhat nonspecific.

An additional direction of research has focused not on the cell surface Aβ interaction but rather the subsequent intracellular signaling response driving acquisition of the reactive phenotype. Although the elucidated pathways have been determined from unique cell systems including primary microglial cultures, microglial cell lines, and monocytic cell lines, there are common aspects of the response. For example, fibrils stimulate a transient increase in activity of a number of tyrosine kinases including Fyn, Lyn, Syk, focal adhesion kinase (FAK), and pyruvate kinase (PYK) in stimulated THP-1 monocytes and primary rodent microglia (Bamberger et al., 2003; Combs et al., 1999, 2001; McDonald et al., 1997, 1998). Subsequent to the increase in tyrosine kinase activities, the cells release calcium from intracellular stores (Combs et al., 1999) and a number of serine threonine kinases are activated. For example, members of the mitogen activated protein (MAP) kinase family, extracellular signal regulated kinases (ERKs), c-Jun N-terminal kinase (JNKs), and p38 (Combs et al., 2001; Giri et al., 2003; McDonald et al., 1998) protein kinase C (PKC) (Combs et al., 1999), and RSK1/2 (McDonald et al., 1998) have all been reported to be activated by fibrillar stimulation of monocytes or microglia. In addition, subsequent changes suggestive of altered transcription occur after fibril stimulation including nuclear factor kappa beta (NFκB) activation, increased c-fos levels, and increased phosphorylation of CREB (Combs et al., 2001; McDonald et al., 1998).

It is not surprising, then, that fibrillar Aβ stimulation leads to increased secretion of several proinflammatory molecules from monocytes and microglia. Fibrils stimulate increased protein secretion and/or mRNA levels of several cytokines from human monocyte cell lines, microglial cell lines, primary rodent microglia and primary human, fetal and adult microglia including TNFα, IL-8, IL-6, MCSF, MIP-1α, IL-1β, matrix metalloproteases 1, 3, 9, 10, and 12 (Combs et al., 2000, 2001, Floden and Combs, 2006; Franciosi et al., 2005; Gasic-Milenkovic et al., 2003; Giri et al., 2003; Lue et al., 2001; Twig et al., 2005; Walker et al., 2001, 2006; Yates et al., 2000). In addition to cytokine secretion, fibrillar Aβ also stimulates secretion of superoxide anion from both human and mouse microglia via increased activity of plasmalemmal NADPH oxidase (Bianca et al, 1999; Wilkinson et al., 2006). Fibrillar stimulation also increases secretion of glutamate (Floden et al., 2005; Noda et al., 1999) and D-serine (Wu et al., 2004) demonstrating that oxidative and excitotoxic species may facilitate microglial mediated neuron death.

However, the identity of the neurotoxic agent(s) generated by Aβ fibril stimulated microglia appears to vary between paradigms ranging from excitotoxins, to cytokines, to oxidative damage dependent death (Banati et al., 1993, 1999; Combs et al., 2001; Floden et al., 2005; Giulian et al., 1995; Ii et al., 1996; Kingham and Pocock, 2001; Li et al., 2004; Monsonego et al., 2003; Tan et al., 2000; Xie et al., 2002b). It is likely that a variety of secreted factors contribute to the eventual loss of neurons that occurs either in the culture paradigms or in vivo following Aβ fibril stimulation of microglia. It will be important to identify which factors, if any, are truly generated by microglia in AD brains to determine the accuracy of in vitro modeling of microglial-dependent inflammatory changes during disease.

4   Oligomeric Aβ is an Activating Stimulus for Microglia

One of the criticisms of the amyloid cascade hypothesis, as originally proposed, has long been a clear lack of correlation between dementia rating and the numbers of fibrillar plaques in the brain (Lue et al., 1999; McLean et al., 1999; Morris et al., 1996). Indeed, it has been suggested that synaptic loss and gliosis precede plaque deposition (Martin et al., 1994). A similar observation has been confirmed in studies employing mouse hAPP transgenic lines. For example, Westerman et al. (2002) observed cognitively normal aged APP$_{SWE}$ mice in spite of high concentrations of insoluble Aβ aggregates leading them to suggest a soluble Aβ form is responsible for neuronal deficit. An additional study using mice overexpressing APP$_{IND}$ and APP$_{SWE,IND}$ demonstrated decreased presynaptic marker immunoreactivity and impaired synaptic transmission prior to plaque deposition (Hsia et al., 1999). Finally, using APP$_{IND}$ and APP$_{SWE, IND}$ as well as APP$_{WT}$ overexpressing cells Mucke et al. (2000) demonstrated that presynaptic marker immunoreactivity correlates inversely with levels of Aβ and plaque load correlates independently of Aβ levels. Collectively these data suggest that the fibrillar, insoluble form of the peptide may not be the most relevant species for mediating neuronal death/dysfunction.

Interestingly, recent data suggests that nonfibrillar Aβ conformations may be more reliable indices of disease progression. An oligomeric form of Aβ has been shown to accumulate in vivo and, more importantly, is elevated in AD brains in correlation with degree of behavioral deficit (Lue et al., 1999; McLean et al., 1999). The oligomeric form of the peptide can vary from dimers to high molecular weight SDS-stable oligomers and also arises in vitro from Aβ secreted into the culture media (Gong et al., 2003; Podlisny et al., 1995; Xia et al., 1997). Importantly, much like their fibrillar derivatives, the oligomers are neurotoxic, stimulate gliosis, produce cognitive dysfunction, and decrease long-term potentiation (LTP) both in vitro and in vivo (Chromy et al., 2003; Cleary et al., 2005; Hu et al., 1998; Klyubin et al., 2005; Lambert et al., 1998; Roher et al., 1996; Walsh et al., 2002; Wang et al., 2002). The low molecular weight dimeric/trimeric multimer of Aβ is reportedly able to mediate reversible inhibition of LTP generation (Klyubin et al, 2005; Walsh et al., 2005), impairment of cognitive dysfunction (Cleary et al., 2005) and microglial-dependent neuron death (Roher et al., 1996). A similar study using monomeric-tetrameric preparations demonstrated robust toxic effects on neuronal cell lines using both Aβ1-40 and Aβ1-42 (Dahlgren et al., 2002). Similarly, higher molecular weight multimers have demonstrated direct neurotoxic effects in paradigms ranging from rodent hippocampal slice cultures (Chong et al., 2006) to cell lines (Chromy et al., 2003; Demuro et al., 2005) to human fetal neuron cultures (Deshpande et al., 2006). Not surprisingly, these high molecular weight multimers have been observed to directly bind to neurons in both diseased brains and rodent hippocampal neuron cultures (Kokubo et al., 2005; Lacor et al., 2004). More recently, a dodecamer, Aβ*56, has been specifically characterized to increase in vivo in the brains of Tg2576 mice correlatively with the appearance of cognitive deficit and induce a reversible spatial memory deficit when microinjected into rat brain (Cleary et al., 2005; Lesne et al., 2006). This plethora of new data has revised the amyloid cascade hypothesis to now state that AD is initiated by neurotoxic stimulation provided by soluble Aβ peptide in its oligomeric rather than fibrillar form (Selkoe, 2002). Unfortunately, the mechanism by which oligomers stimulate neuron loss and glial activation is still unclear.

There is some data demonstrating that increasing oligomer concentrations correlate with microgliosis in vivo in transgenic rodent brains (Gordon et al., 2002;

Koistinaho et al., 2002). In agreement with this observation several in vitro studies have begun characterizing the ability of nonfibrillar Aβ peptides to stimulate microglia. Although the culture paradigms as well as multimeric state have varied between laboratories, a common theme with these studies is that oligomeric stimulation promotes acquisition of a proinflammatory phenotype. As already mentioned, while many studies have characterized putative receptors for fibrillar Aβ, it remains unclear how the oligomeric peptides interact with microglia. Using purified cultures of mouse microglia we have observed increased protein phosphotyrosine levels upon stimulation with the low molecular weight dimer/trimer Aβ1-42 oligomers (unpublished observations). This is similar but not identical to the signaling response initiated by stimulating microglia with fibrillar peptides (Combs et al., 1999). For example, we have not observed any increase in MAP kinase activities upon stimulation with these dimer/trimer oligomers (unpublished observations). There is, however, still a paucity of data describing the extent of the stimulated signaling response in microglia following treatment with not only the low molecular weight dimer/trimer oligomers but also the larger multimers.

On the other hand, the reactive phenotype produced by oligomer stimulation is better characterized. Using rat astrocyte cultures (95–98%astrocytes/2–5% microglia) two different studies demonstrated that the low molecular weight oligomers (White et al., 2005) and soluble Aβ (Hu et al., 1998) stimulate proinflammatory changes including increased protein and mRNA levels of inducible nitric oxide synthase (iNOS) and IL1-β, increased iNOS activity, and increased TNFα secretion. In a similar study Manelli et al. (2006) demonstrated that this mixed glia paradigm produced neurotoxins when cocultured with coverslips of primary rat cortical neurons. Although these data do not prove that the changes in proinflammatory protein expression and neurotoxicity are via oligomer-microglia interaction a study by Roher et al. (1996) demonstrated that rat mixed hippocampal neuron-glia cultures exhibited toxicity when treated with low molecular weight oligomers (dimer/trimer) only when microglia were present.

Other studies have used purified cultures of microglia to characterize the effects of oligomer stimulation on activation. A recent report showed that Aβ1-42 monomer-24mer preparations stimulate rat microglia cultures to secrete IL-1α and interferon-γ (IFN-γ) (Lindberg et al., 2005). A similar study by Takata et al. (2003) showed that rat microglia cultures increase secretion of TNFα, IL-6, and nitric oxide upon stimulation with low molecular weight Aβ1-40 oligomers. Even the Aβ25-35 fragment in its nonfibrillar form has recently demonstrated the ability to activate rat microglia to increase TNFα secretion (Hashioka et al., 2005). Using a dimer/trimer preparation of Aβ1-42, we have observed a similar activating response using cultures of purified mouse microglia. Oligomer stimulation results in increased expression of CD68, increased secretion of IL-6, TNFα, keratinocyte chemoattractant chemokine (KC), and decreased secretion of monocyte chemoattractant protein-1 (MCP-1) (Floden and Combs, 2006; unpublished observations). Moreover, these low molecular weight species are toxic to neurons only in the presence of microglia similar to prior work (Roher et al., 1996; unpublished observations). Therefore, although oligomeric peptides have direct effects on neuron activity and viability these collected studies above suggest that oligomeric peptides, much like their fibrillar counterpart, may mediate a portion of their detrimental effects through microglia activation. It remains to be seen whether different multimeric states have unique stimulatory abilities for microglia.

5 The N-terminal Secreted Fragment of APP, sAPP, is an Activating Stimulus for Microglia

It is now appreciated that additional cleavage products of APP besides the Aβ peptides also mediate distinct, physiologic effects on cells. APP can be processed along two distinct, competing pathways to release a large secreted N-terminal portion of the protein (sAPP). ADAM 10 and TACE are involved in alpha secretase cleavage of APP resulting in generation of a soluble, 612 amino acid, N-terminal fragment of APP (sAPPα) which is released into the extracellular space (Esch et al., 1990; Haass et al., 1991; Sinha and Lieberburg, 1999; Weidemann et al., 1989). The aspartic proteases, BACE1 and BACE2, represent the beta secretase activities responsible for generation of sAPPβ required for the proteolytic processing to generate the Aβ peptides (Bennett et al., 2000; Sinha et al., 1999; Vassar et al., 1999). Much like, the fibrillar form of Aβ, sAPP has a host of effects on neurons. For example, sAPPα has direct protective effects on cultured neurons in response to excitotoxic challenge that is a hundred fold more protective than sAPPβ (Barger and Mattson, 1997; Furukawa et al., 1996). This effect involves increased guanylate cyclase activity and increased NFkB activation as well as decreased NMDA receptor-mediated calcium influx (Barger and Mattson, 1995, 1996; Furukawa et al., 1996; Furukawa and Mattson, 1998). In addition, sAPPα has a demonstrated ability to stimulate increased neurite outgrowth in neuronal cells via a tyrosine kinase stimulated signaling response (Jin et al., 1994; Mook-Jung and Saitoh, 1997).

Almost paradoxically, sAPPα also has a demonstrated ability to robustly stimulate microglial activation. Although the signaling pathway is not completely determined, is has been demonstrated that sAPPα stimulation of microglia involves increased MAP kinase activities. Specifically, treatment of rat microglia with sAPPα leads to increased levels of active ERKs, p38 kinase, and JNKs (Bodles and Barger, 2005). In addition, sAPPα or sAPPβ stimulation of primary microglia as well as the N9 microglia cell line increases NFκB activity (Barger and Harmon, 1997). As might be expected, these changes lead to increased expression or activity of a host of proinflammatory products including iNOS and IL-1β, and reactive oxygen species (Barger and Harmon, 1997; Barger et al., 2000; Bodles and Barger, 2005; Li et al., 2000). Importantly, the production of proinflammatory proteins is dependent upon activity of JNK and p38 kinases and not ERKs since specific inhibitors of JNK and p38 MAP kinases but not ERKs attenuate the sAPPα-induced increase in iNOS protein levels and activity (Bodles and Barger, 2005). Besides cytokine secretion, sAPPα also stimulates microglia to secrete glutamate via the cystine-glutamate antiporter (Barger and Basile, 2001; Ikezu et al., 2003).

Based upon the identity of the secretory products described above, it is not surprising that the secretions from sAPPα or sAPPβ stimulated microglia are toxic to rodent neuron cultures (Barger and Basile, 2001; Barger and Harmon, 1997; Ikezu et al., 2003). The toxicity can be prevented by a superoxide dismutase (SOD) mimetic, MnTBP, specific inhibitors of neuronal nitric oxide synthase (nNOS), specific inhibitors of iNOS, and the NMDA receptor antagonist, MK-801 (Barger and Basile, 2001; Ikezu et al., 2003). Taken together, these data suggest that sAPP-stimulated microglia induce neuron death via combined oxidative and excitotoxic mechanisms. Therefore, although alpha secretase cleaved APP, sAPPα, is a demonstrated neurotrophic factor, it can also drive microglia to acquire a reactive, neurotoxic phenotype. It remains to be seen which of these opposing actions will dominate the in vivo function of sAPP.

6   Full Length APP can Act as a Proinflammatory Receptor on Microglia

We have thus far reviewed the accumulating data describing the ability of APP proteolytic fragments to stimulate microglial activation. Far less information is available regarding the function of full length APP in microglia. This is somewhat surprising since microglia serve as the second major producer of Aβ peptides behind neurons (Banati et al., 1993). It is relevant to discuss microglial APP in the context of this discussion since work by ourselves as well as others has suggested that is behaves as a proinflammatory receptor on monocytes and microglia. It has been known for some time that APP mRNA can be found within microglia of human brains (Schmechel et al., 1988). However, it has also been reported that plaque associated microglia in the AD brain have no detectable APP mRNA (Scott et al., 1993). The more definitive assessment of protein, however, has confirmed that microglia not only express APP but also upregulate protein levels in response to particular stimuli. In vitro cultures of purified rat microglia have verified they can express all isoforms of APP (Haass et al., 1991; LeBlanc et al., 1991). However, basal APP levels are low compared to neurons and very little of the protein is localized to the plasmalemma (Haass et al., 1991; LeBlanc et al., 1991). Not surprisingly, then, in vitro rat microglia studies have demonstrated that very little to no Aβ peptide or sAPP is generated by microglia (Haass et al., 1991; LeBlanc et al., 1991). These data suggest that the holoprotein may function differently in microglia compared to neurons.

However, other in vitro studies have demonstrated that APP protein levels are readily upregulated in microglia upon specific stimulation. For instance, human monocytes differentiated to macrophage in vitro increase their APP protein levels (Bauer et al., 1991). Microglia cultures stimulated with activating ligands like lipopolysaccharide (LPS) or prostaglandin E2 (PGE2) also increase APP, particularly on the cell surface (Pooler et al., 2004; unpublished observations). Using a mouse microglia line, BV-2, Monning et al. (1995) have demonstrated that when microglia express cell surface APP they are fully capable of secreting sAPP fragments. More importantly, this occurs in response to microglial adhesion to extracellular substrates like fibronectin and poly-L-lysine (Monning et al., 1995).

Perhaps the most compelling microglial APP data is derived from a series of in vivo studies demonstrating that APP immunoreactivity increases acutely and transiently within microglia following a variety of insults. For instance, transection of facial or sciatic nerves in rats results in increased microglial APP immunoreactivity within 6 h post lesion not only in the affected nucleus but also in areas of afferent projection (Banati et al., 1993). Lesion of the entorhinal cortex in rats produced a similar, transient profile of increased microglial APP immunoreactivity in the dentate gyrus (Banati et al., 1994). However, microglial APP expression is also responsive to a broader range of insults beyond axotomy. Both an experimental autoimmune encephalomyelitis (EAE) model as well as a transient ischemia model in mice result in an elevation of microglial APP immunoreactivity that lasts for several days-weeks (Banati et al., 1995a, b). Collectively, these data, together with the observation that the structure of APP resembles a cell surface receptor (Kang et al., 1987), suggest that APP has a role in regulating acquisition of a reactive phenotype in microglia.

We have begun work in support of this hypothesis by characterizing the signaling response stimulated by plasmalemmal APP in primary mouse microglia and the human monocytic cell-line THP-1 (Sondag and Combs, 2004, 2006). Utilizing two different stimulation paradigms we have found that APP is associated with a classic tyrosine kinase-based proinflammatory signaling response leading to acquisition of a reactive phenotype in these cells. By plating these cells onto a type I collagen substrate we have modeled $\beta1$ integrin-mediated adhesion-dependent activation. In addition we have used an antibody, 22C11, against the N-terminus of APP to cross-link cell surface APP to simulate ligand binding. Both paradigms stimulate increased protein phosphotyrosine levels in microglia and THP-1 cells indicative of increased tyrosine kinase activity. APP pull-down co-immunoprecipitations have shown that the Src family tyrosine kinase, Lyn, and the tyrosine kinase, Syk, are recruited to a complex with APP upon substrate adhesion or antibody cross-linking (Sondag and Combs, 2004). In addition, substrate adhesion but not antibody cross-linking recruits APP to a multireceptor signaling complex with $\beta1$ integrin along with Syk and Lyn (Sondag and Combs, 2004). Subsequent to increased tyrosine kinase activity, we observed activation of the MAP kinase family following both adhesion and antibody cross-linking (Sondag and Combs, 2004, 2006).

Not surprisingly, adhesion-dependent activation of THP-1 cells stimulates an increase in protein levels of a plethora of proinflammatory markers including COX-2, CD36, iNOS, and IL-1β. However, the more interesting observation is that these changes in protein levels were dependent upon expression of APP and the subsequent increase in tyrosine and MAP kinase activities induced upon ligand binding (Sondag and Combs, 2004). We extended this observation to define the behavior of THP-1 cells and microglia following antibody cross-linking of APP. As with the adhesion studies, the stimulated increase in proinflammatory protein levels was dependent upon recruited tyrosine and MAP kinase activities. Moreover, APP cross-linking increased cytokine secretion by the THP-1 cells and microglia. Most notably, cross-linked cells increased secretion of IL-1β and IL-6 in a tyrosine kinase dependent manner (Sondag and Combs, 2006). Because antibody-mediated receptor cross-linking is expected to influence endocytic events we also determined whether APP was cleaved into Aβ peptides following stimulation. Cross-linking stimulated a selective release of Aβ1–42 compared to Aβ1–40 from the monocytes. However, Aβ1–42 secretion was independent of the increase in tyrosine and MAP kinase activities we had observed since inhibition had no effect on stimulated Aβ1–42 secretion. Therefore, secretase control of APP metabolism was independent of the tyrosine kinase based activation pathway. Our results thus far have suggested that APP has a common function in monocytes and microglia that is important in acquisition of a reactive phenotype. More importantly, it appears that the protein can act as an independent receptor, as in the case of antibody cross-linking, or it can be recruited into a multi-receptor signaling complex, as in the case of adhesion dependent activation. This novel signaling mechanism by which monocytes and microglia generate Aβ peptides could be a relevant contribution to plaque pathology in AD and vascular amyloidosis. Collectively, these results strengthen existing data that suggest microglial-derived APP can contribute to amyloid production in AD (Bauer et al., 1991).

Although we have demonstrated a rather robust role for APP in monocyte/microglial activation as a single receptor or within a multi-receptor complex, it is not clear how APP is involved in activating these cells in vivo. While it is easy to imagine APP participating in adhesion-mediated activation of microglia adhering to extracellular matrix, it is more difficult to envision how APP can behave as an independent proinflammatory receptor. This is largely due the fact that an agonist ligand for the extracellular domain of APP is not yet known. One interesting possibility is that the Aβ peptide itself can behave in an autocrine fashion to interact with APP to mediate clustering and subsequent signal transduction. Interestingly, Aβ has already been demonstrated to bind to the extracellular region of APP (Chung et al., 1999; Lorenzo et al., 2000; Shaked et al., 2006; Van Nostrand et al., 2002; Wagner et al., 2000) offering the possibility of a proinflammatory feed-forward pathway in which Aβ-APP interaction leads to increased APP-dependent proinflammatory signaling that results in further Aβ production. Alternatively, in vitro studies have shown that membrane-bound APP can form homodimers leading to the speculation that full-length APP can be its own ligand acting in a cis (same cell) or trans (opposing cell) fashion (Lu et al., 2003; Rossjohn et al., 1999; Scheuermann et al., 2001; Wang and Ha, 2004). Therefore, although it is well accepted that APP processing to Aβ peptide is an important contribution to plaque formation in AD (Citron et al., 1992, 1994; Jarrett et al., 1993; Masters et al., 1985; Scheuner et al., 1996), it is possible that APP has a multi-faceted role in the progression of this disease particularly as a proinflammatory receptor on microglia.

In conclusion, although reactive microglia are a histological hallmark in the AD brain, their contribution to neuron death and cognitive decline remains unclear. In addition, the stimulus for their reactivity is also not defined. A large collection of data demonstrates that proinflammatory changes occur in not only AD brains but also its animal models. These data offer hope that attenuating microgliosis will offer benefit against disease conditions. However, before this can be approached in a specific fashion it is important to define not only the source of reactivity but also the subtle differences in activation phenotype that surely must exist in vivo. For example, there is a well recognized association of a certain reactive microglial phenotype with mature, dense core plaques and fibrillar Aβ peptides are activating ligands for microglia. However, as illustrated above, it is also clear that nonfibrillar forms of the peptide as well as the secreted N-terminus and full length APP itself, all have the capacity to stimulate microglia to acquire unique, reactive phenotypes. It remains to be seen which of these species, if any, has the most significant role in promoting microglial activation in AD.

A. Aβ fibril stimulation   B. Aβ oligomer stimulation

C. sAPP stimulation   D. Adhesion associated APP Stimulation

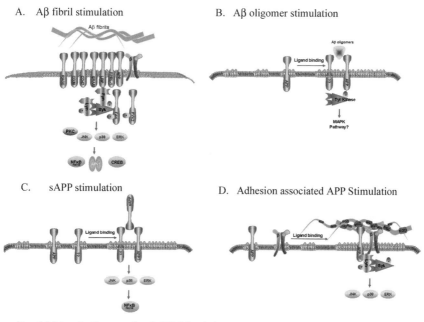

E. Multimerization associated APP Stimulation

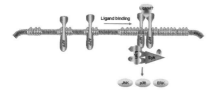

Comparison of Modes of Microglial Actication by APP and its Proteolytic Fragments.

References

Akiyama, H., Arai, T., Kondo, H., Tanno, E., Haga, C. and Ikeda, K. (2000). Cell mediators of inflammation in the Alzheimer disease brain. *Alzheimer Dis Assoc Disord*, *14 Suppl 1*, S47–S53.

Bamberger, M. E., Harris, M. E., McDonald, D. R., Husemann, J. and Landreth, G. E. (2003). A cell surface receptor complex for fibrillar beta-amyloid mediates microglial activation. *J Neurosci*, *23*, 2665–2674.

Banati, R. B., Gehrmann, J., Czech, C., Monning, U., Jones, L. L., Konig, G., Beyreuther, K. and Kreutzberg, G. W. (1993). Early and rapid de novo synthesis of Alzheimer beta A4-amyloid precursor protein (APP) in activated microglia. *Glia*, *9*, 199–210.

Banati, R. B., Gehrmann, J. and Kreutzberg, G. W. (1994). Glial beta-amyloid precursor protein: expression in the dentate gyrus after entorhinal cortex lesion. *Neuroreport, 5,* 1359–1361.

Banati, R. B., Gehrmann, J., Lannes-Vieira, J., Wekerle, H. and Kreutzberg, G. W. (1995a). Inflammatory reaction in experimental autoimmune encephalomyelitis (EAE) is accompanied by a microglial expression of the beta A4-amyloid precursor protein (APP). *Glia, 14,* 209–215.

Banati, R. B., Gehrmann, J., Wiessner, C., Hossmann, K. A. and Kreutzberg, G. W. (1995b). Glial expression of the beta-amyloid precursor protein (APP) in global ischemia. *J Cereb Blood Flow Metab, 15,* 647–654.

Barger, S. W. and Basile, A. S. (2001). Activation of microglia by secreted amyloid precursor protein evokes release of glutamate by cystine exchange and attenuates synaptic function. *J Neurochem, 76,* 846–854.

Barger, S. W. and Harmon, A. D. (1997). Microglial activation by Alzheimer amyloid precursor protein and modulation by apolipoprotein E. *Nature, 388,* 878–881.

Barger, S. W. and Mattson, M. P. (1995). The secreted form of the Alzheimer's beta-amyloid precursor protein stimulates a membrane-associated guanylate cyclase. *Biochem J, 311 (Pt 1),* 45–47.

Barger, S. W. and Mattson, M. P. (1996). Induction of neuroprotective kappa B-dependent transcription by secreted forms of the Alzheimer's beta-amyloid precursor. *Brain Res Mol Brain Res, 40,* 116–126.

Barger, S. W. and Mattson, M. P. (1997). Isoform-specific modulation by apolipoprotein E of the activities of secreted beta-amyloid precursor protein. *J Neurochem, 69,* 60–67.

Barger, S. W., Chavis, J. A. and Drew, P. D. (2000). Dehydroepiandrosterone inhibits microglial nitric oxide production in a stimulus-specific manner. *J Neurosci Res, 62,* 503–509.

Bauer, J., Konig, G., Strauss, S., Jonas, U., Ganter, U., Weidemann, A., Monning, U., Masters, C. L., Volk, B., Berger, M., et al. (1991). In-vitro matured human macrophages express Alzheimer's beta A4-amyloid precursor protein indicating synthesis in microglial cells. *FEBS Lett, 282,* 335–340.

Beher, D., Hesse, L., Masters, C. L. and Multhaup, G. (1996). Regulation of amyloid protein precursor (APP) binding to collagen and mapping of the binding sites on APP and collagen type I. *J Biol Chem, 271,* 1613–1620.

Bennett, B. D., Babu-Khan, S., Loeloff, R., Louis, J. C., Curran, E., Citron, M. and Vassar, R. (2000). Expression analysis of BACE2 in brain and peripheral tissues. *J Biol Chem, 275,* 20647–20651.

Bi, X., Gall, C. M., Zhou, J. and Lynch, G. (2002). Uptake and pathogenic effects of amyloid beta peptide 1–42 are enhanced by integrin antagonists and blocked by NMDA receptor antagonists. *Neuroscience, 112,* 827–840.

Bianca, V. D., Dusi, S., Bianchini, E., Dal Pra, I. and Rossi, F. (1999). Beta-amyloid activates the O-2 forming NADPH oxidase in microglia, monocytes, and neutrophils. A possible inflammatory mechanism of neuronal damage in Alzheimer's disease. *J Biol Chem, 274,* 15493–15499.

Bodles, A. M. and Barger, S. W. (2005). Secreted beta-amyloid precursor protein activates microglia via JNK and p38-MAPK. *Neurobiol Aging, 26,* 9–16.

Boland, K., Manias, K. and Perlmutter, D. H. (1995). Specificity in recognition of amyloid-beta peptide by the serpin-enzyme complex receptor in hepatoma cells and neuronal cells. *J Biol Chem, 270,* 28022–28028.

Borg, J. P., Ooi, J., Levy, E. and Margolis, B. (1996). The phosphotyrosine interaction domains of X11 and FE65 bind to distinct sites on the YENPTY motif of amyloid precursor protein. *Mol Cell Biol, 16,* 6229–6241.

Braak, H. and Braak, E. (1997a). Diagnostic criteria for neuropathologic assessment of Alzheimer's disease. *Neurobiol Aging, 18,* S85–S88.

Braak, H. and Braak, E. (1997b). Frequency of stages of Alzheimer-related lesions in different age categories. *Neurobiol Aging, 18,* 351–357.

Bressler, S. L., Gray, M. D., Sopher, B. L., Hu, Q., Hearn, M. G., Pham, D. G., Dinulos, M. B., Fukuchi, K., Sisodia, S. S., Miller, M. A., Disteche, C. M. and Martin, G. M. (1996). cDNA cloning and chromosome mapping of the human Fe65 gene: interaction of the conserved cytoplasmic domains of the human beta-amyloid precursor protein and its homologues with the mouse Fe65 protein. *Hum Mol Genet, 5*, 1589–1598.

Cai, X. D., Golde, T. E. and Younkin, S. G. (1993). Release of excess amyloid beta protein from a mutant amyloid beta protein precursor. *Science, 259*, 514–516.

Chartier-Harlin, M. C., Crawford, F., Houlden, H., Warren, A., Hughes, D., Fidani, L., Goate, A., Rossor, M., Roques, P., Hardy, J., et al. (1991). Early-onset Alzheimer's disease caused by mutations at codon 717 of the beta-amyloid precursor protein gene. *Nature, 353*, 844–846.

Chong, Y. H., Shin, Y. J., Lee, E. O., Kayed, R., Glabe, C. G. and Tenner, A. J. (2006). ERK1/2 activation mediates Abeta oligomer-induced neurotoxicity via caspase-3 activation and tau cleavage in rat organotypic hippocampal slice cultures. *J Biol Chem, 281*, 20315–20325.

Chromy, B. A., Nowak, R. J., Lambert, M. P., Viola, K. L., Chang, L., Velasco, P. T., Jones, B. W., Fernandez, S. J., Lacor, P. N., Horowitz, P., Finch, C. E., Krafft, G. A. and Klein, W. L. (2003). Self-assembly of Abeta(1–42) into globular neurotoxins. *Biochemistry, 42*, 12749–12760.

Chung, H., Brazil, M. I., Soe, T. T. and Maxfield, F. R. (1999). Uptake, degradation, and release of fibrillar and soluble forms of Alzheimer's amyloid beta-peptide by microglial cells. *J Biol Chem, 274*, 32301–32308.

Citron, M., Oltersdorf, T., Haass, C., McConlogue, L., Hung, A. Y., Seubert, P., Vigo-Pelfrey, C., Lieberburg, I. and Selkoe, D. J. (1992). Mutation of the beta-amyloid precursor protein in familial Alzheimer's disease increases beta-protein production. *Nature, 360*, 672–674.

Citron, M., Vigo-Pelfrey, C., Teplow, D. B., Miller, C., Schenk, D., Johnston, J., Winblad, B., Venizelos, N., Lannfelt, L. and Selkoe, D. J. (1994). Excessive production of amyloid beta-protein by peripheral cells of symptomatic and presymptomatic patients carrying the Swedish familial Alzheimer disease mutation. *Proc Natl Acad Sci U S A, 91*, 11993–11997.

Cleary, J. P., Walsh, D. M., Hofmeister, J. J., Shankar, G. M., Kuskowski, M. A., Selkoe, D. J. and Ashe, K. H. (2005). Natural oligomers of the amyloid-beta protein specifically disrupt cognitive function. *Nat Neurosci, 8*, 79–84.

Combs, C. K., Johnson, D. E., Cannady, S. B., Lehman, T. M. and Landreth, G. E. (1999). Identification of microglial signal transduction pathways mediating a neurotoxic response to amyloidogenic fragments of beta-amyloid and prion proteins. *J Neurosci, 19*, 928–939.

Combs, C. K., Johnson, D. E., Karlo, J. C., Cannady, S. B. and Landreth, G. E. (2000). Inflammatory mechanisms in Alzheimer's disease: inhibition of beta-amyloid-stimulated proinflammatory responses and neurotoxicity by PPARgamma agonists. *J Neurosci, 20*, 558–567.

Combs, C. K., Karlo, J. C., Kao, S. C. and Landreth, G. E. (2001). Beta-amyloid stimulation of microglia and monocytes results in TNFalpha-dependent expression of inducible nitric oxide synthase and neuronal apoptosis. *J Neurosci, 21*, 1179–1188.

Coraci, I. S., Husemann, J., Berman, J. W., Hulette, C., Dufour, J. H., Campanella, G. K., Luster, A. D., Silverstein, S. C. and El-Khoury, J. B. (2002). CD36, a class B scavenger receptor, is expressed on microglia in Alzheimer's disease brains and can mediate production of reactive oxygen species in response to beta-amyloid fibrils. *Am J Pathol, 160*, 101–112.

Dahlgren, K. N., Manelli, A. M., Stine, W. B., Jr., Baker, L. K., Krafft, G. A. and LaDu, M. J. (2002). Oligomeric and fibrillar species of amyloid-beta peptides differentially affect neuronal viability. *J Biol Chem, 277*, 32046–32053.

Demuro, A., Mina, E., Kayed, R., Milton, S. C., Parker, I. and Glabe, C. G. (2005). Calcium dysregulation and membrane disruption as a ubiquitous neurotoxic mechanism of soluble amyloid oligomers. *J Biol Chem, 280*, 17294–17300.

Deshpande, A., Mina, E., Glabe, C. and Busciglio, J. (2006). Different conformations of amyloid beta induce neurotoxicity by distinct mechanisms in human cortical neurons. *J Neurosci, 26*, 6011–6018.

Dickson, D. W., Lee, S. C., Mattiace, L. A., Yen, S. H. and Brosnan, C. (1993). Microglia and cytokines in neurological disease, with special reference to AIDS and Alzheimer's disease. *Glia, 7*, 75–83.

Eikelenboom, P., Hack, C. E., Rozemuller, J. M. and Stam, F. C. (1989). Complement activation in amyloid plaques in Alzheimer's dementia. *Virchows Arch B Cell Pathol Incl Mol Pathol, 56*, 259–262.

El Khoury, J., Hickman, S. E., Thomas, C. A., Cao, L., Silverstein, S. C. and Loike, J. D. (1996). Scavenger receptor-mediated adhesion of microglia to beta-amyloid fibrils. *Nature, 382,* 716–719.

Esch, F. S., Keim, P. S., Beattie, E. C., Blacher, R. W., Culwell, A. R., Oltersdorf, T., McClure, D. and Ward, P. J. (1990). Cleavage of amyloid beta peptide during constitutive processing of its precursor. *Science, 248,* 1122–1124.

Floden, A. M. and Combs, C. K. (2006). Beta-amyloid stimulates murine postnatal and adult microglia cultures in a unique manner. *J Neurosci, 26,* 4644–8.

Floden, A. M., Li, S. and Combs, C. K. (2005). Beta-amyloid-stimulated microglia induce neuron death via synergistic stimulation of tumor necrosis factor alpha and NMDA receptors. *J Neurosci, 25,* 2566–75.

Franciosi, S., Choi, H. B., Kim, S. U. and McLarnon, J. G. (2005). IL-8 enhancement of amyloid-beta (Abeta 1–42)-induced expression and production of pro-inflammatory cytokines and COX-2 in cultured human microglia. *J Neuroimmunol, 159,* 66–74.

Furukawa, K. and Mattson, M. P. (1998). Secreted amyloid precursor protein alpha selectively suppresses N-methyl-D-aspartate currents in hippocampal neurons: involvement of cyclic GMP. *Neuroscience, 83,* 429–438.

Furukawa, K., Sopher, B. L., Rydel, R. E., Begley, J. G., Pham, D. G., Martin, G. M., Fox, M. and Mattson, M. P. (1996). Increased activity-regulating and neuroprotective efficacy of alpha-secretase-derived secreted amyloid precursor protein conferred by a C-terminal heparin-binding domain. *J Neurochem, 67,* 1882–1896.

Gasic-Milenkovic, J., Dukic-Stefanovic, S., Deuther-Conrad, W., Gartner, U. and Munch, G. (2003). Beta-amyloid peptide potentiates inflammatory responses induced by lipopolysaccharide, interferon-gamma and 'advanced glycation endproducts' in a murine microglia cell line. *Eur J Neurosci, 17,* 813–821.

Giri, R. K., Selvaraj, S. K. and Kalra, V. K. (2003). Amyloid peptide-induced cytokine and chemokine expression in THP-1 monocytes is blocked by small inhibitory RNA duplexes for early growth response-1 messenger RNA. *J Immunol, 170,* 5281–5294.

Giulian, D., Haverkamp, L. J., Li, J., Karshin, W. L., Yu, J., Tom, D., Li, X. and Kirkpatrick, J. B. (1995). Senile plaques stimulate microglia to release a neurotoxin found in Alzheimer brain. *Neurochem Int, 27,* 119–137.

Gong, Y., Chang, L., Viola, K. L., Lacor, P. N., Lambert, M. P., Finch, C. E., Krafft, G. A. and Klein, W. L. (2003). Alzheimer's disease-affected brain: presence of oligomeric A beta ligands (ADDLs) suggests a molecular basis for reversible memory loss. *Proc Natl Acad Sci U S A, 100,* 10417–10422.

Gordon, M. N., Holcomb, L. A., Jantzen, P. T., DiCarlo, G., Wilcock, D., Boyett, K. W., Connor, K., Melachrino, J., O'Callaghan, J. P. and Morgan, D. (2002). Time course of the development of Alzheimer-like pathology in the doubly transgenic PS1 + APP mouse. *Exp Neurol, 173,* 183–195.

Griffin, W. S., Sheng, J. G., Roberts, G. W. and Mrak, R. E. (1995). Interleukin-1 expression in different plaque types in Alzheimer's disease: significance in plaque evolution. *J Neuropathol Exp Neurol, 54,* 276–281.

Grundke-Iqbal, I., Iqbal, K., Tung, Y. C., Quinlan, M., Wisniewski, H. M. and Binder, L. I. (1986). Abnormal phosphorylation of the microtubule-associated protein tau (tau) in Alzheimer cytoskeletal pathology. *Proc Natl Acad Sci U S A, 83,* 4913–4917.

Haass, C., Hung, A. Y. and Selkoe, D. J. (1991). Processing of beta-amyloid precursor protein in microglia and astrocytes favors an internal localization over constitutive secretion. *J Neurosci, 11,* 3783–3793.

Hardy, J. A. and Higgins, G. A. (1992). Alzheimer's disease: the amyloid cascade hypothesis. *Science, 256,* 184–185.

Hashimoto, Y., Tsuji, O., Niikura, T., Yamagishi, Y., Ishizaka, M., Kawasumi, M., Chiba, T., Kanekura, K., Yamada, M., Tsukamoto, E., Kouyama, K., Terashita, K., Aiso, S., Lin, A. and Nishimoto, I. (2003). Involvement of c-Jun N-terminal kinase in amyloid precursor protein-mediated neuronal cell death. *J Neurochem, 84,* 864–877.

Hashioka, S., A. Monji, T. Ueda, S. Kanba and H. Nakanishi (2005). Amyloid-beta fibril formation is not necessarily required for microglial activation by the peptides. *Neurochem Int, 47*, 369–376.

Hendriks, L., van Duijn, C. M., Cras, P., Cruts, M., Van Hul, W., van Harskamp, F., Warren, A., McInnis, M. G., Antonarakis, S. E., Martin, J. J., et al. (1992). Presenile dementia and cerebral haemorrhage linked to a mutation at codon 692 of the beta-amyloid precursor protein gene. *Nat Genet, 1*, 218–221.

Hsia, A. Y., Masliah, E., McConlogue, L., Yu, G. Q., Tatsuno, G., Hu, K., Kholodenko, D., Malenka, R. C., Nicoll, R. A. and Mucke, L. (1999). Plaque-independent disruption of neural circuits in Alzheimer's disease mouse models. *Proc Natl Acad Sci U S A, 96*, 3228–3233.

Hu, J., Akama, K. T., Krafft, G. A., Chromy, B. A. and Van Eldik, L. J. (1998). Amyloid-beta peptide activates cultured astrocytes: morphological alterations, cytokine induction and nitric oxide release. *Brain Res, 785*, 195–206.

Hung, A. Y., Koo, E. H., Haass, C. and Selkoe, D. J. (1992). Increased expression of beta-amyloid precursor protein during neuronal differentiation is not accompanied by secretory cleavage. *Proc Natl Acad Sci U S A, 89*, 9439–9443.

Ii, M., Sunamoto, M., Ohnishi, K. and Ichimori, Y. (1996). Beta-amyloid protein-dependent nitric oxide production from microglial cells and neurotoxicity. *Brain Res, 720*, 93–100.

Ikezu, T., Luo, X., Weber, G. A., Zhao, J., McCabe, L., Buescher, J. L., Ghorpade, A., Zheng, J. and Xiong, H. (2003). Amyloid precursor protein-processing products affect mononuclear phagocyte activation: pathways for sAPP- and Abeta-mediated neurotoxicity. *J Neurochem, 85*, 925–934.

Itagaki, S., McGeer, P. L., Akiyama, H., Zhu, S. and Selkoe, D. (1989). Relationship of microglia and astrocytes to amyloid deposits of Alzheimer disease. *J Neuroimmunol, 24*, 173–182.

Jarrett, J. T., Berger, E. P. and Lansbury, P. T., Jr. (1993). The carboxy terminus of the beta amyloid protein is critical for the seeding of amyloid formation: implications for the pathogenesis of Alzheimer's disease. *Biochemistry, 32*, 4693–4697.

Jin, L. W., Ninomiya, H., Roch, J. M., Schubert, D., Masliah, E., Otero, D. A. and Saitoh, T. (1994). Peptides containing the RERMS sequence of amyloid beta/A4 protein precursor bind cell surface and promote neurite extension. *J Neurosci, 14*, 5461–5470.

Kang, J., Lemaire, H. G., Unterbeck, A., Salbaum, J. M., Masters, C. L., Grzeschik, K. H., Multhaup, G., Beyreuther, K. and Muller-Hill, B. (1987). The precursor of Alzheimer's disease amyloid A4 protein resembles a cell-surface receptor. *Nature, 325*, 733–736.

Kibbey, M. C., Jucker, M., Weeks, B. S., Neve, R. L., Van Nostrand, W. E. and Kleinman, H. K. (1993). Beta-amyloid precursor protein binds to the neurite-promoting IKVAV site of laminin. *Proc Natl Acad Sci U S A, 90*, 10150–10153.

Kingham, P. J. and Pocock, J. M. (2001). Microglial secreted cathepsin B induces neuronal apoptosis. *J Neurochem, 76*, 1475–1484.

Klyubin, I., Walsh, D. M., Lemere, C. A., Cullen, W. K., Shankar, G. M., Betts, V., Spooner, E. T., Jiang, L., Anwyl, R., Selkoe, D. J. and Rowan, M. J. (2005). Amyloid beta protein immunotherapy neutralizes Abeta oligomers that disrupt synaptic plasticity in vivo. *Nat Med, 11*, 556–561.

Koenigsknecht, J. and Landreth, G. (2004). Microglial phagocytosis of fibrillar beta-amyloid through a beta1 integrin-dependent mechanism. *J Neurosci, 24*, 9838–9846.

Koistinaho, M., Kettunen, M. I., Goldsteins, G., Keinanen, R., Salminen, A., Ort, M., Bures, J., Liu, D., Kauppinen, R. A., Higgins, L. S. and Koistinaho, J. (2002). Beta-amyloid precursor protein transgenic mice that harbor diffuse A beta deposits but do not form plaques show increased ischemic vulnerability: role of inflammation. *Proc Natl Acad Sci U S A, 99*, 1610–1515.

Kokubo, H., Kayed, R., Glabe, C. G. and Yamaguchi, H. (2005). Soluble Abeta oligomers ultrastructurally localize to cell processes and might be related to synaptic dysfunction in Alzheimer's disease brain. *Brain Res, 1031*, 222–228.

Lacor, P. N., Buniel, M. C., Chang, L., Fernandez, S. J., Gong, Y., Viola, K. L., Lambert, M. P., Velasco, P. T., Bigio, E. H., Finch, C. E., Krafft, G. A. and Klein, W. L. (2004). Synaptic targeting by Alzheimer's-related amyloid beta oligomers. *J Neurosci, 24*, 10191–10200.

Lambert, M. P., Barlow, A. K., Chromy, B. A., Edwards, C., Freed, R., Liosatos, M., Morgan, T. E., Rozovsky, I., Trommer, B., Viola, K. L., Wals, P., Zhang, C., Finch, C. E., Krafft, G. A. and Klein, W. L. (1998). Diffusible, nonfibrillar ligands derived from Abeta1–42 are potent central nervous system neurotoxins. *Proc Natl Acad Sci U S A*, *95*, 6448–6453.

LeBlanc, A. C., Chen, H. Y., Autilio-Gambetti, L. and Gambetti, P. (1991). Differential APP gene expression in rat cerebral cortex, meninges, and primary astroglial, microglial and neuronal cultures. *FEBS Lett*, *292*, 171–178.

Lesne, S., Koh, M. T., Kotilinek, L., Kayed, R., Glabe, C. G., Yang, A., Gallagher, M. and Ashe, K. H. (2006). A specific amyloid-beta protein assembly in the brain impairs memory. *Nature*, *440*, 352–357.

Li, Y., Liu, L., Kang, J., Sheng, J. G., Barger, S. W., Mrak, R. E. and Griffin, W. S. (2000). Neuronal-glial interactions mediated by interleukin-1 enhance neuronal acetylcholinesterase activity and mRNA expression. *J Neurosci*, *20*, 149–155.

Li, M., Pisalyaput, K., Galvan, M. and Tenner, A. J. (2004). Macrophage colony stimulating factor and interferon-gamma trigger distinct mechanisms for augmentation of beta-amyloid-induced microglia-mediated neurotoxicity. *J Neurochem*, *91*, 623–633.

Lindberg, C., Selenica, M. L., Westlind-Danielsson, A. and Schultzberg, M. (2005). Beta-amyloid protein structure determines the nature of cytokine release from rat microglia. *J Mol Neurosci*, *27*, 1–12.

Lorenzo, A. and Yankner, B. A. (1996). Amyloid fibril toxicity in Alzheimer's disease and diabetes. *Ann N Y Acad Sci*, *777*, 89–95.

Lorenzo, A., Yuan, M., Zhang, Z., Paganetti, P. A., Sturchler-Pierrat, C., Staufenbiel, M., Mautino, J., Vigo, F. S., Sommer, B. and Yankner, B. A. (2000). Amyloid beta interacts with the amyloid precursor protein: a potential toxic mechanism in Alzheimer's disease. *Nat Neurosci*, *3*, 460–464.

Lu, D. C., Shaked, G. M., Masliah, E., Bredesen, D. E. and Koo, E. H. (2003). Amyloid beta protein toxicity mediated by the formation of amyloid-beta protein precursor complexes. *Ann Neurol*, *54*, 781–789.

Lue, L. F., Walker, D. G. and Rogers, J. (2001). Modeling microglial activation in Alzheimer's disease with human postmortem microglial cultures. *Neurobiol Aging*, *22*, 945–956.

Lue, L. F., Kuo, Y. M., Roher, A. E., Brachova, L., Shen, Y., Sue, L., Beach, T., Kurth, J. H., Rydel, R. E. and Rogers, J. (1999). Soluble amyloid beta peptide concentration as a predictor of synaptic change in Alzheimer's disease. *Am J Pathol*, *155*, 853–862.

Luterman, J. D., Haroutunian, V., Yemul, S., Ho, L., Purohit, D., Aisen, P. S., Mohs, R. and Pasinetti, G. M. (2000). Cytokine gene expression as a function of the clinical progression of Alzheimer disease dementia. *Arch Neurol*, *57*, 1153–1160.

Manelli, A. M., Bulfinch, L. C., Sullivan, P. M. and Ladu, M. J. (2006). Abeta42 neurotoxicity in primary co-cultures: effect of apoE isoform and Abeta conformation. *Neurobiol Aging*, 1558–1497.

Martin, L. J., Pardo, C. A., Cork, L. C. and Price, D. L. (1994). Synaptic pathology and glial responses to neuronal injury precede the formation of senile plaques and amyloid deposits in the aging cerebral cortex. *Am J Pathol*, *145*, 1358–1381.

Masters, C. L., Multhaup, G., Simms, G., Pottgiesser, J., Martins, R. N. and Beyreuther, K. (1985). Neuronal origin of a cerebral amyloid: neurofibrillary tangles of Alzheimer's disease contain the same protein as the amyloid of plaque cores and blood vessels. *EMBO J*, *4*, 2757–2763.

Matsuda, S., Yasukawa, T., Homma, Y., Ito, Y., Niikura, T., Hiraki, T., Hirai, S., Ohno, S., Kita, Y., Kawasumi, M., Kouyama, K., Yamamoto, T., Kyriakis, J. M. and Nishimoto, I. (2001). c-Jun N-terminal kinase (JNK)-interacting protein-1b/islet-brain-1 scaffolds Alzheimer's amyloid precursor protein with JNK. *J Neurosci*, *21*, 6597–6607.

Mattiace, L. A., Davies, P., Yen, S. H. and Dickson, D. W. (1990). Microglia in cerebellar plaques in Alzheimer's disease. *Acta Neuropathol (Berl)*, *80*, 493–498.

McDonald, D. R., Brunden, K. R. and Landreth, G. E. (1997). Amyloid fibrils activate tyrosine kinase-dependent signaling and superoxide production in microglia. *J Neurosci*, *17*, 2284–2294.

McDonald, D. R., Bamberger, M. E., Combs, C. K. and Landreth, G. E. (1998). Beta-amyloid fibrils activate parallel mitogen-activated protein kinase pathways in microglia and THP1 monocytes. *J Neurosci*, *18*, 4451–4460.

McGeer, P. L., Kamo, H., Harrop, R., McGeer, E. G., Martin, W. R., Pate, B. D. and Li, D. K. (1986). Comparison of PET, MRI, and CT with pathology in a proven case of Alzheimer's disease. *Neurology*, *36*, 1569–1574.

McLean, C. A., Cherny, R. A., Fraser, F. W., Fuller, S. J., Smith, M. J., Beyreuther, K., Bush, A. I. and Masters, C. L. (1999). Soluble pool of Abeta amyloid as a determinant of severity of neurodegeneration in Alzheimer's disease. *Ann Neurol*, *46*, 860–866.

Miyazono, M., Iwaki, T., Kitamoto, T., Kaneko, Y., Doh-ura, K. and Tateishi, J. (1991). A comparative immunohistochemical study of Kuru and senile plaques with a special reference to glial reactions at various stages of amyloid plaque formation. *Am J Pathol*, *139*, 589–598.

Monning, U., Sandbrink, R., Weidemann, A., Banati, R. B., Masters, C. L. and Beyreuther, K. (1995). Extracellular matrix influences the biogenesis of amyloid precursor protein in microglial cells. *J Biol Chem*, *270*, 7104–7110.

Monsonego, A., Imitola, J., Zota, V., Oida, T. and Weiner, H. L. (2003). Microglia-mediated nitric oxide cytotoxicity of T cells following amyloid beta-peptide presentation to Th1 cells. *J Immunol*, *171*, 2216–2224.

Mook-Jung, I. and Saitoh, T. (1997). Amyloid precursor protein activates phosphotyrosine signaling pathway. *Neurosci Lett*, *235*, 1–4.

Moore, K. J., El Khoury, J., Medeiros, L. A., Terada, K., Geula, C., Luster, A. D. and Freeman, M. W. (2002). A CD36-initiated signaling cascade mediates inflammatory effects of beta-amyloid. *J Biol Chem*, *277*, 47373–47379.

Morgan, D., Gordon, M. N., Tan, J., Wilcock, D. and Rojiani, A. M. (2005). Dynamic complexity of the microglial activation response in transgenic models of amyloid deposition: implications for Alzheimer therapeutics. *J Neuropathol Exp Neurol*, *64*, 743–753.

Morris, J. C., Storandt, M., McKeel, D. W., Jr., Rubin, E. H., Price, J. L., Grant, E. A. and Berg, L. (1996). Cerebral amyloid deposition and diffuse plaques in "normal" aging: evidence for presymptomatic and very mild Alzheimer's disease. *Neurology*, *46*, 707–719.

Mrak, R. E. and Griffin, W. S. (2000). Interleukin-1 and the immunogenetics of Alzheimer disease. *J Neuropathol Exp Neurol*, *59*, 471–476.

Mucke, L., Masliah, E., Yu, G. Q., Mallory, M., Rockenstein, E. M., Tatsuno, G., Hu, K., Kholodenko, D., Johnson-Wood, K. and McConlogue, L. (2000). High-level neuronal expression of Abeta 1–42 in wild-type human amyloid protein precursor transgenic mice: synaptotoxicity without plaque formation. *J Neurosci*, *20*, 4050–4058.

Mullan, M., Crawford, F., Axelman, K., Houlden, H., Lilius, L., Winblad, B. and Lannfelt, L. (1992). A pathogenic mutation for probable Alzheimer's disease in the APP gene at the N-terminus of beta-amyloid. *Nat Genet*, *1*, 345–347.

Murrell, J., Farlow, M., Ghetti, B. and Benson, M. D. (1991). A mutation in the amyloid precursor protein associated with hereditary Alzheimer's disease. *Science*, *254*, 97–99.

Noda, M., Nakanishi, H. and Akaike, N. (1999). Glutamate release from microglia via glutamate transporter is enhanced by amyloid-beta peptide. *Neuroscience*, *92*, 1465–1474.

O'Banion, M. K., Chang, J. W., Kaplan, M. D., Yermakova, A. and Coleman, P. D. (1997). Glial and neuronal expression of cyclooxygenase-2: relevance to Alzheimer's disease. *Adv Exp Med Biol*, *433*, 177–180.

Okamoto, T., Takeda, S., Murayama, Y., Ogata, E. and Nishimoto, I. (1995). Ligand-dependent G protein coupling function of amyloid transmembrane precursor. *J Biol Chem*, *270*, 4205–4208.

Pike, C. J., Burdick, D., Walencewicz, A. J., Glabe, C. G. and Cotman, C. W. (1993). Neurodegeneration induced by beta-amyloid peptides in vitro: the role of peptide assembly state. *J Neurosci*, *13*, 1676–1687.

Podlisny, M. B., Ostaszewski, B. L., Squazzo, S. L., Koo, E. H., Rydell, R. E., Teplow, D. B. and Selkoe, D. J. (1995). Aggregation of secreted amyloid beta-protein into sodium dodecyl sulfate-stable oligomers in cell culture. *J Biol Chem*, *270*, 9564–9570.

Pooler, A. M., Arjona, A. A., Lee, R. K. and Wurtman, R. J. (2004). Prostaglandin E2 regulates amyloid precursor protein expression via the EP2 receptor in cultured rat microglia. *Neurosci Lett*, *362*, 127–130.

Roher, A. E., Chaney, M. O., Kuo, Y. M., Webster, S. D., Stine, W. B., Haverkamp, L. J., Woods, A. S., Cotter, R. J., Tuohy, J. M., Krafft, G. A., Bonnell, B. S. and Emmerling, M. R. (1996). Morphology and toxicity of Abeta-(1–42) dimer derived from neuritic and vascular amyloid deposits of Alzheimer's disease. *J Biol Chem*, *271*, 20631–20635.

Rossjohn, J., Cappai, R., Feil, S. C., Henry, A., McKinstry, W. J., Galatis, D., Hesse, L., Multhaup, G., Beyreuther, K., Masters, C. L. and Parker, M. W. (1999). Crystal structure of the N-terminal, growth factor-like domain of Alzheimer amyloid precursor protein. *Nat Struct Biol*, *6*, 327–331.

Sabo, S. L., Ikin, A. F., Buxbaum, J. D. and Greengard, P. (2001). The Alzheimer amyloid precursor protein (APP) and FE65, an APP-binding protein, regulate cell movement. *J Cell Biol*, *153*, 1403–1414.

Sasaki, A., H. Yamaguchi, A. Ogawa, S. Sugihara and Y. Nakazato (1997). Microglial activation in early stages of amyloid beta protein deposition. *Acta Neuropathol (Berl)*, *94*, 316–322.

Scheuermann, S., Hambsch, B., Hesse, L., Stumm, J., Schmidt, C., Beher, D., Bayer, T. A., Beyreuther, K. and Multhaup, G. (2001). Homodimerization of amyloid precursor protein and its implication in the amyloidogenic pathway of Alzheimer's disease. *J Biol Chem*, *276*, 33923–33929.

Scheuner, D., Eckman, C., Jensen, M., Song, X., Citron, M., Suzuki, N., Bird, T. D., Hardy, J., Hutton, M., Kukull, W., Larson, E., Levy-Lahad, E., Viitanen, M., Peskind, E., Poorkaj, P., Schellenberg, G., Tanzi, R., Wasco, W., Lannfelt, L., Selkoe, D. and Younkin, S. (1996). Secreted amyloid beta-protein similar to that in the senile plaques of Alzheimer's disease is increased in vivo by the presenilin 1 and 2 and APP mutations linked to familial Alzheimer's disease. *Nat Med*, *2*, 864–870.

Schmechel, D. E., Goldgaber, D., Burkhart, D. S., Gilbert, J. R., Gajdusek, D. C. and Roses, A. D. (1988). Cellular localization of messenger RNA encoding amyloid-beta-protein in normal tissue and in Alzheimer disease. *Alzheimer Dis Assoc Disord*, *2*, 96–111.

Schulz, J. G., Megow, D., Reszka, R., Villringer, A., Einhaupl, K. M. and Dirnagl, U. (1998). Evidence that glypican is a receptor mediating beta-amyloid neurotoxicity in PC12 cells. *Eur J Neurosci*, *10*, 2085–2093.

Scott, S. A., Johnson, S. A., Zarow, C. and Perlmutter, L. S. (1993). Inability to detect beta-amyloid protein precursor mRNA in Alzheimer plaque-associated microglia. *Exp Neurol*, *121*, 113–118.

Selkoe, D. J. (2001). Alzheimer's disease: genes, proteins, and therapy. *Physiol Rev*, *81*, 741–766.

Selkoe, D. J. (2002). Alzheimer's disease is a synaptic failure. *Science*, *298*, 789–791.

Selkoe, D. J. (2005). Defining molecular targets to prevent Alzheimer disease. *Arch Neurol*, *62*, 192–195.

Shaked, G. M., Kummer, M. P., Lu, D. C., Galvan, V., Bredesen, D. E. and Koo, E. H. (2006). Abeta induces cell death by direct interaction with its cognate extracellular domain on APP (APP 597–624). *FASEB J*, *20*, 1254–1256.

Sheng, J. G., Mrak, R. E. and Griffin, W. S. (1995). Microglial interleukin-1 alpha expression in brain regions in Alzheimer's disease: correlation with neuritic plaque distribution. *Neuropathol Appl Neurobiol*, *21*, 290–301.

Sheng, J. G., Mrak, R. E. and Griffin, W. S. (1997). Neuritic plaque evolution in Alzheimer's disease is accompanied by transition of activated microglia from primed to enlarged to phagocytic forms. *Acta Neuropathol (Berl)*, *94*, 1–5.

Sheng, J. G., Griffin, W. S., Royston, M. C. and Mrak, R. E. (1998). Distribution of interleukin-1-immunoreactive microglia in cerebral cortical layers: implications for neuritic plaque formation in Alzheimer's disease. *Neuropathol Appl Neurobiol*, *24*, 278–283.

Sinha, S. and Lieberburg, I. (1999). Cellular mechanisms of beta-amyloid production and secretion. *Proc Natl Acad Sci U S A*, *96*, 11049–11053.

Sinha, S., Anderson, J. P., Barbour, R., Basi, G. S., Caccavello, R., Davis, D., Doan, M., Dovey, H. F., Frigon, N., Hong, J., Jacobson-Croak, K., Jewett, N., Keim, P., Knops, J., Lieberburg, I., Power, M., Tan, H., Tatsuno, G., Tung, J., Schenk, D., Seubert, P., Suomensaari, S. M., Wang, S., Walker, D., Zhao, J., McConlogue, L. and John, V. (1999). Purification and cloning of amyloid precursor protein beta-secretase from human brain. *Nature*, *402*, 537–540.

Sondag, C. M. and Combs, C. K. (2004). Amyloid precursor protein mediates proinflammatory activation of monocytic lineage cells. *J Biol Chem*, *279*, 14456–14463.

Sondag, C. M. and Combs, C. K. (2006). Amyloid precursor protein cross-linking stimulates beta amyloid production and pro-inflammatory cytokine release in monocytic lineage cells. *J Neurochem*, *97*, 449–461.

Strauss, S., Bauer, J., Ganter, U., Jonas, U., Berger, M. and Volk, B. (1992). Detection of interleukin-6 and alpha 2-macroglobulin immunoreactivity in cortex and hippocampus of Alzheimer's disease patients. *Lab Invest*, *66*, 223–230.

Suzuki, N., Cheung, T. T., Cai, X. D., Odaka, A., Otvos, L., Jr., Eckman, C., Golde, T. E. and Younkin, S. G. (1994). An increased percentage of long amyloid beta protein secreted by familial amyloid beta protein precursor (beta APP717) mutants. *Science*, *264*, 1336–1340.

Takata, K., Kitamura, Y., Umeki, M., Tsuchiya, D., Kakimura, J., Taniguchi, T., Gebicke-Haerter, P. J. and Shimohama, S. (2003). Possible involvement of small oligomers of amyloid-beta peptides in 15-deoxy-delta 12,14 prostaglandin J2-sensitive microglial activation. *J Pharmacol Sci*, *91*, 330–333.

Tan, J., Town, T. and Mullan, M. (2000). CD45 inhibits CD40L-induced microglial activation via negative regulation of the Src/p44/42 MAPK pathway. *J Biol Chem*, *275*, 37224–37231.

Tarr, P. E., Roncarati, R., Pelicci, G., Pelicci, P. G. and D'Adamio, L. (2002). Tyrosine phosphorylation of the beta-amyloid precursor protein cytoplasmic tail promotes interaction with Shc. *J Biol Chem*, *277*, 16798–16804.

Twig, G., Graf, S. A., Messerli, M. A., Smith, P. J., Yoo, S. H. and Shirihai, O. S. (2005). Synergistic amplification of beta-amyloid- and interferon-gamma-induced microglial neurotoxic response by the senile plaque component chromogranin A. *Am J Physiol Cell Physiol*, *288*, C169–C175.

Van Nostrand, W. E., Melchor, J. P., Keane, D. M., Saporito-Irwin, S. M., Romanov, G., Davis, J. and Xu, F. (2002). Localization of a fibrillar amyloid beta-protein binding domain on its precursor. *J Biol Chem*, *277*, 36392–36398.

Vassar, R., Bennett, B. D., Babu-Khan, S., Kahn, S., Mendiaz, E. A., Denis, P., Teplow, D. B., Ross, S., Amarante, P., Loeloff, R., Luo, Y., Fisher, S., Fuller, J., Edenson, S., Lile, J., Jarosinski, M. A., Biere, A. L., Curran, E., Burgess, T., Louis, J. C., Collins, F., Treanor, J., Rogers, G. and Citron, M. (1999). Beta-secretase cleavage of Alzheimer's amyloid precursor protein by the transmembrane aspartic protease BACE. *Science*, *286*, 735–741.

Wagner, M. R., Keane, D. M., Melchor, J. P., Auspaker, K. R. and Van Nostrand, W. E. (2000). Fibrillar amyloid beta-protein binds protease nexin-2/amyloid beta-protein precursor: stimulation of its inhibition of coagulation factor XIa. *Biochemistry*, *39*, 7420–7427.

Walker, D. G., Lue, L. F. and Beach, T. G. (2001). Gene expression profiling of amyloid beta peptide-stimulated human post-mortem brain microglia. *Neurobiol Aging*, *22*, 957–966.

Walker, D. G., Link, J., Lue, L. F., Dalsing-Hernandez, J. E. and Boyes, B. E. (2006). Gene expression changes by amyloid beta peptide-stimulated human postmortem brain microglia identify activation of multiple inflammatory processes. *J Leukoc Biol*, *79*, 596–610.

Walsh, D. M., Klyubin, I., Fadeeva, J. V., Cullen, W. K., Anwyl, R., Wolfe, M. S., Rowan, M. J. and Selkoe, D. J. (2002). Naturally secreted oligomers of amyloid beta protein potently inhibit hippocampal long-term potentiation in vivo. *Nature*, *416*, 535–539.

Walsh, D. M., Townsend, M., Podlisny, M. B., Shankar, G. M., Fadeeva, J. V., El Agnaf, O., Hartley, D. M. and Selkoe, D. J. (2005). Certain inhibitors of synthetic amyloid beta-peptide (Abeta) fibrillogenesis block oligomerization of natural Abeta and thereby rescue long-term potentiation. *J Neurosci*, *25*, 2455–2462.

Wang, Y. and Ha, Y. (2004). The X-ray structure of an antiparallel dimer of the human amyloid precursor protein E2 domain. *Mol Cell*, *15*, 343–353.

Wang, H. Y., Lee, D. H., Davis, C. B. and Shank, R. P. (2000). Amyloid peptide Abeta(1–42) binds selectively and with picomolar affinity to alpha7 nicotinic acetylcholine receptors. *J Neurochem*, *75*, 1155–1161.

Wang, H. W., Pasternak, J. F., Kuo, H., Ristic, H., Lambert, M. P., Chromy, B., Viola, K. L., Klein, W. L., Stine, W. B., Krafft, G. A. and Trommer, B. L. (2002). Soluble oligomers of beta amyloid (1–42) inhibit long-term potentiation but not long-term depression in rat dentate gyrus. *Brain Res*, *924*, 133–140.

Weidemann, A., Konig, G., Bunke, D., Fischer, P., Salbaum, J. M., Masters, C. L. and Beyreuther, K. (1989). Identification, biogenesis, and localization of precursors of Alzheimer's disease A4 amyloid protein. *Cell*, *57*, 115–126.

Westerman, M. A., Cooper-Blacketer, D., Mariash, A., Kotilinek, L., Kawarabayashi, T., Younkin, L. H., Carlson, G. A., Younkin, S. G. and Ashe, K. H. (2002). The relationship between Abeta and memory in the Tg2576 mouse model of Alzheimer's disease. *J Neurosci*, *22*, 1858–1867.

White, J. A., Manelli, A. M., Holmberg, K. H., Van Eldik, L. J. and Ladu, M. J. (2005). Differential effects of oligomeric and fibrillar amyloid-beta 1–42 on astrocyte-mediated inflammation. *Neurobiol Dis*, *18*, 459–465.

Wilkinson, B., Koenigsknecht-Talboo, J., Grommes, C., Lee, C. Y. and Landreth, G. (2006). Fibrillar beta-amyloid-stimulated intracellular signaling cascades require Vav for induction of respiratory burst and phagocytosis in monocytes and microglia. *J Biol Chem*, *281*, 20842–20850.

Williamson, T. G., Nurcombe, V., Beyreuther, K., Masters, C. L. and Small, D. H. (1995). Affinity purification of proteoglycans that bind to the amyloid protein precursor of Alzheimer's disease. *J Neurochem*, *65*, 2201–2208.

Wu, S. Z., Bodles, A. M., Porter, M. M., Griffin, W. S., Basile, A. S. and Barger, S. W. (2004). Induction of serine racemase expression and D-serine release from microglia by amyloid beta-peptide. *J Neuroinflammation*, *1*, 2.

Xia, W., Zhang, J., Kholodenko, D., Citron, M., Podlisny, M. B., Teplow, D. B., Haass, C., Seubert, P., Koo, E. H. and Selkoe, D. J. (1997). Enhanced production and oligomerization of the 42-residue amyloid beta-protein by Chinese hamster ovary cells stably expressing mutant presenilins. *J Biol Chem*, *272*, 7977–7982.

Xiang, Z., Haroutunian, V., Ho, L., Purohit, D. and Pasinetti, G. M. (2006). Microglia activation in the brain as inflammatory biomarker of Alzheimer's disease neuropathology and clinical dementia. *Dis Markers*, *22*, 95–102.

Xie, L., Helmerhorst, E., Taddei, K., Plewright, B., Van Bronswijk, W. and Martins, R. (2002a). Alzheimer's beta-amyloid peptides compete for insulin binding to the insulin receptor. *J Neurosci*, *22*, RC221.

Xie, Z., M. Wei, T. E. Morgan, P. Fabrizio, D. Han, C. E. Finch and V. D. Longo (2002b). Peroxynitrite mediates neurotoxicity of amyloid beta-peptide1-42- and lipopolysaccharide-activated microglia. *J Neurosci*, *22*, 3484–3492.

Yan, S. D., Chen, X., Fu, J., Chen, M., Zhu, H., Roher, A., Slattery, T., Zhao, L., Nagashima, M., Morser, J., Migheli, A., Nawroth, P., Stern, D. and Schmidt, A. M. (1996). RAGE and amyloid-beta peptide neurotoxicity in Alzheimer's disease. *Nature*, *382*, 685–691.

Yates, S. L., Burgess, L. H., Kocsis-Angle, J., Antal, J. M., Dority, M. D., Embury, P. B., Piotrkowski, A. M. and Brunden, K. R. (2000). Amyloid beta and amylin fibrils induce increases in proinflammatory cytokine and chemokine production by THP-1 cells and murine microglia. *J Neurochem*, *74*, 1017–1025.

Yazawa, H., Yu, Z. X., Takeda, Le, Y., Gong, W., Ferrans, V. J., Oppenheim, J. J., Li, C. C. and Wang, J. M. (2001). Beta amyloid peptide (Abeta42) is internalized via the G-protein-coupled receptor FPRL1 and forms fibrillar aggregates in macrophages. *FASEB J*, *15*, 2454–2462.

3
Pericytes

Martin Krüger and Ingo Bechmann

1 Overview

Pericytes were first described by Rouget in 1873 and since then much has been speculated on their nature and role(s). They are located between the inner and outer vascular basement membrane of arterioles, capillaries, and venules, and are thus part of the vascular wall. In standard hematoxylin-eosin or Nissl-stained sections, pericytes are difficult to recognize, but they can be identified under the electron microscope and in semithin sections, where the vascular basement membranes are visible. Moreover several antibodies demark their unique morphology and therefore, allow clear-cut identification at the light microscopic level.

Much of the confusion regarding their origin, function, and phenotype derives from the fact that pericytes are often mixed up with adjacent populations of cells located within the vascular wall, the perivascular space, or the parenchyma. Moreover, they are difficult to study in vivo and therefore, their function is difficult to address. Here, we will briefly describe the topography of the vascular, perivascular, and juxtavascular compartment, and summarize current concepts of pericyte biology.

2 Definition by Topography and Morphology

In the brain, much of the confusion derives from the complex anatomy of the blood-brain interface, where three compartments must be distinguished:

- The vascular wall consisting of endothelial and smooth muscle cells
- The perivascular space harboring leptomeningeal cells, macrophages and other antigen-presenting cells
- The juxtavascular compartment which is delineated by the glia limitans consisting of endfeet of astrocytes and juxtavascular microglia

The best way to describe pericytes is already given by their name which emphasizes that they are wrapped around microvessels. Their long processes are oriented along the longitudinal axis of the blood vessel, while smaller radial arms appear to engirdle

T.E. Lane et al. (eds.), *Central Nervous System Diseases and Inflammation.*
© Springer 2008

the vascular wall (Zimmermann, 1923; Rhodin, 1968; Rucker et al., 2000). Ultrastructural analysis reveals that pericytes are completely covered by the outermost basement membrane of the vascular wall and in vitro data suggest that both that both endothelial cells and pericytes contribute to its synthesis (Cohen et al., 1980; Mandarino et al., 1993). In capillaries, the outer membrane fuses with the membrane of the glia limitans ("fused gliovascular membrane"), thereby occluding the perivascular (Virchow–Robin) space (Fig. 3.1). In histological sections of these

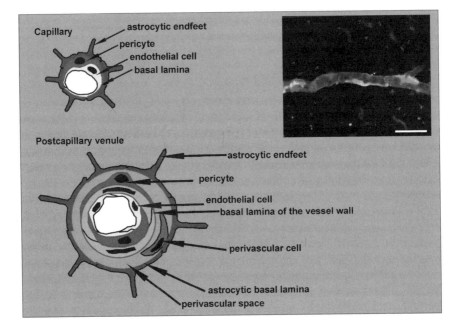

Fig. 3.1 Left side: **Topography of the blood-brain interface: Difference between capillaries and venule**
Upper panel: By definition, capillaries lack a "media" of smooth muscle cells. The pericyte (green) is thus separated from the endothelial cells (beige) by a basement membrane (red) only. The outer vascular basement membrane and the one on top of the glia limitans (blue) are fused to a common "gliovascular" membrane. Note the intimate contact of the astrocytic endfeet and the pericytes and endothelial cells which may underlie certain aspects of blood-brain barrier differentiation.
Lower panel: Less pronounced barrier function of the endothelium may be explained by the following differences between venules and capillaries:
Astrocytic endfeet cannot contact the pericytes and the endothelium due to the perivascular space and the smooth muscle cells of the "media" (orange); the latter also separates the pericytes from the endothelium.
Pericytes are often mixed up with perivascular cells (pink), a heterogeneous population of leptomeningeal cells, macrophages, and other leukocytes.
The brain parenchyma proper (neuropil) is delineated by the astrocytic endfeet forming the glia limitans (blue).
It is not clear if all pericytes are located at the outermost portion of the vascular wall (upper green cell). Some may engulf the endothelium or be part of the media (lower green cell).
Right side: Pericytes (green) oriented along the longitudinal axis of a capillary. The fused gliovascular membrane is labeled with an antibody to laminin (red). Scale bar: 20 μm. (*See also color plates*).

areas, it is often difficult to distinguish pericytes from adjacent perivascular cells or even juxtavascular microglia. Moreover, even if perivascular spaces are evident, cells located therein are often addressed as "pericytes." The underlying anatomical misconception and terminological inconsistency often renders it difficult to compare or integrate data from different studies (Bechmann et al., 2007).

By definition, capillaries in contrast to pre- and postcapillary vessels lack a "media" consisting of smooth muscle cells. Therefore, pericytes and endothelial cells of capillaries are only separated by a basement membrane with an intercellular distance of less than 20 nm (Sims, 1986). There is also evidence for direct contact zones provided by gap junctions allowing direct signaling and transport of ions and small molecules (Cuevas et al., 1984). In non-capillary vessels, pericytes and endothelial cells are separated by several layers of muscle cells excluding such direct interaction. This may underlie the different roles pericytes appear to have in the different parts of the vascular tree.

The density of pericytes differs between the tissues they are situated in. The brain exhibits the highest density of all tissues. The ratio between pericytes and endothelial cells is 1:3 in the brain, 1:5 in the retina, but only 1:100 in striated muscles. However, their density and morphology also differs along the vascular tree. Pericytes of pre-capillary and capillary sections have been reported to be more elongated and slender, while those of post-capillary venules are rather stubby with shorter processes. (Shepro and Morel, 1993). The varying density has been attributed to different hydrostatic pressure (Sims, 2000), but may also be a consequence of other features such as the distance to the endothelium and/or the glia limitans. Vice versa, the density of pericytes may impact on endothelial tightness and thus, influence its barrier function (Sims, 1986).

3   Phenotype and Origin

3.1   Detection

Unfortunately, the detection of pericytes at the light microscopic level is complex, while their clear-cut detection using electron microscopy is time-consuming and not applicable for quantitative analysis. Semithin sections provide an alternative, but are difficult to combine with immunocytochemistry.

Nevertheless, pericytes do express surface antigens for their identification, albeit none of them being specific [CD13, α-smooth muscle actin (SMA), desmin, vimentin, NG-2]. These molecules may be induced in neighboring cells by experimental manipulations. Moreover, there are evident differences between pericytes of capillaries and larger vessels in regard to their expression of contractile elements. If present, their use for identification is inflicted by the simultaneous expression in the media (see below).

An antigen suitable for the detection of pericytes in their environment was established by the group of Dermietzel who proved its specificity at the ultrastructural level and identified the pericytic ectoenzyme aminopeptidase N (pAPN) as the recognized antigen (Kunz et al., 1994). This aminopeptidase occurs at day E18 of brain angiogenesis (Dermietzel and Krause, 1991) and belongs to the family of matrix metalloproteinases which are involved in zinc dependent cleavage of extracellular matrix molecules as well as non matrix substrates such as growth factors or neuropeptides (Sato, 2004). Interestingly, regions devoid of a tight endothelium also lack pAPN expression (Kunz et al., 1994). This was further investigated during experimental autoimmune encephalomyelitis (EAE), an animal model of multiple sclerosis. During neuroinflammation, diminished pAPN expression correlated to increased permeability of the blood–brain barrier (BBB) (Kunz et al., 1995). Furthermore, in the process of inflammation invasive pAPN positive cells, which were not related to microvessels, could be seen in the white matter of the spinal cord. These infiltrating cells were also positive for ED-1, a marker of activated rat macrophages (Dijkstra, 1985), and OX-17, which is also strongly up-regulated on macrophages and microglia during EAE (Hickey et al., 1985). Unfortunately, it is still unclear whether these infiltrates resembled pericytes emigrated from the vascular wall or blood-borne macrophages.

3.2   *Members of the Monocytes/Macrophage Family?*

Pericytes were repeatedly reported to express markers of the monocyte/macrophage lineage such as ED1 or CD11b (Graeber et al., 1989; Balabanov et al., 1996). In vitro, Dore-Duffy and colleagues showed that they respond to IFN-γ stimulation with up-regulation of the major histocompatibility class II antigen (MHC-II) (Dore-Duffy and Balabanov, 1998). Their phagocytic potential has been explored using antibody-coated zymosan and fluorochrome-conjugated polystyrene beads (Balabanov et al., 1996). However, as clear-cut identification is lacking in these studies, it is possible that the reported observations relate to perivascular cells which are separated from pericytes by a basement membrane only (Fig. 3.1). In fact, perivascular cells have been shown to exhibit phagocytosis, to act as antigen-presenting cells, and to be supplemented by haematogenous cells (Hickey and Kimura, 1988; Hickey et al., 1992; Bechmann et al., 2001; Priller et al., 2001, Greter et al., 2005). The latter has also been established for pericytes (Kokovay et al., 2005, Ozerdem et al., 2005), but the respective precursor and thus the lineage pericytes belong to remain enigmatic. Thus, the question of whether pericytes comprise a specialized subtype or immature form of macrophages which can be recruited to the neuropil under certain conditions is open.

4   Putative Functions of Pericytes

4.1   *Regulation of Homeostasis and BBB Integrity*

Originally, astrocytes were thought to maintain and induce BBB properties of endothelial cells (Stewart and Wiley, 1981; Arthur et al., 1987). However, several groups showed that the development of endothelial tight junctions does not directly depend on astrocytes (Jaeger and Blight, 1997; Felts and Smith, 1996). In GFAP-deficient mice an increased pericyte coverage in the microvasculature was observed (Balabanov and Dore-Duffy, 1998) and interpreted as a possible counter-regulation for vascular leakage. Indeed, addition of pericytes to co-cultured monolayers of endothelial cells caused increased barrier function for hydrophilic molecules and enhanced the transendothelial resistance (Dente et al., 2001).

Transforming growth factor (TGF)-β is an established factor inducing vascular stability as well as vessel maturation. Early studies demonstrated a crucial role of pericytes for the activation of this cytokine (Sato and Rifkin, 1989; Antonelli-Orlidge et al., 1989). TGF-β and TGF-βR2 knockout mice developed defects of the vascular wall which have been attributed to the lack of inhibition of endothelial proliferation and migration (Oshima et al., 1996, Li et al., 1999).

The two angiopoietins Ang-1 and Ang-2 and their receptor, the tyrosine kinase Tie-2 also impact on vessel development. Due to complementary phenotypes in Ang-1 or Tie-2 deficient and Ang-2 over-expressing mice an antagonistic model concerning the effects of these ligands has been suggested (Sato et al., 1995; Suri et al., 1996; Maisonpierre et al., 1997). Here, Ang-1 acts as an agonist and Ang-2 as an antagonist of Tie-2 receptors. During development and in adult tissue the endothelium expresses Tie-2 whereas perivascular cells and pericytes are responsible for the production of Ang-1 (Suri et al., 1996; Davis et al., 1996). The binding of Ang-1 to the receptor Tie-2 on the one hand fostered vascular stabilization. This was done in platelet derived growth factor (PDGF)-β deficient mice that develop microaneurysms and endothelial leakage due to the absence of mural cells. Ang-1 readily restored the vascular structure and function (Uemura et al., 2002). The reduction of BBB leakage via Ang-1 has also been shown in ischemic models (Zhang et al., 2002). On the other hand and quite contrary to Ang-1, Ang-2 is merely a destabilizing factor which is restricted to endothelial cells in areas of vascular remodeling and binds to Tie-2 without inducing any signal transduction (Maisonpierre et al., 1997). Hypoxia also induces Ang-2 without changing the expression patterns of Ang-1 (Mandriota and Pepper, 1998). Therefore Ang-2 has a rather pro-angiogenic effect whereas Ang-1 and its source, the pericytes, act in a stabilizing way.

Pericyte recruitment to sites of angiogenesis is another requirement for proper vascular development. This recruitment is driven by PDGF-&betabdot; and the corresponding receptor PDGFR-β. PDGFR-β is expressed on pericytes, while its ligand is produced by sprouting endothelial cells (Hellstrom et al., 1999; Lindahl et al., 1997). Mice lacking PDGF-&betabdot; or its receptor develop identical phenotypes. The vessels appear to be depleted of pericyte causing neonatal lethality, microvascular leakage and hemorrhage (Soriano, 1994; Leveen et al., 1994). This is supported by in vitro co-culture studies demonstrating that proliferating endothelial cells produce PDGF-β which acts as a mitogen and chemoattractant for mural precursors (Hirshi et al., 1998, 1999). Thus, the developing vasculature remains unstable and immature until pericytes are recruited (Benjamin et al., 1998).

4.2 The Role in Angiogenesis and Neovascularization

A regulatory function of brain pericytes during angiogenesis and neovascularization is also evident in models of traumatic brain injury and brain hypoxia, a strong
simulus for angiogenesis (Dore-Duffy et al., 1999). Pericytes were shown to be the
first population to respond to brain hypoxia in cats. While endothelial cells and
astrocytes were not visibly changed, morphologic alterations of pericytes were evident as soon as 2 h after hypoxia. As an initial step of migration, the abluminal surface of pericytes developed characteristic "spikes" pointing towards the parenchyma.
Simultaneously, the basement membrane covering the endothelium at the luminal
side began to thicken (Gonul et al., 2002). Vascular endothelial growth factor
(VEGF) is produced by pericytes under hypoxic conditions and my drive the onset
of migration (Yamagishi et al., 1999).

Similar findings have been reported after traumatic brain injury at the ultrastructural level. Thickened luminal basement membranes and ruffled borders at the
abluminal side were observed, followed by elongation and the disappearance of the
basal lamina at the leading edge of migrating cells. Forty percent of the population
per section were found to be migrating (Dore-Duffy et al., 2000).

Migration normally involves the expression of urokinase plasminogen activator
(uPA) and its receptor (uPAR) on migrating cells (Washington et al., 1996; Blasi,
1999). These molecules are expressed at the mRNA level and the protein is found
on the leading tips of migratory pericytes. Pericytes remaining in position exhibited
signs of cytoplasmic and nuclear degeneration. Interestingly, pericytic processes
were reported to interdigitate with synaptic complexes and this observation has
been interpreted as a sign of synaptic stripping which may protect the neurons from
excess glutamate levels The dynamic of this event remains to be visualized; an
alternate explanation for this observation is that de-establishing of synapses is a
solely neuronal process providing space for the processes of immigrating pericytes
and other non-neuronal cells. However, the migration of pericytes was confirmed
by calculating the ratio of pericytes versus endothelial cells, which dropped from
1:5 to 1:10–12 (Dore-Duffy et al., 2000).

VEGF has also been used to investigate the interaction between pericytes and
the endothelium. In the chicken chorioallantoic membrane assay the effect of
VEGF on vessel formation has been studied. In this setting, VEGF induced CD31-
positive angiogenic sprouts to display α-SMA and desmin expressing cells of pericytic phenotype. As CD31 is an established marker for mature and embryonic
endothelial cells, this may indicate that endothelial cells and pericytes have a common precursor (Hagedorn et al., 2004). Maybe. All newly formed capillaries were
covered with pericytes and their number on the vessel could be increased with
VEGF treatment (Benjamin et al., 1998).

Finally, the role of pericytes in angiogenesis has also become evident by targeting
NG-2. This antigen (high molecular weight melanoma associated antigen
HMW-MAA, called nerve/glial antigen 2, NG 2) is expressed by immature pericytes.
Blockage by antibodies and depletion of the gene coding for NG-2 abrogated

vascular growth in various models of induced angiogenesis (Ruiter et al., 1993; Ozerdem and Stallcup, 2003, 2004).

4.3 Pericytes as Regulators of Microvascular Blood Flow?

Since 1873, when Rouget regarded pericytes as contractile elements (Rouget 1874), the discussion concerning this ability has never ceased. Pericytes are likely to respond to vasoactive molecules such as nitric oxide, prostacyclin, angiotensin II, and endothelin-1, because they express respective receptors (Dehouck et al., 1997, Chakravarthy and Gardiner, 1999, Healy and Wilk, 1993). However, a prerequisite of contraction is the expression of contractile filaments of actin and myosin. Cerebral pericytes express both smooth muscle and non-smooth muscle isoforms of actin and myosin (Herman and D'Amore, 1985). In embryonic chicken, but not in rat and mouse, all pericytes express α-SMA (Hellstrom et al., 1999). In mice, pericytes surrounding capillaries with a diameter of less than 10 μm do not show α-SMA activity, whereas pericytes of larger vessels like arterioles or post-capillary venules are regularly immune positive throughout the brain (Alliott et al., 1999; Nehls and Drenckhahn, 1991). This may indicate functional heterogeneity, which is also supported by the strikingly different morphologies described above. In fact, considering the established interaction between endothelial cells and pericytes, it seems trivial to state that the distance from the endothelium, which varies along the vascular tree, strongly impacts on the functional state of pericytes.

Blood flow is widely believed to be regulated in precapillary arterioles, but 65% of the noradrenergic innervation of CNS blood vessels terminate in vicinity to capillaries (Cohen et al., 1997). A recent study demonstrated that pericytes induce the constriction of capillaries induced by ATP and noradrenalin. Glutamate suppressed, while GABA fostered this process which could be observed through live-imaging of retinal and cerebellar slices obtained from young rats (Peppiatt et al., 2006).

5 Concluding Remarks

Studying pericytes is inflicted by the lack of a specific marker. This is apparent in studies addressing their immune function as a closer look often reveals that the cells were not unequivocally identified as pericytes. In vitro studies, besides the problem of an extremely altered environment also leave open the question whether pure culture can be established. Nevertheless, several lines of evidence point to a role in angiogenesis and control over endothelial growth. It seems also clear that they are capable of leaving the vascular wall to infiltrate the neuropil (Dore-Duffy et al., 2000). Moreover, it has been established that pericytes are constantly replaced by haematogenous cells (Kokovay et al., 2005), but the question of a common lineage of endothelial cells and pericytes on the one hand, and pericytes and perivascular cells and juxtavascular microglia on the other, remains open. The biggest success of the last years may be the formal demonstration of their impact on capillary contraction (Peppiatt et al., 2006). Thus, Rouget's old idea eventually turned out to be true, albeit in an unexpected segment of the vascular tree. As we learn to observe pericytes on duty, more surprises are to come.

References

Alliott F, Rutin J, Leenen PJM, Pessac B (1999). Pericytes and periendothelial cells of brain parenchyma vessels co-express aminopeptidase n, aminopeptidase a and nestin. *J Neurosci Res*, 58, 367–378.

Antonelli-Orlidge A, Saunders KB, Smith SR, D'Amore PA (1989). An activated form of transforming growth factor beta is produced by cocultures of endothelial cells and pericytes. *Proc Natl Acad Sci U S A*, 86, 4544–4548.

Arthur FE, Shivers RR, Bowman PD (1987). Astrocyte-mediated induction of tight junctions in brain capillary endothelium: an efficient in vitro model. *Brain Res*, 433, 155–159.

Balabanov R, Dore-Duffy P (1998). Role of the CNS microvascular pericyte in the blood–brain barrier. *J Neurosci Res*, 53, 637–44.

Balabanov R, Washington R, Wagnerova J, Dore-Duffy P (1996). CNS microvascular pericytes express macrophage-like function, cell surface integrin alpha M, and macrophage marker ED-2. *Microvasc Res*, 52, 127–42.

Bechmann I, Priller J, Kovac A, Bontert M, Wehner T, Klett FF, Bohsung J, Stuschke M, Dirnagl U, Nitsch R (2001). Immune surveillance of mouse brain perivascular spaces by blood-borne macrophages. *Eur J Neurosci*, 14, 1651–1658.

Bechmann I, Goldmann J, Kovac A, Kwidzinski E, Simburger E, Naftolin F, Dirnagl U, Nitsch R, Priller J (2005). Circulating monocytic cells infiltrate layers of anterograde axonal degeneration where they transform into microglia. *FASEB J*, 19(6), 647–649.

Bechmann I, Galea I, Perry VH (2007) What is the blood–brain barrier (not)? *Trends Immunol*, 28(1), 5–11.

Benjamin LE, Hemo I, Keshet E (1998). A plasticity window for blood vessel remodeling is defined by pericyte coverage of the preformed endothelial network and is regulated by PDGF-B and VEGF. *Development*, 125, 1591–1598.

Blasi F (1999). UPA, uPA, PAI-1: key intersection of proteolytic adhesive and chemotactic highways? *Immunol Today*, 18, 415–417.

Chakravarthy U, Gardiner TA (1999). Endothelium-derived agents in pericyte function/dysfunction. *Prog Retin Eye Res*, 18, 511–527.

Cohen MP, Frank RN, Khalifa AA (1980). Collagen production by cultured retinal capillary pericytes. *Invest Ophthalmol Vis Sci*, 19, 90–94.

Cohen Z, Molinatti G, Hamel E (1997), Astroglial and vascular interactions of noradrenaline terminals in the rat cerebral cortex. *J Cereb Blood Flow Metab*, 17, 894–904.

Cuevas P, Gutierrez-Diaz JA, Reimers D, Dujovny M, Diaz FG, Ausman JI (1984). Pericyte endothelial gap junctions in human cerebral capillaries. *Anat Embryol (Berl)*, 170, 155.

Davis S, Aldrich TH, Jones PF, Acheson A, Compton DL, Jain V, Ryan TE, Bruno J, Radziejewski C, Maisonpierre PC, Yancopoulos GD (1996). Isolation of angiopoietin-1, a ligand for the TIE2 receptor, by secretion-trap expression cloning. *Cell*, 87, 1161–1169.

Dehouck MP, Vigne P, Torpier G, Breittmayer JP, Cecchelli R, Frelin C (1997). Endothelin-1 as a mediator of endothelial cell-pericyte interactions in bovine brain capillaries. *J Cereb Blood Flow Metab*, 17, 464–469.

Dente CJ, Steffes CP, Speyer C, Tyburski JG (2001). Pericytes augment the capillary barrier in in vitro cocultures. *J Surg Res*, 97, 85–91.

Dermietzel R, Krause D (1991). Molecular anatomy of the blood–brain barrier as defined by immunocytochemistry. *Int Rev Cytol*, 127, 57–109.

Dijkstra CD, Dopp EA, Joling P, Kraal G (1985). The heterogeneity of mononuclear phagocytes in lymphoid organs: distinct macrophage subpopulations in the rat recognized by monoclonal antibodies ED1, ED2 and ED3. *Immunology*, 54(3), 589–99.

Dore-Duffy P, Balabanov R (1998). The role of CNS microvascular pericyte in leukocyte polarization of cytokine-secreting phenotype. *J Neurochem*, 70, 72.

Dore-Duffy P, Balabanov R, Beaumont T, Hritz MA, Harik SI, LaManna JC (1999). Endothelial activation following hypobaric hypoxia. *Microvasc Res*, 57, 75–85.

Dore-Duffy P, Owen C, Balabanov R, Murphy S, Beaumont T, Rafols JA (2000). Pericyte migration from the vascular wall in response to traumatic brain injury. *Microvasc Res*, 60, 55–69.

Felts PA, Smith KJ (1996). Blood–brain barrier permeability in astrocyte-free regions of the central nervous system remyelinated by Schwann cells. *Neuroscience*, 75, 643–655.

Gonul E, Duz B, Kahraman S, Kayali H, Kubar A, Timurkaynak E (2002). Early response to brain hypoxia in cats: an ultrastructural study. *Microvasc Res*, 64, 116–119.

Graeber MB, Streit WJ, Kreutzberg GW (1989). Identity of ED-2 positive perivascular cells in rat brain. *J Neurosci Res*, 22, 103.

Greter M, Heppner FL, Lemos MP, Odermatt BM, Goebels N, Laufer T, Noelle RJ, Becher B. (2005). Dendritic cells permit immune invasion of the CNS in an animal model of multiple sclerosis. *Nat Med*, 11(3), 328–334.

Hagedorn M, Balke M, Schmidt A, Bloch W, Kurz H, Javerzat S, Rousseau B, Wilting J, Bikfalvi A (2004). VEGF coordinates interaction of pericytes and endothelial cells during vasculogenesis and experimental angiogenesis. *Dev Dyn*, 230, 23–33.

Healy DP, Wilk S (1993). Localization of immunoreactive glutamyl aminopeptidase in rat brain. II. Distribution and correlation with angiotensin II. *Brain Res*, 606, 295–303.

Hellstrom M, Kalen M, Lindahl P, Abramsson A, Betsholtz C (1999). Role of PDGF-B and PDGFR-beta in recruitment of vascular smooth muscle cells and pericytes during embryonic blood vessel formation in the mouse. *Development*, 126, 3047–3055.

Herman IM, D'Amore PA (1985). Microvascular pericytes contain muscle and nonmuscle actins. *J Cell Biol*, 101, 43–52.

Hickey WF, Kimura H (1988). Perivascular microglial cells of the CNS are bone marrow-derived and present antigen in vivo. *Science*, 239, 290–292.

Hickey WF, Osborne JP, Kirby WM (1985). Expression of Ia antigen molecules by astrocytes during acute experimental allergic encephalomyelitis in the Lewis rat. *Cell Immunol*, 91, 528–535.

Hickey WF, Vass K, Lassmann H (1992). Bone marrow-derived elements in the central nervous system: an immunohistochemical and ultrastructural survey of rat chimeras. *J Neuropathol Exp Neurol*, 51, 246–256.

Hirshi KK, Rohovsky SA, D'Amore PA (1998). PDGF, TGF-beta, and heterotypic cell-cell interactions mediate endothelial cell-induced recruitment of 10T1/2 cells and their differentiation to a smooth muscle fate. *J Cell Biol*, 141, 805–814.

Hirshi KK, Rohovsky SA, Beck LH, Smith SR, D'Amore PA (1999). Endothelial cells modulate the proliferation of mural cell precursors via platelet-derived growth factor-BB and heterotypic cell contact. *Circ Res*, 84, 298–305.

Jaeger CB, Blight AR (1997). Spinal cord compression injury in guinea pigs: structural changes of endothelium and its perivascular cell associations after blood–brain barrier breakdown and repair. *Exp Neurol*, 144, 381–399.

Kokovay E, Li L, Cunningham LA (2005). Angiogenic recruitment of pericytes from bone marrow after stroke. *J Cereb Blood Flow Metabol*, 26, 545–555.

Kunz J, Krause D, Kremer M, Dermietzel R (1994). The 140-kDa protein of blood–brain barrier-associated pericytes is identical to aminopeptidase N. *J Neurochem*, 62, 2375–2386.

Kunz J, Krause D, Gehrmann J, Dermietzel R (1995). Changes in the expression pattern of blood–brain barrier-associated pericytic aminopeptidase N (pAP N) in the course of acute experimental autoimmune encephalomyelitis. *J Immunol*, 59, 41–55.

Leveen P, Pekny M, Gebre-Medhin S, Swolin B, Larsson E, Betsholtz C (1994). Mice deficient for PDGF B show renal, cardiovascular, and hematological abnormalities. *Genes Dev*, 8, 1875–1887.

Li DY, Sorensen LK, Brooke BS, Urness LD, Davis EC, Taylor DG, Boak BB, Wendel DP (1999). Defective angiogenesis in mice lacking endoglin. *Science*, 284, 1534–1537.

Lindahl P, Johansson BR, Leveen P, Betsholtz C (1997). Pericyte loss and microaneurysm formation in PDGF-B-deficient mice. *Science*, 277, 242–245.

Mandarino LJ, Sundarraj N, Finlayson J, Hassell HR (1993). Regulation of fibronectin and laminin synthesis by retinal capillary endothelial cells and pericytes in vitro. *Exp Eye Res*, 57, 609–621.

Mandriota SJ, Pepper MS (1998). Regulation of angiopoietin-2 mRNA levels in bovine microvascular endothelial cells by cytokines and hypoxia. *Circ Res*, 83, 852–859.

Maisonpierre PC, Suri C, Jones PF, Bartunkova S, Wiegand SJ, Radziejewski C, Compton D, McClain J, Aldrich TH, Papadopoulos N (1997). Angiopoietin-2, a natural antagonist for Tie2 that disrupts in vivo angiogenesis. *Science*, 277, 55–60.

Nehls V, Drenckhahn D (1991). Heterogeneity of microvascular pericytes for smooth muscle type alpha-actin. *J Cell Biol*, 113, 147–154.

Oshima M, Oshima H, Taketo MM (1996). TGF-beta receptor type II deficiency results in defects of yolk sac hematopoiesis and vasculogenesis. *Dev Biol*, 179(1), 297–302.

Ozerdem U, Stallcup WB (2003). Early contribution of pericytes to angiogenic sprouting and tube formation. *Angiogenesis*, 6, 241–249.

Ozerdem U, Stallcup WB (2004). Pathological angiogenesis is reduced by targeting pericytes via the NG2 proteoglycan. *Angiogenesis*, 7, 269–276.

Ozerdem U, Alitalo K, Salven P, Li A (2005). Contribution of bone marrow-derived pericyte precursor cells to corneal vasculogenesis. *Invest Ophthalmol Vis Sci*, 46(10), 3502–3506.

Peppiatt CM, Howarth C, Mobbs P, Attwell D (2006), Bidirectional control of CNS capillary diameter by pericytes. *Nature*, 443, 700–704.

Priller J, Flugel A, Wehner T, Boentert M, Haas CA, Prinz M, Fernandez-Klett F, Prass K, Bechmann I, de Boer BA, Frotscher M, Kreutzberg GW, Persons DA, Dirnagl U (2001), Targeting gene-modified hematopoietic cells to the central nervous system: use of green fluorescent protein uncovers microglial engraftment. *Nat Med*, 7, 1356–1361.

Rhodin JAG (1968). Ultrastructure of mammalian venous capillaries, venules and small collecting venules. *J Ultrastruct Res*, 25, 452–500.

Rouget C (1874). Note sur le développement de la tunique contractile des vaisseaux. Compt. *rend. acad. sci.* Paris, 59, 559–562.

Rucker HK, Wynder HJ, Thomas WE (2000). Cellular mechanisms of CNS pericytes. *Brain Res Bull*, 51, 363–369.

Ruiter DJ, Schlingemann RO, Westphal JR, Denijn M, Rietveld FJ, De Waal RM (1993). Angiogenesis in wound healing and tumor metastasis. *Behring Inst Mitt*, 1993, 258–272.

Sato Y (2004). Role of aminopeptidases in angiogenesis. *Biol Pharm Bull*, 27(6), 772–776.

Sato Y, Rifkin DB (1989). Inhibition of endothelial cell movement by pericytes and smooth muscle cells: activation of a latent transforming growth factor-beta 1-like molecule by plasmin during co-culture. *J Cell Biol*, 109, 309–315.

Sato TN, Tozawa Y, Deutsch U, Wolburg-Buchholz K, Fujiwara Y, Gendron-Maguire M, Gridley T, Wolburg H, Risau W, Qin Y (1995). Distinct roles of the receptor tyrosine kinases Tie-1 and Tie-2 in blood vessel formation. *Nature*, 376, 70–74.

Shepro D, Morel NM (1993). Pericyte physiology. *FASEB J*, 7, 1031–1038.

Sims DE (1986). The pericyte – a review. *Tissue Cell*, 18, 153–174.

Sims DE (2000). Diversity within pericytes. *Clin Exp Pharmacol Physiol*, 27, 842–846.

Soriano P (1994). Abnormal kidney development and hematological disorders in PDGF beta-receptor mutant mice. *Genes Dev*, 8, 1888–1896.

Stewart PA, Wiley MJ (1981). Developing nervous tissue induces formation of blood–brain barrier characteristics in invading endothelial cells: a study using quail – chick transplantation chimeras. *Dev Biol*, 84, 183–192.

Suri C, Jones PF, Patan S, Bartunkova S, Maisonpierre PC, Davis S, Sato TN, Yancopoulos GD (1996). Requisite role of angiopoietin-1, a ligand for the TIE2 receptor, during embryonic angiogenesis. *Cell*, 87, 1171–1180.

Uemura A, Ogawa M, Hirashima M, Fujiwara T, Koyama S, Takagi H, Honda Y, Wiegand SJ, Yancopoulos GD, Nishikawa SI (2002). Recombinant angiopoietin-1 restores higher-order architecture of growing blood vessels in mice in the absence of mural cells. *J Clin Invest*, 110, 1615–1617.

Washington RA, Becher B, Balabanov R, Antel J, Dore-Duffy P (1996). Expression of the activation marker urokinase plasminogen-activator receptor in cultured human central nervous system microglia. *J Neurosci Res*, 45, 392–399.

Yamagishi S, Yonekura H, Yamamoto Y (1999). Vascular endothelial growth factor acts as a pericyte mitogen under hypoxic conditions. *Lab Invest*, 79, 501–509.

Zhang ZG, Zhang L, Croll SD, Chopp M (2002). Angiopoietin-1 reduces cerebral blood vessel leakage and ischemic lesion volume after focal cerebral embolic ischemia in mice. *Neuroscience*, 113, 683–687.

Zimmermann KW (1923). Der feinere Bau der Blutkapillaren. *Z Anat Entwicklungsgesch*, 68, 29–109.

4
Imaging Microglia in the Central Nervous System: Past, Present and Future

Dimitrios Davalos and Katerina Akassoglou

The development of in vivo imaging technology in mice has been a powerful tool to study mechanisms of physiology and pathology in both the nervous and the immune systems. Recent studies have revolutionized our understanding on how glial cells, T cells and neurons interact with each other and respond to damage in the nervous system. This chapter aims to summarize the advances of in vivo imaging as they relate to microglial activation and present the challenges of utilizing this technology directly in animal models of neuroimmunologic disease.

1 In vivo Imaging

Imaging approaches have always been implemented by biologists in their efforts to describe cellular morphology and structure and analyze cellular functions and interactions. A combination of advances in both microscopy and surgery has allowed the in vivo imaging of individual cellular responses as well as complex biological processes, as they occur in real time and within their natural microenvironment, in both health and disease. In particular, the development of two photon-excited fluorescence laser scanning microscopy (Denk et al., 1990) that uses low-energy light and penetrates deep in the intact living tissue has allowed the high-resolution imaging of cells located several hundred microns below the surface of various organs of living animals, with minimal photobleaching and photodamage (Helmchen and Denk, 2005; Svoboda and Yasuda, 2006). This has been facilitated tremendously by the generation of transgenic animals which express fluorescent proteins by specific cell types (Feng et al., 2000; Tsien, 1998), a technology that made the direct observation of cells in vivo possible in both physiological and pathological settings (Germain et al., 2006; Helmchen and Denk, 2005; Misgeld and Kerschensteiner, 2006; Svoboda and Yasuda, 2006).

Prior to the development of in vivo imaging, studies were performed by either standard histopathological techniques or in vitro imaging of individual cultured cells or ex vivo preparations of tissues. Histopathological techniques have provided a wealth of information on the anatomical and structural features of the nervous system and have revealed several aspects of nervous system pathophysiology.

T.E. Lane et al. (eds.), *Central Nervous System Diseases and Inflammation.*
© Springer 2008

However, the analysis of a "snapshot" of a biological event might obscure the appreciation and understanding of ongoing cellular processes, responses to environmental stimuli and transient cell-cell interactions. In vitro imaging of individual cultured cells that allows their manipulation by transfection of exogenous genes and the controlled addition of stimulatory agents or inhibitors has elucidated in detail both cellular and molecular mechanisms. Studies of cells within their microenvironment have been conducted in the context of acutely extracted tissue. For example, imaging in brain slices has expanded our understanding of dendritic calcium dynamics and the biochemical signals regulated by neural activity, as well as the protein-protein interactions in neuronal micro-compartments such as axons, dendrites and their spines (Svoboda et al., 1997; Yasuda et al., 2006). However, in several cases cell viability and behavior within the brain slices may depend upon the preparation and preservation conditions of the tissue in vitro.

2 Imaging of Microglia In vivo

2.1 Microglial Subtypes

Ramon y Cajal was the first who introduced a cellular "third element" besides the neurons and neuroglia in the central nervous system (CNS) in 1913. Del Rio–Hortega (1932) further identified oligodendroglia and microglia within that "third element" and discussed their morphological and ontogenic differences for the first time (Kaur et al., 2001). Microglia are unique among other cells of the nervous system due to (a) their origin, (b) the multiple microglia subtypes present in the healthy brain and (c) their differentiation upon injury that is accompanied by dramatic changes in both cellular morphology and gene expression.

Regarding their origin, microglia have been the subject of numerous studies by many investigators using several experimental systems and animal models, yet they remain among the least understood of all the cell types in the mammalian brain. Their exact origin and subclassification are matters of active debate, due to the lack of microglia-specific markers, their morphological polymorphism and the antigenic plasticity of microglial populations (Gonzalez-Scarano and Baltuch, 1999). It is currently thought that pial macrophages and peripheral monocytes infiltrate the brain during the early developmental stages (Kaur et al., 2001), where they differentiate into a precursor form, the amoeboid microglia.

Besides the amoeboid microglia, there are three additional microglial subtypes. The parenchymal or ramified microglia, represent the resting state of the cells and are found throughout the healthy CNS. They bear long processes with many branches that define each cell's territory in a non-overlapping manner. The reactive microglia, are rounded cells without processes, present around various types of traumatic injury. Finally the perivascular microglia are in close association with the vasculature, and are believed to be important for communicating with components of the BBB and with peripheral immune cells in the blood circulation (Grossmann et al., 2002). Although these are distinct microglial subtypes, it is widely believed that they can all arise by interconversion within a common set of cells (Stence et al., 2001) depending on the presence of activating signals from the surrounding tissue and the severity of the tissue injury. It is well established that following brain damage microglia become activated (Gonzalez-Scarano and Baltuch, 1999; Kreutzberg, 1996; Raivich et al., 1999; Thomas, 1992). Their activation triggers a stereotypical series of both morphological and molecular alterations which are believed to be responsible for the functional differences of the emerging microglial subtypes (Davis et al., 1994; Petersen and Dailey, 2004; Stence et al., 2001). The acquired functions include cell proliferation, migration, phagocytosis and up-regulation of typical immunological markers and antigen-presenting cell capabilities (Aloisi, 2001; Carson, 2002).

2.2   Studies of Microglial Morphology and Activation In vitro or Ex vivo

Microglia are involved in pathologies of the brain and the spinal cord, acute as well as chronic. They are equipped with a broad spectrum of membrane receptors and can release a wide variety of biologically active substances if challenged. Their impressive firepower puts them in the front line against a wide range of insults of their immune-privileged environment. On the other hand, when their involvement fails to produce the desirable results, they can exacerbate a pathological condition and accelerate the progression of a disease, doing more harm than good to the already weakened CNS. Due to the diversity of their abilities and the controversy about their actual contribution to the diseased brain, microglia are an attractive subject for a wide range of studies of the brain pathophysiology. At the same time, studies of the role of microglia represent a challenge, since they require both analysis at the molecular level as well as characterization of the morphological changes that these cells undergo in health and disease. These studies have been primarily done in dissociated cell cultures or in brain slices.

Dissociated cell culture studies have demonstrated that amoeboid microglia can perform highly active movements (Booth and Thomas, 1991; Haapaniemi et al., 1995; Takeda et al., 1998; Tomita et al., 1996; Ward et al., 1991). In situ, the presumed sessile (Stoll and Jander, 1999) ramified microglia withdraw all of their "resting" processes and become highly motile upon activation, and able to migrate toward sites of injury, proliferate and extend new "reactive" processes in order to phagocytose dead or injured cells and cellular debris (Czapiga and Colton, 1999; Stence et al., 2001). Interestingly, the transformation between the two different states is described as a process that is "not immediate", but instead requires nearly complete resorption of existing branches before the cells can exhibit protrusive motility and locomotion. Moreover, time-lapse observations of "resting" and "activated" processes showed significant differences in their behavior: Following activation, the preexisting ramified branches seemed incapable of extension and only underwent a slow retraction, sometimes until their complete withdrawal (in < 2 h) (Stence et al., 2001). The function of microglia in organotypic slice cultures has been shown to vary significantly with time, animal species and medium contents (Czapiga and Colton, 1999), whereas diverse microglial motility behaviors have been described during the clearance of dead cells in hippocampal slices over time (Petersen and Dailey, 2004). Since microglia are the brain's prime responders to damage, the method of preparation and conditioning of the brain slice is perceived as injury that inherently activates microglia. Using in vivo imaging to study these cells in their intact environment reveals important information about their role in the CNS and presents a powerful approach to investigate their interactions with the surrounding tissue in the injured or diseased brain.

2.3   Studies of Microglial Activation In vivo

2.3.1   Challenging the Concept of "Resting" Microglia

Microglia display a highly branched morphology in the unperturbed cerebral cortex, with each cell soma decorated by long processes with fine termini. For decades, this morphological phenotype has been considered completely sessile; its functional potential was found limited, compared to the activated type, which in the adult was only observed following tissue trauma (Davis et al., 1994; Stoll and Jander, 1999; Streit et al., 1988). Based solely on similarities in receptor and cytokine expression profiles with phagocytic macrophages, microglia were characterized as "resting" and expected to perform tissue surveillance in the CNS. Two recent studies by Davalos et al. and Nimmerjahn et al. have challenged the term of "resting" microglia. By combining transcranial two-photon imaging (Grutzendler et al., 2002) and CX3CR1$^{GFP/+}$ mice (Jung et al., 2000) in which brain microglia selectively express the enhanced green fluorescent protein (EGFP), they showed in real time a physical demonstration of their tissue surveillance function. It is important to emphasize that the use of two-photon microscopy is ideal for the study of microglia, since it minimizes photodamage to the living tissue which could by itself cause activation of microglia. In these studies, ramified microglia demonstrated a highly motile behavior of their higher order processes while their main branches and cell bodies remained at relatively fixed positions in the tissue (Fig. 4.1). Through cycles of small extensions and retractions of only their finer termini, microglia were able to patrol their territory without perturbing the densely packed and mostly stable neuronal network (Grutzendler et al., 2002). This thorough scanning of the extracellular space of the brain allows microglia to sample the tissue's integrity on a continuous basis and be in a constant state of readiness to respond to any challenge, whenever it should occur.

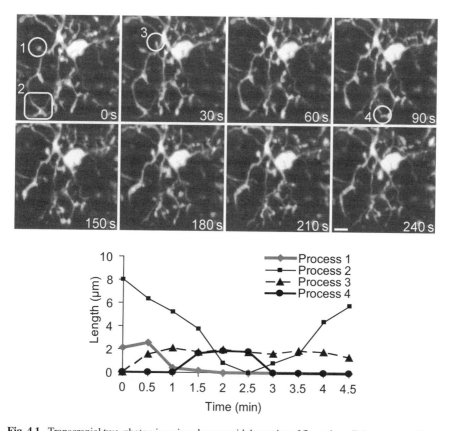

Fig. 4.1 Transcranial two-photon imaging shows rapid dynamics of fine microglial processes. (*Top*) Time-lapse imaging of the same microglial branches demonstrates rapid extension and retraction of fine microglial processes over seconds. Circles and rounded box indicate four representative processes that change in length and shape over time. (*Bottom*) Length changes of the four processes marked in (**a**) as a function of time. Scale bar, 5 μm. Reproduced from Davalos et al. (2005). For timelapse imaging of baseline microglial dynamics see Supplementary Video 1 online at: http://www.nature.com/neuro/journal/v8/n6/suppinfo/nn1472_S1.html.

2.3.2 Microglial Responses to Localized Trauma In vivo

As mentioned above, studies in brain slices are limited by the fact that the slicing procedure can inherently activate microglia, induce their transformation into a morphologically distinct and highly reactive state (Koshinaga et al., 2000; Petersen and Dailey, 2004; Stence et al., 2001) and thereby obscure potentially important dynamic processes (Davalos et al., 2005). In order to challenge these cells in a more localized manner in vivo, both Davalos et al. and Nimmerjahn et al. took advantage of the focal properties of the two photon laser and introduced a very confined injury in the mouse cortex. They both observed a very rapid response of neighboring microglial cells within only minutes of laser-induced injury (Fig. 4.2). The microglial processes moved directly towards the site of injury and appeared to surround and contain it while keeping their cell bodies at their original positions. Mechanical damage of the brain tissue introduced with a glass electrode through a small craniotomy demonstrated the traumatic nature of the laser ablation, while an ablation of two sites very close to each other showed that individual microglial processes of even the same cell can differentially respond to challenges in their close proximity (Davalos et al., 2005). When the laser ablation was performed on a blood vessel in the brain, microglia responded again in a very rapid manner and appeared to contain the hemorrhage by wrapping their processes around the damaged vessel wall (Nimmerjahn et al., 2005). Although it is unclear if the actual damage of the vessel wall and the cellular processes around it caused the microglial response towards the ablated vessel, it is indeed an attractive hypothesis that the disruption of the blood-brain barrier (BBB) and/or the release of blood factor(s) in the brain parenchyma can potentially activate and attract microglial processes.

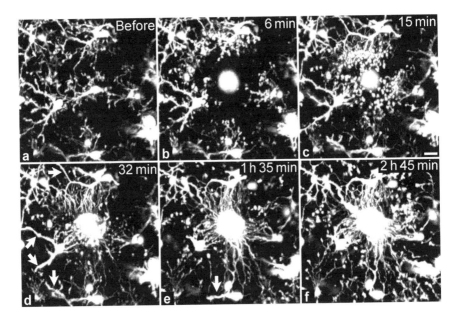

Fig. 4.2 Microglial processes move rapidly towards the site of injury induced by the two-photon laser. (**a–f**) After creating a localized ablation inside the cortex (~15 μm in diameter) with a two-photon laser (**b**), nearby microglial processes respond immediately with bulbous termini (**b**) and extend toward the ablation until they form a spherical containment around it (**c–f**). At the same time, the same cells retract those of their processes that lay in directions opposite to the site of injury (arrows in **d** and **e**). Scale bar, 10 μm. Reproduced from Davalos et al. (2005). For timelapse imaging of the rapid microglial responses towards a localized injury in the living mouse brain see Supplementary Video 2 online at: http://www.nature.com/neuro/journal/v8/n6/suppinfo/nn1472_S1.html.

2.3.3   Challenging the Concept of Microglial "Reactive Transformation"

It is important to note that in all injury paradigms performed in both studies, the onset of microglial responses showed significant differences to activation previously described in ex vivo setups. There was no microglial proliferation or migration observed toward the sites of injury for at least 10 h (Davalos et al., 2005), nor did pre-existing microglial processes need to retract in order for new motile ones to form and respond to the damage. On the contrary, the so far believed inactive microglial processes showed signs of morphological activation within seconds after they were challenged, and started to extend toward each site of injury without delay. Moreover, the ability of even individual microglial processes to identify, extend and eventually contain any tissue damage in their vicinity occurs without a morphological transformation to a reactive state and in a much faster timescale than ever described before (Davalos et al., 2005). It seems that small-size damage that can be contained locally, does not require the massive mobilization described in brain slice experiments and thus the behavior of these cells in the in vivo setting is quite different from what was previously predicted.

Finally, the studies of microglia in the living brain revealed no morphological differences between microglial subtypes in the healthy, intact or locally injured mouse cortex. Individual microglial cells exhibited similar dynamic behaviors in all the in vivo experiments, namely baseline motility, laser ablation, mechanical injury of the brain parenchyma or the vasculature (Davalos et al., 2005; Nimmerjahn et al., 2005). The distinction between perivascular and other parenchymal microglia may also be simply conditional; although some cell bodies appeared in closer proximity (or even in direct contact) with the vasculature compared to others, there was no apparent behavioral distinction between cells situated closer or further away from blood vessels, in both baseline and following injury microglial responses.

2.3.4   Studying Signaling Mechanisms in Microglia In vivo

Many studies performed on microglia in culture have attempted to correlate their molecular properties with their functional roles, often leading to conflicting conclusions. Since gene expression profiles in cultured microglia differ significantly among cells that were extracted by using different isolation methods, and often even more so when compared to microglia in vivo, the interpretation of these results towards a functional prediction warrants extreme caution (Melchior et al., 2006). The surprising features of the observed microglial behaviors in vivo created the need to readdress the role of these cells by studying their molecular properties and signaling mechanisms in their natural habitat, namely directly inside the living brain. Questions raised focused upon the mechanism that regulates their tissue surveillance function and the molecules involved in attracting microglial processes towards sites of damage. Several molecules have been shown to activate microglia in vitro, induce morphological changes and attract them in a concentration gradient dependent manner. However, most studies have addressed these issues on a time scale of several hours to days in accordance with typical immunological response paradigms. The constant motility of microglial processes in unperturbed conditions implies a mechanism that is acutely regulated within the brain tissue.

Nimmerjahn et al. explored the possibility that neuronal or synaptic activity could be regulating the baseline microglial dynamics. By using the Na^+-channel blocker tetrodotoxin (TTX) to reduce neuronal activity they found no significant effect on microglial activity. On the other hand, enhancing synaptic activity by using the GABA-receptor blocker bicuculline had a slightly stimulating effect on microglial processes. However, since bicuculline also elicits seizure activity, the motility-enhancing effect that was observed could have also been due to the incipient cortical damage (Raivich, 2005). Davalos et al. reasoned that the directional convergence of microglial processes towards the trauma implies the presence of a gradient of one or more highly diffusible and abundant molecules that can mediate this phenomenon. Using a combination of inhibitory and activating approaches while imaging microglial responses in vivo they demonstrated that extracellular ATP and activation of P2Y receptors are necessary for the rapid microglial response towards injury. Interestingly, the baseline activity of microglial processes was shown to be also affected by pharmacological inhibitors of purinergic signaling, indicating that there could be a uniform mechanism in place that controls microglial motility under both normal and injury conditions. Moreover, these results appear to suggest a role of the surrounding tissue in preserving and regenerating the signals that fuel the observed microglial response to damage. Although neurons, oligodendrocytes and endothelial cells are likely to release large amounts of ATP upon injury and thereby contribute to microglial reaction, they provided evidence to suggest that astrocytes may play an important role in mediating the rapid and widespread microglial response (Davalos et al., 2005). The study of this phenomenon directly inside the intact brain, where all the potential players are present and able to perform their roles as they are faced with an experimental challenge, allowed the identification of a very intriguing cell-cell interaction between astrocytes and microglia in an orchestrated attempt to respond to the incurred tissue damage (Davalos et al., 2005).

3 Challenges/Perspective: Imaging Microglia in Animal Models of CNS Disease

Ever since transgenic animals expressing fluorescent proteins were combined with two-photon microscopy our understanding of the brain's pathophysiology has undergone a fundamental reform. In vivo imaging allows the observation of cellular processes in real time within the intact tissue microenvironment. Application of in vivo imaging on microglia revealed their baseline motility, their responses to stimuli, injury, as well as their potential for interactions with neighboring cells. In addition, direct injection of microglial activators and inhibitors in the brain parenchyma allowed the elucidation of molecular mechanisms that mediate microglial functions.

The technical ability to relocate and re-image a specific area in the living brain revealed the progression of microglia responses, which is particularly important in understanding the mechanisms of disease in the nervous system.

In vivo imaging can be further applied to study the involvement of microglia in multiple animal models of disease pathogenesis. Microglia have been shown to be involved in a variety of diseases, such as Alzheimer's disease (AD), multiple sclerosis (MS), HIV and other infections of the central nervous system. Overall, their prolonged intervention has been found to be exacerbating rather than improving the tissue condition (Gonzalez-Scarano and Baltuch, 1999; Stoll and Jander, 1999). Breakthrough studies have applied in vivo two-photon microscopy in AD, ischemia and seizures (for review see (Misgeld and Kerschensteiner, 2006)). For example in AD, the contribution of amyloid plaques to neurodegeneration has been documented in a pioneering study using mice with fluorescently labeled neurons and transcranial two-photon imaging (Tsai et al., 2004). In this study it was shown that dendrites passing through or near amyloid deposits undergo spine loss, and nearby axons develop large varicosities, leading to neurite breakage and permanent disruption of neuronal connections (Tsai et al., 2004). The ability to repetitively image the brain in vivo allowed for the first time to study the responses of neurons to a toxic stimulus over time. Although correlations between neuronal damage and beta-amyloid had been previously made with conventional histopathology, the use of two-photon microscopy allowed the assessment of the dynamics between neurodegeneration and a toxic stimulus in the CNS. It has been proposed that similar studies could be performed to observe and correlate microglial activation in AD and other neurodegenerative models (Misgeld and Kerschensteiner, 2006).

Microglial activation is a hallmark of neuroinflammatory diseases such as MS. MS lesions are heterogeneous and characterized by infiltration of immune cells into the CNS parenchyma, destruction of myelin and axonal damage (Lassmann et al., 2001; 2007). Microglia play a central role in this processes, due to their ability to phagocytose myelin via the CD11b/CD18 integrin receptor (van der Laan et al., 1996) and secrete proinflammatory cytokines that are actively involved in demyelination (Platten and Steinman, 2005). Recently, the blood protein fibrinogen was identified as the CD11b/CD18 ligand that drives microglial activation and phagocytosis (Adams et al., 2007). Genetic or pharmacologic inhibition of the binding of fibrinogen to the microglial CD11b/CD18 integrin suppressed microglial activation and relapsing paralysis in experimental autoimmune encephalomyelitis (EAE), an established animal model of MS (Adams et al., 2007). These studies proposed fibrinogen as a "danger signal" that triggers microglial activation after BBB disruption (Adams et al., 2007). Interestingly, in human MS microglial activation in areas of demyelinating plaques correlates with areas of deposition of fibrinogen due to BBB disruption (Gay et al., 1997), which is one of the earliest histopathologic alterations in MS lesions (Vos et al., 2005). Recent studies have demonstrated that microglia not only act as phagocytes, but they are also involved in the onset of inflammatory demyelination in CNS autoimmune disease (Heppner et al., 2005). Overall, these studies suggest that signals, such as fibrinogen, that may initiate microglial activation could be crucial for the onset of disease pathogenesis in the CNS.

Although several imaging studies have been applied in different animal models of neurodegenerative disease, in vivo imaging of neuroinflammation using two-photon microscopy has not been performed. Imaging in animal models of inflammatory demyelination such as EAE, has been either limited to the vasculature with the use of intravital microscopy (Vajkoczy et al., 2001), or performed ex vivo using tissue explants (Kawakami et al., 2005; Nitsch et al., 2004). Inflammatory demyelination is a complex pathological setting where microglia can exchange signals with both neuronal and immune cells (Carson, 2002). Given the immunologic properties of microglia and their complex interactions with cells on both sides of the BBB, imaging their functions in EAE can reveal crucial information regarding their dual identity as immune cells that reside within the central nervous system. For example, imaging microglia in EAE could identify the link between microglial activation and the onset of inflammatory demyelination. Overall, establishing an in vivo imaging technique in animal models of inflammatory demyelination could reshape our understanding of the pathophysiological role of microglia within demyelinating lesions and reveal the cell-cell interactions that could develop during an inflammatory attack that leads to nervous system damage.

From an imaging perspective, two-photon microscopy in demyelinated areas presents several challenges. First, myelin – and as a result inflammatory demyelination – is located in deep layers of the brain. In EAE, a "hot spot" for demyelination in the brain is the cerebellar white matter. Therefore, by contrast to animal models of AD where the pathology is mostly localized in the mouse cortex, inflammation in animal models of MS is localized deep into the brain tissue. Given the depth limitations of even two-photon microscopy, it would be extremely challenging to apply this technique for imaging demyelination (Svoboda and Yasuda, 2006). Second, although demyelinating lesions in the spinal cord would be an ideal site to study microglial responses in vivo, limited work has been performed on imaging the living spinal cord to date, due mainly to the movement artifacts generated by the heartbeat and the breathing of the animal. Regarding the brain, alternative approaches, such as intravital microscopy or regeneratively amplified two-photon microscopy (Helmchen and Denk, 2005; Theer et al., 2003) could possibly be efficacious in imaging microglia located within the demyelinated white matter. Regarding the spinal cord, the development of novel techniques that would allow the stable and repetitive imaging of the mouse spinal cord using two-photon microscopy would be essential to decipher the role of microglia within the demyelinated lesions. Therefore, due to the distinct characteristics of the location of neuroimmunologic lesions, their imaging requires further advances in both microscopy and surgery. Future research should take into account the specific features of inflammatory demyelination and aim towards the development of novel methodologies to access inflammatory foci in the CNS in ways that would allow their study by two-photon microscopy. Performing in vivo imaging of microglia in the presence of neuronal and inflammatory stimuli can further correlate microglial functions with the progression of inflammation, demyelination and neuronal damage. Moreover, the application of such novel imaging approaches to study neuroinflammatory lesions can be crucial for identifying and testing potential pharmacological targets and regulate microglial activation within demyelinating lesions.

References

Adams, R.A., Bauer, J., Flick, M.J., Sikorski, S.L., Nuriel, T., Lassmann, H., Degen, J.L., and Akassoglou, K. (2007). The fibrin-derived gamma377-395 peptide inhibits microglia activation and suppresses relapsing paralysis in central nervous system autoimmune disease. *J Exp Med 204*, 571–582.

Aloisi, F. (2001). Immune function of microglia. *Glia 36*, 165–179.

Booth, P.L. and Thomas, W.E. (1991). Dynamic features of cells expressing macrophage properties in tissue cultures of dissociated cerebral cortex from the rat. *Cell Tissue Res 266*, 541–551.

Carson, M.J. (2002). Microglia as liaisons between the immune and central nervous systems: functional implications for multiple sclerosis. *Glia 40*, 218–231.

Czapiga, M. and Colton, C.A. (1999). Function of microglia in organotypic slice cultures. *J Neurosci Res 56*, 644–651.

Davalos, D., Grutzendler, J., Yang, G., Kim, J.V., Zuo, Y., Jung, S., Littman, D.R., Dustin, M.L., and Gan, W.B. (2005). ATP mediates rapid microglial response to local brain injury in vivo. *Nat Neurosci 8*, 752–758.

Davis, E.J., Foster, T.D., and Thomas, W.E. (1994). Cellular forms and functions of brain microglia. *Brain Res Bull 34*, 73–78.

Denk, W., Strickler, J.H., and Webb, W.W. (1990). Two-photon laser scanning fluorescence microscopy. *Science 248*, 73–76.

Feng, G., Mellor, R.H., Bernstein, M., Keller-Peck, C., Nguyen, Q.T., Wallace, M., Nerbonne, J.M., Lichtman, J.W., and Sanes, J.R. (2000). Imaging neuronal subsets in transgenic mice expressing multiple spectral variants of GFP. *Neuron 28*, 41–51.

Gay, F.W., Drye, T.J., Dick, G.W., and Esiri, M.M. (1997). The application of multifactorial cluster analysis in the staging of plaques in early multiple sclerosis. Identification and characterization of the primary demyelinating lesion. *Brain 120(Pt 8)*, 1461–1483.

Germain, R.N., Miller, M.J., Dustin, M.L., and Nussenzweig, M.C. (2006). Dynamic imaging of the immune system: progress, pitfalls and promise. *Nat Rev Immunol 6*, 497–507.

Gonzalez-Scarano, F. and Baltuch, G. (1999). Microglia as mediators of inflammatory and degenerative diseases. *Annu Rev Neurosci 22*, 219–240.

Grossmann, R., Stence, N., Carr, J., Fuller, L., Waite, M., and Dailey, M.E. (2002). Juxtavascular microglia migrate along brain microvessels following activation during early postnatal development. *Glia 37*, 229–240.

Grutzendler, J., Kasthuri, N., and Gan, W.B. (2002). Long-term dendritic spine stability in the adult cortex. *Nature 420*, 812–816.

Haapaniemi, H., Tomita, M., Tanahashi, N., Takeda, H., Yokoyama, M., and Fukuuchi, Y. (1995). Non-amoeboid locomotion of cultured microglia obtained from newborn rat brain. *Neurosci Lett 193*, 121–124.

Helmchen, F. and Denk, W. (2005). Deep tissue two-photon microscopy. *Nat Methods 2*, 932–940.

Heppner, F.L., Greter, M., Marino, D., Falsig, J., Raivich, G., Hovelmeyer, N., Waisman, A., Rulicke, T., Prinz, M., Priller, J., et al. (2005). Experimental autoimmune encephalomyelitis repressed by microglial paralysis. Nat Med *11*, 146–152.

Jung, S., Aliberti, J., Graemmel, P., Sunshine, M.J., Kreutzberg, G.W., Sher, A., and Littman, D. R. (2000). Analysis of fractalkine receptor CX(3)CR1 function by targeted deletion and green fluorescent protein reporter gene insertion. Mol Cell Biol *20*, 4106–4114.

Kaur, C., Hao, A.J., Wu, C.H., and Ling, E.A. (2001). Origin of microglia. *Microsc Res Tech 54*, 2–9.

Kawakami, N., Nagerl, U.V., Odoardi, F., Bonhoeffer, T., Wekerle, H., and Flugel, A. (2005). Live imaging of effector cell trafficking and autoantigen recognition within the unfolding autoimmune encephalomyelitis lesion. *J Exp Med 201*, 1805–1814.

Koshinaga, M., Katayama, Y., Fukushima, M., Oshima, H., Suma, T., and Takahata, T. (2000). Rapid and widespread microglial activation induced by traumatic brain injury in rat brain slices. *J Neurotrauma 17*, 185–192.

Kreutzberg, G.W. (1996). Microglia: a sensor for pathological events in the CNS. *Trends Neurosci 19*, 312–318.

Lassmann, H., Brück, W., and Lucchinetti C. (2001). Heterogeneity of multiple sclerosis pathogenesis: implications for diagnosis and therapy. *Trends Mol Med 7*, 115–121.

Lassmann, H., Brück, W., and Lucchinetti C. (2007). The immunopathology of multiple sclerosis: an overview. *Brain Pathol 17*, 210–218.

Melchior, B., Puntambekar, S.S., and Carson, M.J. (2006). Microglia and the control of autoreactive T cell responses. *Neurochem Int 49*, 145–153.

Misgeld, T. and Kerschensteiner, M. (2006). In vivo imaging of the diseased nervous system. *Nat Rev Neurosci 7*, 449–463.

Nimmerjahn, A., Kirchhoff, F., and Helmchen, F. (2005). Resting microglial cells are highly dynamic surveillants of brain parenchyma in vivo. *Science 308*, 1314–1318.

Nitsch, R., Pohl, E.E., Smorodchenko, A., Infante-Duarte, C., Aktas, O., and Zipp, F. (2004). Direct impact of T cells on neurons revealed by two-photon microscopy in living brain tissue. *J Neurosci 24*, 2458–2464.

Petersen, M.A., and Dailey, M.E. (2004). Diverse microglial motility behaviors during clearance of dead cells in hippocampal slices. *Glia 46*, 195–206.

Platten, M. and Steinman, L. (2005). Multiple sclerosis: trapped in deadly glue. *Nat Med 11*, 252–253.

Raivich, G. (2005). Like cops on the beat: the active role of resting microglia. *Trends Neurosci 28*, 571–573.

Raivich, G., Bohatschek, M., Kloss, C.U., Werner, A., Jones, L.L., and Kreutzberg, G.W. (1999). Neuroglial activation repertoire in the injured brain: graded response, molecular mechanisms and cues to physiological function. *Brain Res Brain Res Rev 30*, 77–105.

Stence, N., Waite, M., and Dailey, M.E. (2001). Dynamics of microglial activation: a confocal time-lapse analysis in hippocampal slices. *Glia 33*, 256–266.

Stoll, G. and Jander, S. (1999). The role of microglia and macrophages in the pathophysiology of the CNS. *Prog Neurobiol 58*, 233–247.

Streit, W.J., Graeber, M.B., and Kreutzberg, G.W. (1988). Functional plasticity of microglia: a review. Glia *1*, 301–307.

Svoboda, K. and Yasuda, R. (2006). Principles of two-photon excitation microscopy and its applications to neuroscience. *Neuron 50*, 823–839.

Svoboda, K., Denk, W., Kleinfeld, D., and Tank, D.W. (1997). In vivo dendritic calcium dynamics in neocortical pyramidal neurons. Nature *385*, 161–165.

Takeda, H., Tomita, M., Tanahashi, N., Kobari, M., Yokoyama, M., Takao, M., Ito, D., and Fukuuchi, Y. (1998). Hydrogen peroxide enhances phagocytic activity of ameboid microglia. *Neurosci Lett 240*, 5–8.

Theer, P., Hasan, M.T., and Denk, W. (2003). Two-photon imaging to a depth of 1000 micron in living brains by use of a Ti:Al2O3 regenerative amplifier. *Opt Lett 28*, 1022–1024.

Thomas, W.E. (1992). Brain macrophages: evaluation of microglia and their functions. *Brain Res Brain Res Rev 17*, 61–74.

Tomita, M., Fukuuchi, Y., Tanahashi, N., Kobari, M., Takeda, H., Yokoyama, M., Ito, D., and Terakawa, S. (1996). Swift transformation and locomotion of polymorphonuclear leukocytes and microglia as observed by VEC-DIC microscopy (video microscopy). *Keio J Med 45*, 213–224.

Tsai, J., Grutzendler, J., Duff, K., and Gan, W.B. (2004). Fibrillar amyloid deposition leads to local synaptic abnormalities and breakage of neuronal branches. *Nat Neurosci 7*, 1181–1183.

Tsien, R.Y. (1998). The green fluorescent protein. *Annu Rev Biochem 67*, 509–544.

Vajkoczy, P., Laschinger, M., and Engelhardt, B. (2001). Alpha4-integrin-VCAM-1 binding mediates G protein-independent capture of encephalitogenic T cell blasts to CNS white matter microvessels. *J Clin Invest 108*, 557–565.

van der Laan, L.J., Ruuls, S.R., Weber, K.S., Lodder, I.J., Dopp, E.A., and Dijkstra, C.D. (1996). Macrophage phagocytosis of myelin in vitro determined by flow cytometry: phagocytosis is mediated by CR3 and induces production of tumor necrosis factor-alpha and nitric oxide. *J Neuroimmunol 70*, 145–152.

Vos, C.M., Geurts, J.J., Montagne, L., van Haastert, E.S., Bo, L., van der Valk, P., Barkhof, F., and de Vries, H.E. (2005). Blood-brain barrier alterations in both focal and diffuse abnormalities on postmortem MRI in multiple sclerosis. *Neurobiol Dis 20*, 953–960.

Ward, S.A., Ransom, P.A., Booth, P.L., and Thomas, W.E. (1991). Characterization of ramified microglia in tissue culture: pinocytosis and motility. *J Neurosci Res 29*, 13–28.

Yasuda, R., Harvey, C.D., Zhong, H., Sobczyk, A., van Aelst, L., and Svoboda, K. (2006). Supersensitive Ras activation in dendrites and spines revealed by two-photon fluorescence lifetime imaging. *Nat Neurosci 9*, 283–291.

5

Cytokines in CNS Inflammation and Disease

Malú G. Tansey and Tony Wyss-Coray

Abstract A growing number of increasingly sophisticated studies over the past two decades have implicated key immune regulatory factors in CNS function and disease. Chief among those factors are cytokines, which were originally described to regulate communication between immune cells. The purpose of this chapter is to review what is known regarding normal expression of key cytokines and their receptors in the CNS as well as their roles in development and disease. We will focus in particular on tumor necrosis factor α (TNF), transforming growth factor β (TGF-β), interleukin 1 (IL-1), IL-6, IL-10 and IL-12. We will discuss how these cytokines modulate not only normal CNS function but how abnormal cytokine signaling may contribute to major acute and chronic CNS diseases. We will rely heavily on studies in experimental mouse models that overproduce or ablate these cytokines and discuss the potential therapeutic benefits from 'proof-of-concept' targeting studies of specific cytokines in the treatment of CNS diseases.

1 Introduction

For many decades it was believed that the blood–brain barrier prevented access of immune cells to the brain and as a result the immune and central nervous system were relatively independent of each other. However, it has become increasingly clear that the permeability of this so-called barrier can be regulated under normal conditions and may in fact increase or become dysregulated in disease states. The capacity of the brain to activate an inflammatory reaction involving central production of cytokines in response to an immune challenge in the periphery has been demonstrated in recent years. In addition, inflammatory challenges may function as triggers for uncovering pre-existing vulnerabilities or may exacerbate previous functional deficits, giving rise to clinical symptoms of neurological or psychiatric conditions. Surprisingly, certain molecular mechanisms involved in modulating neuronal plasticity and cellular immunity may be shared in common as evidenced by the involvement of specific immune molecules in mediating intercellular communication in the nervous system (Boulanger et al., 2001; Chun, 2001). The cellular processes of astrocytes make intimate contacts with essentially all areas of the brain and are functionally coupled to neurons, oligodendrocytes, and other astrocytes via both contact-dependent and non-contact-dependent pathways to maintain the homeostatic environment, thus promoting the proper functioning of the neuronal network. Inflammation in the CNS disrupts this process either transiently or permanently and cytokines have important roles in reprogramming gene expression in glia during injury and recovery (John et al., 2003). Within the CNS, cytokines mediate inflammatory processes that modulate blood–brain barrier permeability or promote apoptosis of neurons, oligodendrocytes and astrocytes as well as damage to myelinated axons. Moreover, direct neurotoxic and neuroprotective effects of cytokines may be independent of their immunoregulating properties. The immune reactions initiated by viruses, bacteria and parasites may contribute to latent vulnerabilities which could become manifest with future stressors or challenges. Collectively, whether cytokine action has beneficial or harmful outcomes in the brain depends on the dynamics, cellular source, degree of compart-mentalization of cytokine release, the pathophysiological context, and the presence of co-expressed factors.

2   Tumor Necrosis Factor-α

2.1   Signaling Pathways and Expression of TNF and its Receptors in Normal CNS

TNF belongs to a superfamily of ligands with pivotal roles in immune system function (Wajant et al., 2003; Shen and Pervaiz, 2006); many of these ligands have been implicated in the etiology of several acquired and genetic diseases (Locksley et al., 2001; MacEwan, 2002). TNF is synthesized as a monomeric 26 kD type-2 transmembrane protein that assembles into trimers and is cleaved by TACE metal-loprotease to a soluble circulating form (Aggarwal et al., 2000b; Idriss and Naismith, 2000). Both forms of TNF are biologically active and can be synthesized in the brain by microglia, astrocytes and subsets of neurons (Lieberman et al., 1989; Morganti-Kossman et al., 1997; Chung et al., 2005). TNF receptors, commonly referred to as type 1 receptor (R1, p55, Tnfrsf1a) and type 2 receptor (R2, p75, Tnfrsf1b) bind the two forms of TNF with different affinities and are constitutively expressed on neurons and glia in the CNS (Benveniste and Benos, 1995). R1 is activated equally well by soluble and membrane-bound TNF; whereas R2 is pref-erentially activated by transmembrane (tm) TNF and to a lesser extent by soluble TNF (Grell et al., 1998; Aggarwal et al., 2000a). TNFR1 and R2 have a cysteine-rich extracellular domain with 28% shared homology and have completely distinct transmembrane and cytoplasmic domains with no homology between them (Bodmer et al., 2002). Since the original identification of TNFRs about 15 years ago, modern molecular techniques have revealed a complex array of potential pro-teins that may interact with TNFRs to modulate activation and interplay of a number of well characterized signaling pathways (for reviews see (Locksley et al.,

2001; MacEwan, 2002; Liu, 2005)). The death domain (DD) present in the cytoplasmic tail of R1 recruits TNFR1-associating death domain (TRADD) protein and at least 20 different proteins including FADD, TRAF2 and receptor interacting protein kinase 1 (RIP) to form a cascade leading to activation of sphingomyelinase-ceramide, caspases, NFkB transcription, ASK1/c-Jun kinase (JNK), and p38 MAPK pathways. Many of these pathways exert effects on mitochondrial function and redox balance and can lead to activation of apoptosis or alternatively, new gene transcription for promotion of cell survival, proliferation and differentiation. Depending on cellular context, R1 and R2 can activate many of the same downstream pathways and act synergistically or in opposing fashion (Haridas et al., 1998; Aggarwal et al., 2000a; Liu et al., 2000; Liu, 2005) but less is known regarding integration of signaling pathways through R2. In the nervous system, the kinetics and upstream mediators for activation of the NFkB pathway by R1 and R2 have been shown to be different and to result in opposing effects on cortical neuron survival (Marchetti et al., 2004) and hippocampal neurons (Yang et al., 2002; Heldmann et al., 2005). The RIP adaptor protein, which associates with both TNFRs, may be a molecular switch that allows R2 signaling to alternate between anti-apoptotic NFkB activation and death induction through caspase mechanisms (Kelliher et al., 1998; Pimentel-Muinos and Seed, 1999).

TNF-immunoreactive neurons have been reported in the hypothalamus, in the caudal raphe nuclei, in the bed nucleus of the stria terminalis, and along the ventral pontine and medullary surface. They are also found in areas involved in autonomic and neuroendocrine regulation, such as the hypothalamus, amygdala, parabrachial nucleus, dorsal vagal complex, nucleus ambiguous, and the thoracic sympathetic preganglionic cell column (Breder et al., 1993). TNF receptor expression has been detected in brainstem, cortex, cerebellum, thalamus and basal ganglia (Kinouchi et al., 1991). TNFR1 is expressed in many cell types, whereas TNFR2 is expressed less broadly and primarily by cells of the immune system including microglia (Dopp et al., 1997). However, R2 expression has also been reported in cardiac myocytes, endothelial cells (Aggarwal et al., 2000a), dopaminergic (McGuire et al., 2001), cortical (Marchetti et al., 2004), and hippocampal (Bernardino et al., 2005; Heldmann et al., 2005) neurons. CNS functions which have been ascribed to TNF include activation of microglia and astrocytes (Selmaj et al., 1990; Merrill, 1991), regulation of endothelial cell permeability at the blood–brain barrier (Sedgwick et al., 2000), generation of the febrile response (Leon, 2002), enhancement of slow-wave sleep (Deboer et al., 2002) and synaptic strength (Beattie et al., 2002). TNF exerts potent actions on glutamatergic synaptic transmission and modulates synaptic plasticity (Pickering et al., 2005). TNF and its downstream targets appear to regulate hippocampal neuron development as mice doubly deficient in R1 and R2 (double knockout; dko) have decreased arborization of the apical dendrites of the CA1 and CA3 regions and accelerated dentate gyrus development (Golan et al., 2004). Membrane-tethered TNF is critical for lymphoid organ development (Ruuls et al., 2001) but its role in cellular responses related to brain function is less well understood. It appears to be important for proliferation of oligodendrocytes and nerve remyelination (Arnett et al., 2001) and proliferation of hippocampal neuroblasts after stroke (Heldmann et al., 2005). Inflammatory signals activated by TNF are primarily mediated through soluble TNF binding to R1 (Pasparakis et al., 1996; Ruuls et al., 2001; Quintana et al., 2005); although some have proposed that under ischemic conditions R2 can contribute to inflammatory responses (Akassoglou et al., 2003). In cells that co-express R1 and R2, the functional outcome of a TNF stimulus in vivo (neurotoxic versus neuroprotective) may be determined by the R1/ R2 expression ratio (MacEwan, 2002); if this is true in the brain, it may be possible to design new pharmacological and/or gene therapy-based approaches to preferentially upregulate R2 activity and/or expression to achieve neuroprotection in brain regions where TNF normally exerts neurotoxic effects.

2.2   Role of TNF in CNS Disease

The potent cytotoxic effects implicate TNF in the pathophysiology of inflammatory and autoimmune diseases. TNF levels in the brain and/or cerebrospinal fluid (CSF) become elevated in a wide range of CNS disorders, including ischemia (Feuerstein et al., 1994; Liu et al., 1994), trauma (Goodman et al., 1990), multiple sclerosis (Hofman et al., 1989; Selmaj et al., 1991; Sharief and Hentges, 1991; Rieckmann et al., 1995; Raine et al., 1998), Alzheimer's disease (AD)(Fillit et al., 1991; Paganelli et al., 2002; Alvarez et al., 2007), and Parkinson's disease (PD)(Boka et al., 1994; Mogi et al., 1994, 1999, 2000; Hirsch et al., 1998; Bessler et al., 1999; Hunot et al., 1999; Hasegawa et al., 2000; Nagatsu et al., 2000a, b; Nagatsu and Sawada, 2005). While there is no evidence to support a role for TNF in the etiology of any of these diseases, it has been suggested that TNF-dependent mechanisms act to modify, and typically accelerate disease progression.

Polymorphisms in inflammatory genes for interleukin 1α, interleukin 1β, interleukin-6, TNF, α 2-macroglobulin, and α 1-antichymotrypsin have been investigated both for CNS disease association and as markers of disease progression. In particular, the TNF gene has received much attention possibly because it resides within the MHC gene cluster. Individual studies have reported polymorphisms in the TNF promoter that may affect susceptibility to different CNS diseases including cerebral malaria (Knight et al., 1999), stroke (Karahan et al., 2005), vascular dementia (McCusker et al., 2001), Alzheimer's disease (McGuire et al., 1994, 1999; Knight et al., 1999; Perry et al., 2001a, b), and Parkinson's Disease (Nishimura et al., 2001). Several studies suggest that heritable differences in TNF production could be linked to certain HLA haplotypes within human populations that are over-represented in several autoimmune and inflammatory diseases (Wilson et al., 1993). However, consistent and reliable results on the functional signficance of the genetic variations have not been obtained. It is therefore possible that many of the reported associations between TNF alleles and susceptibility to disease merely reflect linkage with MHC genes or chance association.

Much of our understanding of TNF and its receptors in the CNS has come from detailed evaluation of the phenotypes of a number of genetic mouse models developed in the last 15 years and their phenotypes after exposure to various toxins or pathogens

(Table 5.1) (reviewed in Probert and Akassoglou, 2001; Corti and Ghezzi, 2004; Kollias, 2005). Consistent with a role of TNF in modulating synaptic plasticity, hippocampal brain slices from TNFR-deficient mice do not display long-term depression induced by low-frequency stimulation of Schaffer collateral axons (Albensi and Mattson, 2000). Whole animal studies in which TNF knockouts were compared to normal animals indicated TNF deficient animals performed better in spatial memory and learning tasks (Morris Water Maze) (Golan et al., 2004). Conversely, two mouse lines overexpressing hTNF show significant impairment in spatial learning (Aloe et al., 1999b). One obvious caveat in the interpretation of these studies is that the 'substrate' of learning and memory is not the same in knockout and wild-type mice since TNF deficiency affects hippocampal development. At pathophysiological levels, TNF has been shown to have inhibitory effects via the p38 mitogen activated kinase pathway on hippocampal long-term potentiation (LTP) (Cunningham et al., 1996; Butler et al., 2004), a long-lasting increase in synaptic efficacy involving glutamate receptor activation and increased intracellular calcium levels and thought to be an underlying mechanism of learning and memory formation (Bliss and Collingridge, 1993). Elevated levels of TNF, also through a p38 MAPK-dependent pathway, may further contribute to LTP impairment through upregulation of RGS7 (a regulator of G-protein signaling) expression (Benzing et al., 2002). Lastly, studies using TNF over-expressing mice demonstrated an indirect role of TNF in influencing survival of basal forebrain cholinergic neurons via direct regulation of the levels of nerve growth factor (NGF) (Aloe et al., 1993), a key survival factor for this and other neuronal populations. Genetic ablation of TNF or TNF receptors in rodent models of PD, which show neurotoxin induced loss of dopaminergic neurons yielded variable results (Table 5.1) (Bruce et al., 1996; Albensi and Mattson, 2000; Rousselet et al., 2002; Ferger et al., 2004; Leng et al., 2005; Sriram et al., 2006a, b). However, because lack of TNF signaling during development results in arrested dendritic cell development (Pasparakis et al., 1996) and stunted microglial responsiveness in adult animals (Sriram et al., 2006b) it is difficult to implicate TNF directly in neurodegeneration based on these studies.

2.2.1 Multiple Sclerosis

Given that TNF was known to regulate immune function, its role in the autoimmune dysregulation characteristic of multiple sclerosis has been extensively investigated. TNF and its receptors are upregulated in active MS lesions and TNF levels in the CSF of MS patients correlate with disease severity (Hofman et al., 1989; Selmaj et al., 1991; Sharief and Hentges, 1991; Raine et al., 1998). Strong evidence that TNF is important in the MS disease process was derived from experimental rodent models of MS. In particular, TNF blockade was shown to prevent or treat the development of experimental autoimmune encephalomyelitis (EAE) in rodents (Ruddle et al., 1990; Baker et al., 1994; Selmaj et al., 1995; Korner et al., 1997). As indicated in Table 5.1, a role for TNF in the induction phase of EAE via modulation of leukocyte traffic into the CNS parenchyma (Korner et al., 1997; Kassiotis et al., 1999)

Table 5.1 CNS phenotypes resulting from genetic modulation of TNF pathway genes (listed are only models with CNS phenotypes)

Mouse model[a]	Transgene (promoter/target cell)[b]	TNF receptor status[c]	CNS phenotype
TNF tg	Overexpression of mTNF (GFAP/astrocytes)	Wildtype	Lymphocytic meningoencephalomyelitis and paralysis (Campbell et al., 1997) ↓ Kainate-induced seizures (Balosso et al., 2005)
Tg6074	Overexpression of mTNF-hβ-globin (TNF/subset of neurons)	Wildtype	↑ Inflammation, oligodendrocyte apoptosis, demyelination: model for chronic MS (Probert et al., 1995) ↑ Grooming in the novel object investigation test ↓ Rearing in novel olfactory cues test; delayed passive avoidance acquisition ↑ Thermal response in hot-plate test (Fiore et al., 1996) Altered cholinergic neuron survival due to ↓ NGF (Aloe et al., 1999a); impaired learning and memory (MWM) (Aloe et al., 1999b)
Tg6074 × TNFR1 ko	Same as above	TNFR1 ko	None compared to above (Akassoglou et al., 1998)
Tg6074 × TNFR2 ko	Same as above	TNFR2 ko	↔ From Tg6074 with wildtype TNFRs (Akassoglou et al., 1998)
MBP-TNF	Overexpression of mTNF (MBP/oligodendrocytes)	Wildtype	No spontaneous pathology but ↑ EAE with MBP adjuvant progressing to chronic demyelination w/macrophage and microglia activation (Taupin et al., 1997)
TgK742	Overexpression of hTNF (NF-L/neurons)	Wildtype	Meningeal inflammation (Akassoglou et al., 1997)
TgK21	Overexpression of htmTNF (GFAP/astrocytes)	Wildtype	↑ Inflammation, oligodendrocyte apoptosis, demyelination: model for acute MS (Akassoglou et al., 1997, 1999)
TgK21 × TNFR1 ko	Same as above	TNFR1 ko	No CNS pathology compared to TgK21 with wildtype TNFRs (Akassoglou et al., 1997, 1999)
TgK21 × TNFR2 ko	Same as above	TNFR2 ko	↔ From TgK21 with wildtype TNFRs (Akassoglou et al., 1997, 1999)
TNF ko	Constitutive deletion of TNF gene expression (all cells)	Wildtype	Resistant to MPTP toxicity (Ferger et al., 2004)
TNFR1 ko	Constitutive deletion of TNFR1 (all cells)	Wildtype TNFR2	Not resistant to MPTP (Rousselet et al., 2002; Sriram et al., 2002; Leng et al., 2005)

(continued)

Table 5.1 (continued)

TNFR2 ko	Constitutive deletion of TNFR2 (all cells)	Wildtype TNFR1	Not resistant to MPTP (Rousselet et al., 2002; Sriram et al., 2002; Leng et al., 2005); prolonged kainate-induced seizures (Balosso et al., 2005)
TNFR1 × TNFR2 dko	Constitutive deletion of both TNFR1 and R2 (all cells)	None present	Partially resistant to MPTP toxicity (Sriram et al., 2002); Not resistant to MPTP toxicity; abnormal dopamine metabolism (Rousselet et al., 2002) ↑ Vulnerability to focal cerebral ischemia and excitotoxicity; Defective LTD (Bruce et al., 1996; Albensi and Mattson, 2000; Guo et al., 2004); prolonged kainate-induced seizures (Balosso et al., 2005)

Tg transgenic; *CNS* central nervous system; *MS* multiple sclerosis; *EAE* experimental autoimmune encephalomyelitis; *MBP* myelin basic protein; *MPTP* 1-Methyl-4-phenyl-1,2,3,6-tetrahydropyridine; *LTD* long-term depression; *NF-L* neurofilament L; *NGF* nerve growth factor; *ko* knockout; *dko* double knockout; *TACE* TNF alpha-converting enzyme; *MWM* Morris water maze

[a]Details on mouse models including expression pattern and levels, the presence of mutations, etc. are described in the cited papers

[b]Transgene or targeted deletion of indicated form (*m* mouse or *h* human) of TNF or TNF receptors.

[c]Status of TNF receptors in mouse model

↔, no change; ↑, increased; ↓, decreased

and a role for TNFR1 in demyelination were demonstrated (Probert et al., 2000) using TNF genetic models. Similarly, an important role for tmTNF and its preferred receptor TNFR2 in oligodendrocyte precursor proliferation and remyelination was demonstrated using TNF genetic models in the cuprizone toxin model of MS (Arnett et al., 2001). In fact, these data offered a mechanistic explanation for the unfortunate failure of lenercept, an Fc-fused p55/TNFR1, in phase I clinical trials with MS patients whose symptoms worsened between bouts of relapsing-remitting episodes due to the lack of TNF-mediated remyelination (Wiendl and Hohlfeld, 2002).

2.2.2 Traumatic Brain Injury (TBI)

A modulatory role for TNF following traumatic brain injury as well as the impact of TNF neutralization on behavioral deficits has been extensively investigated in a number of different models, in particular fluid percussion injury (Fan et al., 1996; Knoblach et al., 1999; Vitarbo et al., 2004; Marklund et al., 2005), controlled cortical injury (Scherbel et al., 1999), and spinal cord injury (Harrington et al., 2005) with a number of different outcomes depending on the timing of the intervention.

2.2.3 Stroke (Cerebral Infarction) and Excitotoxic Injury

TNF can potentiate glutamate excitotoxicity directly by upregulating expression of both NMDA (Zou and Crews, 2005) and AMPA on synapses (Hermann et al., 2001; Beattie et al., 2002; Leonoudakis et al., 2004) and indirectly by inhibiting glial glutamate transporters on astrocytes (Choi, 1988). However, in animal models, the importance of TNF in hippocampal repair after ischemic injury is supported by the finding that stroke-induced hippocampal neurogenesis can be abolished in the presence of a TNF neutralizing antibody (Heldmann et al., 2005), presumably due to inhibition of TNFR2 signaling by tmTNF. Consistent with this finding, neuronal damage caused by focal cerebral ischemia and epileptic seizures were exacerbated in mice lacking both TNF receptors, suggesting that TNF serves a neuroprotective function in hippocampus under ischemic and excitotoxic conditions (Bruce et al., 1996; Guo et al., 2004). Similarly, intrahippocampal injection of TNF or transgenic targeted overexpression of TNF in astrocytes has been shown to inhibit susceptibility to kainate-induced seizures whereas mice lacking TNFR2 receptors (or both TNFR1 and R2) displayed increased susceptibility and prolonged seizure activity, suggesting that the protective effect of TNF is mediated by TNFR2 (Balosso et al., 2005). Data from these studies suggests that use of drugs that target TNF pathways to treat stroke or traumatic brain injury may have deleterious effects on hippocampal repair and neurogenesis.

2.2.4 Alzheimer's Disease

Interest in identifying polymorphisms in the TNF or TNF receptor genes linked to AD was largely fueled by the presence of this cytokine at amyloidogenic plaques and by results of genome-wide screening of families affected with late-onset AD. While a few individual studies find associations between polymorphisms in the TNF gene or the TNFR2 receptor gene with late-onset AD in families with no individuals possessing the APOE ε4 allele (Collins et al., 2000), others find no significant associations of three polymorphisms in the TNFR1 gene to AD (Perry et al., 2001b). Meta-analyses of genetic association studies will be required to assess overall genetic effect of genetic susceptibility loci and other cytokine genes on AD risk (Cacabelos et al., 2005; Wyss-Coray, 2006). Since polymorphisms in cytokine genes have already been linked to peripheral inflammatory disorders, such as juvenile rheumatoid arthritis, myasthenia gravis, and periodontitis, associations between cytokine gene polymorphisms and several chronic degenerative diseases may eventually be demonstrated (McGeer and McGeer, 2001). Dysregulated levels of TNF and other cytokines has been reported in AD patients and mouse models of AD, raising the possibility that they have disease-modifying effects and could be targeted in therapies. Higher serum TNF and TNF/IL-1β ratio have been detected in patients with severe AD compared to mild-moderate AD (Paganelli et al., 2002) whereas other studies have found no significant differences between studied groups (Blasko et al., 2006). Further investigations are warranted to validate these findings and assess their functional significance. In mouse models of AD-like pathology, elevated TNF and MCP-1 transcript levels were reported in entorhinal cortex of 3-month old 3 × Tg AD mice (Janelsins et al., 2005) coincident with accumulation of intraneuronal Aβ in these brain regions (Billings et al., 2005). Since these mice carry three transgenes encoding mutant proteins linked to Familial Alzheimer's Disease (FAD), these findings suggest that a sensitized genetic background may trigger an early chronic neuroinflammatory response that may involve (but not be limited to) TNF-dependent JNK activation leading to increased γ-secretase activity and enhanced progression of AD-like plaque and tau pathology (Janelsins et al., 2005). Indeed, chronic exposure to systemic lipopolysaccharide (LPS) was shown to hasten pathology in these mice (Kitazawa et al., 2005). In other mouse models of AD-like pathology such as the Tg2576 mice, elevated TNF levels are detectable around amyloid plaques (Mehlhorn et al., 2000; Munch et al., 2003) and exposure to systemic LPS worsens their pathology (Sly et al., 2001). Taken together, these findings strongly suggest that TNF may be an important modulator of AD-associated pathology via multiple cellular mechanisms including modulation of microglial phenotypes and regulation of the various proteases that process APP through TNF-dependent molecular signaling pathways. Consistent with this idea, a prospective, single-center, open-label, clinical pilot study involving 15 patients with mild-to-severe AD receiving twice weekly perispinal extrathecal administration of the TNF inhibitor etanercept for 6 months reported improved cognitive performance with treatment (Tobinick et al., 2006). These promising findings will need to be confirmed and investigated further in randomized, placebo-controlled clinical trials.

2.2.5 Parkinson's Disease

The levels of several cytokines, including TNF, IL-1β, and IFNγ are significantly increased in the substantia nigra pars compacta (SNpc) of PD patients compared to normal controls (Hirsch et al., 1998), particularly in the area of maximal destruction where the vulnerable melanin-containing dopamine-producing neurons reside. Although the genes for various cytokines, chemokines and acute phase proteins have been surveyed and individual reports demonstrate that certain single nucleotide polymorphisms in the TNF promoter that drive transcriptional activity are over-represented in a cohort of early onset Parkinson's Disease patients (Nishimura et al., 2001), these findings have not been confirmed in replicative studies or reported in other populations. Once these become available, meta-analyses of multiple such association studies will be needed to assess the overall genetic effect of TNF gene polymorphisms.

In experimental models of PD, significantly elevated levels of TNF mRNA and protein can be detected in the rodent midbrain substantia nigra within hours of in vivo administration of two neurotoxins widely used to model parkinsonism in rodents, 6-hydroxydopamine (6-OHDA) (Nagatsu et al., 2000b) and 1-methyl-4-phenyl-1,2,3,6-tetrahydropyridine (MPTP) (Rousselet et al., 2002; Sriram et al., 2002; Ferger et al., 2004). Consistent with a role of TNF in contributing to dopaminergic neuron death in chronic parkinsonism, plasma TNF levels were shown to remain elevated in MPTP-treated non-human primates 1 year after administration of the neurotoxin (Barcia et al., 2005). In addition, mice deficient in TNF or both TNF receptors have been reported to have altered dopamine metabolism and reduced survival of dopaminergic terminals (Rousselet et al., 2002) or reduced sensitivity to MPTP-induced neurotoxicity (Sriram et al., 2002; Ferger et al., 2004) (Table 5.1). Additional evidence that inflammation, and possibly TNF, is involved in nigral DA neuron degeneration comes from two recently developed endotoxin models of PD. In the first model, chronic low dose lipopolysaccharide (LPS) infusion into SNpc of rats results in delayed, selective and progressive loss of nigral DA neurons (Gao et al., 2002b). In the second model, exposure of pregnant rats to LPS and thus, in utero exposure of embryos to the endotoxin, caused a loss of DA neurons in postnatal brains (Ling et al., 2002). Most importantly, chronic infusion of dominant negative TNF inhibitor proteins into SNpc of adult rats protected nigral DA neurons from LPS and 6-OHDA induced degeneration (McCoy et al., 2006). Given that TNF receptors are expressed in nigrostriatal dopamine neurons (Tartaglia et al., 1993; Boka et al., 1994) and these neurons are selectively vulnerable to TNF-induced toxicity (Aloe and Fiore, 1997; Ling et al., 1998; McGuire et al., 2001; Gayle et al., 2002; Carvey et al., 2005), these early genetic studies and the more recent chronic inflammation models of PD strongly implicate TNF-dependent mechanisms and downstream targets in neurotoxin- and endotoxin-induced loss of nigral DA neurons and suggest that high TNF levels in the midbrain may increase susceptibility for PD in humans.

2.3   TNF as a Pharmacological Target

Many molecules have been shown to inhibit TNF synthesis or bioactivity with varying specificity. Endogenous inhibitors are particularly important for limiting TNF production and include soluble TNF receptors, glucocorticoids, prostaglandins, and cAMP (reviewed in Corti and Ghezzi, 2004). In addition, 'anti-inflammatory' responses initiated by IL-10 and IL-4 as well as vagus nerve-mediated central cholinergic activation through muscarinic receptors counteract TNF action in the periphery independent of muscarinic receptors present on macrophages (Pavlov and Tracey, 2005; Pavlov et al., 2006). Engineered Fc-fused versions of soluble TNF receptors like lenercept (Fc-TNFR1) and etanercept (Fc-TNFR2) were subsequently developed and modified further by PEGylation to increase their half-life in the circulation and anti-TNF antibodies have been humanized (e.g. infliximab, adalimumab). Many of these TNF antagonists have been used successfully in patients with inflammatory diseases like rheumatoid arthritis and Crohn's disease (reviewed in Szymkowski, 2005) and their success in the clinic underscores the deleterious effects of TNF overproduction in the periphery. However, the use of drugs like Enbrel (etanercept) and Remicade (infliximab) has been associated with an increased number of infections (Bongartz et al., 2006) and demyelinating disease (Arnett et al., 2001; Sukal et al., 2006) due to their ability to block not only soluble TNF but also tmTNF function which is critical in host defense, innate immunity, and nerve myelination. The newest generation of specific TNF inhibitors was designed by engineering a native agonistic protein (monomeric TNF) into a dominant negative one (DN-TNF) that does not bind TNFRs but still exchanges with native soluble TNF to form heterotrimers with attenuated activity; systemic administration of DN-TNFs reduced TNF-induced hepatotoxicity and collagen-induced arthritis (Steed et al., 2003), and intranigral administration of DN-TNFs attenuated loss of midbrain dopaminergic neurons in rat models of Parkinson's Disease (McCoy et al., 2006). These findings suggest that soluble-TNF selective and tm-TNF sparing DN-TNFs may represent a 'safer' second generation anti-TNF therapy for systemic and CNS applications.

Furthermore, since none of the currently available selective TNF inhibitors cross the blood–brain barrier, further modifications or development of alternative delivery modes will be needed if they are to be used to treat CNS diseases. The effectiveness of anti-TNF biologics has prompted investigations to identify small molecules that might be administered orally to act as TNF inhibitors. Several classes of drugs, including inhibitors of Nuclear Factor-kappa B (NFkB), TNF-α-converting enzyme (TACE) metalloprotease, p38 MAPK kinase inhibitors, and thalidomide, have been reported to act in this manner but none are specific for TNF. Consistent with a role of inflammation in the pathophysiology of PD, a prospective study of hospital workers found that daily use of non-steroidal anti-inflammatory drugs (NSAIDs) for a period greater than 2 years lowered the risk of developing PD by 46% (Chen et al., 2003), strongly suggesting that neuroin-flammation contributes to dopamine neuron loss and development of PD in humans. These findings raise the possibility that anti-inflammatory therapy could delay or prevent onset of PD. Thus, general anti-oxidants (Ling et al., 1999; Gao et al., 2002a, 2003; Isacson, 2002; Lin et al., 2003) as well as anti-inflammatory agents (Gao et al., 2003; Sairam et al., 2003; Cleren et al., 2005; Lund et al., 2005; Marchetti and Abbracchio, 2005; Youdim and Buccafusco, 2005) are being intensely investigated for their ability to offer neuroprotection to dopamine neurons in experimental models of PD and other models of neurodegeneration including Huntington's and AD. Early promising findings include the report that chronic extrathecal administration of the TNF inhibitor etancercept was able to improve cognitive performance in patients with mild-to-severe AD (Tobinick et al., 2006). These findings will need to be confirmed and investigated further in randomized, placebo-controlled clinical trials. On the other side of the spectrum, data from studies with TNF antibodies in animal models of stroke suggest that use of drugs that target TNF pathways to treat stroke or traumatic brain injury may have dele-terious effects on hippocampal repair and neurogenesis; it is clear that further research is needed to verify these findings. Perhaps the most rational approach will be directing the inflammatory machinery via selective regional targeting of inflammatory mediators with neurotoxic effects rather than suppressing microglia activation in general (Wyss-Coray and Mucke, 2002; Marchetti and Abbracchio, 2005), especially in light of the clear and unequivocal data that indicate certain inflammatory responses in the brain are beneficial and necessary for neural repair after injury.

3 Transforming Growth Factor-β

3.1 *Signaling Pathway and Expression in Normal CNS*

TGF-β1 is a member of the TGF-β superfamily which consists of several groups of highly conserved cytokines, growth factors, and morphogens. These include TGF-βs, bone morphogenic proteins (BMPs), activins, growth inhibitory factors, and nodal; all proteins with key functions during development and in maintaining tissue homeostasis (reviewed in Massagué et al., 2000; Derynck and Zhang, 2003; ten Dijke and Hill, 2004). The TGF-β subfamily includes three isoforms in mammals, TGF-β1, 2 and 3 which act in a highly contextual manner and depending on cell type and environment, may promote cell survival or induce apoptosis, stimulate cell proliferation or induce differentiation, and initiate or resolve inflammation. Accurate regulation of TGF-β bioactivity and signaling is key to control these opposing functions and essential to health and normal aging. Consequently, disruption of TGF-βs or TGF-β signaling molecules results in severe defects and usually embryonic lethality in mice (Letterio, 2000; Weinstein et al., 2000). Gene mutations leading to dysfunction of TGF-β signaling in humans have been linked to certain forms of cancer (Massagué et al., 2000; Bierie and Moses, 2006), Camurati–Engelmann and related developmental disorders (Loeys et al., 2005; Janssens et al., 2006), and hereditary hemorrhagic telangiectases (HHTs; (Fernandez et al., 2006)). HHT frequently affects the brain leading to cerebral hemorrhaging and will be discussed below.

The biological actions of TGF-βs are mediated by a receptor complex consisting of the TGF-β type 1 (ALK5) and type 2 (TβRII) serine/threonine kinase receptor subunits (reviewed in Massagué et al., 2000; Derynck and Zhang, 2003; ten Dijke and Hill, 2004). Endoglin and betaglycan are co-receptors or type-3 receptors that can modulate signaling (ten Dijke and Hill, 2004). Ligand binding results in recruitment and phosphorylation of a receptor-regulated Smad (R-Smad) protein which is facilitated by the Smad anchor for receptor activation (SARA; (Tsukazaki et al., 1998)). R-Smads 2 and 3 transmit signals for the TGF-β and activin pathways while Smad1, Smad5, and Smad8 do the same for the BMP pathway. Once phosphorylated, these R-Smads associate with Smad4, a co-Smad which is used by TGF-β, activin, and BMP pathways and together, the Smad heterodimers translocate into the nucleus where they bind to the Smad DNA-binding element (SBE) to regulate gene transcription (Itoh et al., 2000). The SBE is present in an estimated 400 genes (Yang et al., 2003) and specificity is obtained by binding of the Smad complex to other transcription factors. In the cytosol, inhibitory Smads such as Smad7 for TGF-β and activin signaling, and Smad6 for BMP signaling compete with R-Smad binding to the type I receptor and inhibit signaling by targeting the receptor complex to the ubiquitin degradation pathway.

Besides this canonical Smad-dependent signaling, TGF-β1 has been shown to engage the type 1 receptor ALK1 together with TβRII and signal via Smad 1. This pathway was first described in endothelial cells where canonical signaling promotes basement membrane synthesis and differentiation, while ALK1 dependent signaling promotes proliferation and angiogenesis (Oh et al., 2000; Goumans et al., 2002, 2003). Interestingly, ALK1 is also expressed by neurons (Konig et al., 2005) and ALK1 dependent signaling after injury has been proposed to mediate neuroprotective effects of TGF-βs by activating NF-kB (Konig et al., 2005). In addition, TGF-βs can activate and crosstalk with several other signaling pathways including the p38 MAP kinase pathway, and the JNK or NF-kB pathways (Derynck and Zhang, 2003).

In the normal CNS, TGF- β1, 2, and 3, and their receptors are expressed within neurons, astrocytes, and microglia (Flanders et al., 1998; Buckwalter and Wyss-Coray, 2004). TGF-β1 appears restricted to meningeal cells, choroid plexus epithelial cells, and glial cells, while TGF-β2 and -β3 are expressed in both glia and neurons (Unsicker et al., 1991). Similarly, TGF-β receptor immunostaining has been described in the CNS (Krieglstein et al., 1995). Neurons show immunoreactivity for both TβRI and TβRII in the human brain (Lippa et al., 1998; Tesseur et al., 2006b). TβRII immunoreactivity was prominent in the substantia nigra reticula and globus pallidus, whereas the striatum was only moderately stained, and the substantia nigra pars compacta lacked immunoreactivity. Immunoreactivity was associated with both neurons and glial cells, but not with myelinated fiber tracts (Andrews et al., 2006). Expression patterns of Smad proteins have also been characterized in mouse embryos (Flanders et al., 2001; de Sousa Lopes et al., 2003) or in adult mice (Huang et al., 2000). Evidence for an important role of TGF-β signaling in the mammalian brain comes from transgenic reporter mice in which luciferase is expressed under control of a heterologous SBE promoter (Lin et al., 2005; Luo et al., 2006). The brain, and within the brain the hippocampus, showed the highest baseline reporter activity of all major organs in the mouse indicating significant constitutive TGF-β signaling in the CNS. In support of these findings phosporylated, active forms of Smad2 were detected predominantly in large pyramidal neurons in the hippocampus and to a lesser extent in the neocortex (Lin et al., 2005; Luo et al., 2006).

What the role of such constitutive TGF-β signaling is remains to be shown but studies of the TGF-β signaling pathway in *Aplysia Californica* and *Drosophila melanogaster*, provide evidence for a role in synaptic function, synaptogenesis, and retrograde signaling (reviewed in Kalinovsky and Scheiffele, 2004; Sanyal et al., 2004). In *Aplysia*, TGFβ1 induces long term facilitation at sensory-motor synapses (Zhang et al., 1997), which can be blocked by expression of a dominant-negative TβRII. TGF-β1 also induced phosphorylation and redistribution of the presynaptic protein synapsin and modulates synaptic function (Zhang et al., 1997; Chin et al., 2002). In *Drosophila*, TGF-β receptors and dSmad2 are required for neuronal remodeling (Zheng et al., 2003). Mutations in *spinster*, a multipass transmembrane protein that is a negative regulator of synaptic growth, resulted in enhanced TGF-β signaling, and caused synaptic overgrowth, which could be reverted by downregulation of TGF-β signaling (Sweeney and Davis, 2002).

3.2   Role of TGF-β1 in Disease

Consistent with its central role in injury responses, angiogenesis and vascular maintenance, immune regulation and cell proliferation, TGF-β signaling has been implicated in brain injury and stroke, neurodegeneration, cerebrovascular disorders, and multiple sclerosis. The TGF-β1 promoter is activated by wounding, ischemia, and cellular stress (Kim et al., 1990; Roberts and Sporn, 1996). Accordingly, TGF-β1 increases rapidly after injury as well as with age (Finch et al., 1993). The main sources of TGF-β1 in the injured brain are astrocytes and microglia (Finch et al., 1993). TGF-β2 and TGF-β3 are regulated mainly by hormonal and developmental signals and may be more important in CNS development (Flanders et al., 1998) (Dennler et al., 2002; Shi and Massague, 2003).

3.2.1   Brain Injury and Stroke

TGF-β1 protects neurons against various toxins and injurious agents in cell culture and in vivo (Flanders et al., 1998; Unsicker and Krieglstein, 2002; Buckwalter and Wyss-Coray, 2004). For example, intracarotid infusion of TGF-β1 in rabbits reduces cerebral infarct size when given at the time of ischemia (Gross et al., 1993). Rat studies also showed that TGF-β1 protects hippocampal neurons from death when given intrahippocampally or intraventricularly 1 h prior to transient global ischemia (Henrich-Noack et al., 1996). Mice infected with adenovirus that overexpressed TGF-β1 5 days prior to transient ischemia also had smaller infarctions than control animals (Pang et al., 2001). Astroglial overexpression of TGF-β1 in transgenic mice protected against neurodegeneration induced with the acute neurotoxin kainic acid or degeneration associated with chronic lack of apolipoprotein E expression (Table 5.2) (Brionne et al., 2003). Similarly, viral expression of TGF-β1 protects neurons from excitotoxic death in mice (Boche et al., 2003). In agreement with this protective effect, TGF-β1 knockout mice displayed spontaneous neuronal death with prominent clusters of TUNEL-positive cells in different parts of the brain (Brionne et al., 2003). In addition, unmanipulated 3-week-old TGF-β1 knockout mice had significantly fewer synaptophysin positive synapses in the neocortex and hippocampus compared to wildtype littermate controls, and showed increased susceptibility to kainic acid-induced neurotoxicity (Brionne et al., 2003). Primary neurons lacking TGF-β1 (Brionne et al., 2003) or expressing a defective, dominant-negative TβRII (Table 5.2) (Tesseur et al., 2006b) do not survive as well as wildtype cells supporting a critical role for neuronal TGF-β signaling in cellular maintenance and survival.

It is not clear how TGF-β1 protects neurons, but several mechanisms have been postulated. For example, TGF-β1 decreases Bad, a pro-apoptotic member of the Bcl-2 family, and contributes to the phosphorylation, and thus inactivation, of Bad by activation of the Erk/MAP kinase pathway (Zhu et al., 2002). TGF-β1 also increases production of the anti-apoptotic protein Bcl-2 (Prehn et al., 1994) possibly

Table 5.2 CNS phenotypes resulting from genetic modulation of other cytokine and their receptors (listed are only models with CNS phenotypes)

Mouse model[a]	Manipulation	Affected cell type (transgene promoter)	CNS phenotype
TGF-β1 ko	Deletion of TGF-β1	All cells	NIH: ↑ CNS inflammation ↑ Neuronal apoptosis ↑ Synaptic deficits ↑ Neurodegeneration (Brionne et al., 2003)
TGF-β1 tg	Overexpression of TGF-β1	Astrocytes (GFAP)	High levels result in hydrocephalus (Galbreath et al., 1995; Wyss-Coray et al., 1995) ↑ Age- and injury-related neuroprotection (Brionne et al., 2003) ↑ Fibrosis of blood vessels (Wyss-Coray et al., 1995; Wyss-Coray et al., 2001) ↑ Inflammation (Wyss-Coray et al., 1997a) ↑ Amyloid in blood vessels (Wyss-Coray et al., 1997b) ↓ Amyloid in parenchyma (Wyss-Coray et al., 2001) ↓ Brain tissue perfusion (Gaertner et al., 2005) ↓ Adult neurogenesis (Buckwalter et al., 2006a)
TGF-β1 tg	Tet-O-TGF-β1 × CamKII-tTA	Neurons (CamKII)	↑ Age- and injury-related neuroprotection ↑ Fibrosis of blood vessels ↑ Amyloid in blood vessels (Ueberham et al., 2005)
TGF-β2 ko	Deletion of TGF-β2	All cells	Lethal around birth: craniofacial malformations, neural tube defects (Andrews et al., 2006) Age-related nigrostriatal dopamine deficits (Andrews et al., 2006)
TβRII deficient	Tet-O-TβRIIΔk × Prnp-tTA	Neurons	Age-related neurodegeneration and Aβ deposition (Tesseur et al., 2006a)
Endoglin deficient	Heterozygous ko allele	All cells	Sporadic telangiectases with cerebral hemorrhages (Bourdeau et al., 1999, 2001)
IL-1R ko	Deletion of IL-1 receptor 1	All cells	↓ Sleep; deficits in synaptic plasticity; impaired spatial memory (Allan and Rothwell, 2001)
IL-1α/IL-1β dko	Deletion of IL-1α and IL-1β	All cells	↓ Ischemia-induced neuronal apoptosis; single ligand ko susceptible (Seripa et al., 2003)

(continued)

Table 5.2 (continued)

Mouse model[a]	Manipulation	Affected cell type (transgene promoter)	CNS phenotype
GILRA2, GILRA4	Overexpression of human IL-1ra	Astrocytes (GFAP)	↓ IL-1β-induced febrile response (Lundkvist et al., 1999)
IL-18 ko	Deletion of IL-18	All cells	Lack of IL-6 production in CNS (Sekiyama et al., 2005)
IL-6 ko	Deletion of IL-6	All cells	Normal open-field behavior (exploratory, anxiety and depression-related behavior)
			Resistance to MOG-induced EAE (Swiergiel and Dunn, 2006)
			↓ Response to conditioning lesion in peripheral nerve (Cafferty et al., 2004)
IL-6 tg	Overexpression	Astrocytes (GFAP)	Severe neurologic disease with runting, ataxia, seizures; neurodegeneration, astrocytosis
			↑ Acute phase proteins (Campbell et al., 1993)
			Cognitive decline (Campbell et al., 1997)
IL-10 ko	Deletion of IL-10	All cells	↑ EAE severity (Bettelli et al., 1998; Samoilova et al., 1998a)
			↑ Damage after spinal cord injury (Abraham et al., 2004) and ischemia (Grilli et al., 2000)
			↑ Susceptibility to prion disease (Thackray et al., 2004)
IL-12 ko	Deletion of IL-12p35 subunit	All cells	↑ EAE severity (Becher et al., 2002; Cua et al., 2003)
			↓ Neurodegeneration and inflammation after excitotoxic brain injury (Chen et al., 2004)
IL-12/IL-23 ko	Deletion of p40 subunit (shared by IL-12 and IL-23)	All cells	↓ EAE severity, resistant (Segal et al., 1998; Becher et al., 2002; Cua et al., 2003)
IL-12 tg	Overexpression of IL-12p35 and p40	Astrocytes (GFAP)	Severe inflammation and neurodegeneration in the cerebellum, neocortex, and hippocampus (Pagenstecher et al., 2000)

CNS central nervous system; *EAE* experimental autoimmune encephalomyelitis; *MBP* myelin basic protein; *GFAP* glial fibrillary acidic protein; *ko* knockout; *dko* double knockout; *tg* transgenic

[a]Details on mouse models including expression pattern and levels, the presence of mutations, etc. are described in the cited papers

↑, increased; ↓, decreased

via ALK1 and NFkB signalling (Konig et al., 2005). In addition, TGF-β1 synergizes with neurotrophins and may be necessary for at least some of the effects of neurotrophins, fibroblast growth factor-2, and glial cell-line derived neurotrophic factor (reviewed in Unsicker and Krieglstein, 2000; Unsicker and Krieglstein, 2002). In addition, TGF-β1 increases laminin expression (Wyss-Coray et al., 1995) and is necessary for normal laminin protein levels in the brain (Brionne et al., 2003). Laminin is thought to provide critical support for neuronal differentiation and survival and may be important for learning and memory (Venstrom and Reichardt, 1993; Luckenbill-Edds, 1997). Lastly, TGF-β1 may decrease inflammation in the infarction area, attenuating secondary neuronal damage (Pang et al., 2001).

3.2.2 Neurodegeneration

Abnormal TGF-β signalling has been described in AD, PD, and other neurodegenerative conditions of the CNS (Flanders et al., 1998; Buckwalter and Wyss-Coray, 2004). Many reports describe changes in TGF-β expression in AD brains, CSF, and serum (van der Wal et al., 1993; Peress and Perillo, 1995; Wyss-Coray et al., 1997b; Luterman et al., 2000; Wyss-Coray et al., 2001; Grammas and Ovase, 2002; Tarkowski et al., 2002). TGF-β1 immunoreactivity is increased in or close to amyloid plaques (van der Wal et al., 1993; Peress and Perillo, 1995) and around cerebral blood vessels (Wyss-Coray et al., 1997b; Wyss-Coray et al., 2001; Grammas and Ovase, 2002). Brain TGF-β1 mRNA levels vary greatly between individuals but are overall increased in AD brains compared with controls and correlate positively with the extent of cerebrovascular amyloid deposition or cerebral amyloid angiopathy (CAA) (Wyss-Coray et al., 1997b; Wyss-Coray et al., 2001). Interestingly, TGF-β1 mRNA levels are inversely correlated with parenchymal Aβ deposition (Wyss-Coray et al., 2001), consistent with the observation that cerebrovascular and parenchymal Aβ deposition are inversely correlated (Wyss-Coray et al., 2001; Tian et al., 2003). Additional support for a role of TGF-β1 in CAA comes from studies of mice that model AD and also overproduce TGF-β1 (Table 5.2) (Wyss-Coray et al.). While the human amyloid precursor protein (hAPP) transgenic mice develop amyloid plaques mainly in the brain parenchyma, hAPP/TGF-β1 doubly transgenic mice accumulate amyloid overwhelmingly in blood vessels. It is conceivable that this CAA phenotype is the result of abnormal TGF-β signaling in endothelial or smooth muscle cells and pericytes, leading to excessive accumulation of basement membrane proteins in the vasculature (Wyss-Coray et al., 2000; Grammas and Ovase, 2002).

Interestingly, TGF-β1 overexpression in hAPP mice not only promoted CAA but also resulted in 60–75% reduction in the number of parenchymal plaques and overall Aβ levels (Wyss-Coray et al.). At least some of this reduction in Aβ accumulation may involve innate immune mechanisms including activation of microglial cells (Wyss-Coray et al., 2002) and increased production of complement C3. Other mechanisms of action may involve the neuroprotective functions of

TGF-β signaling. Thus, reduced TGF-β signaling in neurons of hAPP mice resulted in a twofold increase in amyloid accumulation and increased neurode-generation (Tesseur et al., 2006b). As brains from patients with AD had promi-nently reduced levels of TβRII compared to those without disease (Tesseur et al., 2006b) it is possible that deficiency in TGF-β signaling may contribute to AD as well.

3.2.3 Multiple Sclerosis

TGF-β1 can exert either anti- or pro-inflammatory effects and initiate or resolve inflammatory responses through the regulation of chemotaxis, activation, and survival of immune cells (Luethviksson and Gunnlaugsdottir, 2003; Li et al., 2006). This dual nature has become evident recently in EAE, a model for MS. TGF-β1 and its receptors are expressed in the CNS inflammatory lesions of MS patients and in various animal models of the disease (Johns et al., 1991; Issazadeh et al., 1995b; De Groot et al., 1999) suggesting a role in the disease process. In patients with MS, higher levels of TGF-β1 mRNA have been associated with reduced disability (Link et al., 1994) and MRI evidence of disease (Bertolotto et al., 1999) while serum levels of TGF-β1 are elevated both at the time of relapses and in response to IFN-γ treatment (Nicoletti et al., 1998). Consistent with these potentially protective effects, peripheral treatment with exogenous TGF-β1 reduced the incidence and severity of inflammation and demyelination in the CNS of rodents with EAE (Johns et al., 1991; Kuruvilla et al., 1991; Racke et al., 1991) and administration of TGF-β neutralizing antibodies worsened disease (Johns and Sriram, 1993). In contrast, astrocyte restricted overexpression of TGF-β1 in brains of transgenic mice results in more severe EAE, with earlier disease onset and larger mononuclear cell infiltrates in the CNS (Table 5.2) (Wyss-Coray et al., 1997a). In a related experiment, overexpression of TGF-β1 in hAPP transgenic mice resulted in increased T cell accumulation in the brain after immunization with Aβ peptide, which forms the characteristic amyloid plaques in this model for AD (Buckwalter et al., 2006b). Thus, the effect of TGF-β1 on cerebral accumulation of T cells is not restricted to myelin-specific cells. Even in the absence of immunization, T cells accumulate with aging in unmanipulated TGF-β1 transgenic mice (Buckwalter et al., 2006b).

While these studies suggest that local expression of TGF-β1 in the target organ is responsible for its pro-inflammatory effects recent studies show that TGF-β1 is also necessary for the priming and expansion of autoreactive T cells. Strikingly, transgenic mice with CD4 T cells expressing a dominant negative form of the TβRII and thus, not responsive to TGF-β1, failed to generate Th-17 cells and did not develop EAE (Veldhoen et al., 2006a). Indeed, TGF-β1 is a crucial factor for the differentiation of Th-17 cells (Bettelli et al., 2006; Mangan et al., 2006; Veldhoen et al., 2006b), a T cell subset producing IL-17 and believed to be autoreactive and disease promoting (Langrish et al., 2005; Park et al., 2005; Veldhoen et al., 2006a). In summary, the effects of TGF-β1 in autoimmune encephalitis and MS seem highly dependent on temporal and spatial expression making this pathway a potentially challenging therapeutic target in this disease.

3.2.4 Hydrocephalus

Hydrocephalus and brain fibrosis are common outcomes after whole brain inflammation due to bacterial meningitis, subarachnoid hemorrhage, or severe traumatic brain injury and TGF-β1 may have an important role in its progression (Crews et al., 2004). High CSF levels of TGF-β1 in patients with subarachnoid hemorrhage confer an increased risk of developing chronic hydrocephalus (Kitazawa and Tada, 1994; Takizawa et al., 2001). Indeed, TGF-β1 injected into the lateral ventricles produces hydrocephalus in mice (Tada et al., 1994) and TGF-β1 mice with high levels of expression in astrocytes had persistent communicating hydrocephalus at birth, enlargement of cerebral hemispheres, and thinning of the overlying cerebral cortex (Galbreath et al., 1995; Wyss-Coray et al., 1995). Histological analysis showed decreased stratification of neuronal cell layers and leukomalacia-like areas in addition to excess basement membrane synthesis around blood vessels and in the meninges. Given the extensive fibrosis of the meninges in TGF-β1 mice hydrocephalus may be the result of decreased CSF reabsorption at fibrotic arachnoid villi (Crews et al., 2004).

3.2.5 Cerebrovascular Disorders

TGF-β1 has a key role in formation and maintenance of blood vessels which may be relevant for cerebrovascular disease. In general, low levels of TGF-β1 appear necessary for endothelial cell proliferation and angiogenesis, but higher levels result in increased synthesis of basement membrane proteins and differentiation (Oh et al., 2000; Goumans et al., 2002, 2003). Thus, studies in TGF-β1 knockout and other mice demonstrated an essential role for TGF-β1 in vasculogenesis and angiogenesis during development (Table 5.2) (Dickson et al., 1995) (Kulkarni et al., 1993) as well as in maintaining vascular integrity in adults (Madri et al., 1992; Pepper, 1997). Two TGF-β receptors on endothelial cells, endoglin and ALK1, mediate at least part of these effects and targeted mutations of endoglin (Li et al., 1999) or ALK1 (Oh et al., 2000) lead to lethal defects in vasculogenesis and angiogenesis in mice. In humans, hereditary mutations in the genes encoding for endoglin or ALK1 cause hereditary hemorrhagic telangiectasia (HHT) (McAllister et al., 1994; Johnson et al., 1996), a disease characterized by recurrent epistaxis, telangiectasias, hemorrhages, and arteriovenous malformations. Many of these malformations occur in the brain and mice heterozygous for endoglin show that haploinsufficiency leads to HHT-like disease in the CNS (Satomi et al., 2003).

On the other hand, too much TGF-β1 is detrimental for the vasculature as well, possibly by inducing excessive extracellular matrix deposition. Transgenic mice overexpressing TGF-β1 in astrocytes demonstrated an age- and dose-dependent accumulation of thioflavin S–positive perivascular amyloid deposits and degeneration of vascular cells (Wyss-Coray et al., 2000). Nevertheless, the amyloid deposits which did not contain appreciable amounts of endogenous Aβ had an appearance similar to those found in brains of AD patients with concomitant CAA. Analysis of the progression of the cerebrovascular changes in TGF-β1 mice showed a significant accumulation of basement membrane proteins perlecan and fibronectin in microvessels of young adult mice that preceded the formation of thioflavin S positive amyloid (Wyss-Coray et al., 2000). Consistent with these findings, inducible neuronal expression of TGF-β1 led to the formation of thioflavin S–positive cerebrovascular deposits which remained even after transgene expression was turned off (Ueberham et al., 2005). These pro-fibrotic effects are in line with the pathological role TGF-β1 has in fibrosis in the lung (Kumar et al., 1995; Pittet et al., 2001) and the kidney (Border and Noble, 1997). Cerebrovascular fibrosis observed in normal aging or hypertension (Suthanthiran et al., 2000) may well be the result of increased vascular TGF-β signaling and could be a factor in the formation of cerebrovascular amyloid in humans.

3.3   The TGF-β Pathway as a Pharmacological Target

Due to their potent actions not only in neurological disease but in peripheral diseases as well TGF-β1 and the TGF-β signaling pathway are attractive drug targets. However, because TGF-β1 can exert opposing actions on cellular processes depending on contextual factors a better understanding of the regulation of TGF-β1 activation and signaling may be necessary to target regulators or downstream modulators of TGF-β signaling. The therapeutic benefit of TGF-β protein has been tested in animal models or humans to promote wound healing, bone formation and a number of other conditions although delivery remains a major problem [for review see (Flanders and Burmester, 2003)]. As discussed above, TGF-βs delivered via stereotactic or viral means reduced ischemic damage in a number of stroke models (Gross et al., 1993; Henrich-Noack et al., 1996; Pang et al., 2001). TGF-β1 has also been shown to reduce autoimmune encephalomyelitis in a rodent model for multiple sclerosis if administered in the periphery (Racke et al., 1991) or via genetically engineered TGF-β1 expressing T cells (Chen et al., 1998). A small molecule compound with TGF-β-like activity has been described to target histone deacetylase and may be developed to harness some of these beneficial effects of TGF-β (Glaser et al., 2002).

To target the detrimental actions of the TGF-β pathway a number of small molecule inhibitors of the kinase activity of the TGF-β type I receptor ALK5 have been developed. These drugs are tested to inhibit fibrosis, scarring and tumor-associated angiogenesis among other processes (Callahan et al., 2002; Hjelmeland et al., 2004; Kim et al., 2004; Uhl et al., 2004; Moon et al., 2006). Based on preclinical studies ALK5 inhibitors seem particularly promising in the treatment of gliomas (Uhl et al., 2004).

4 Interleukin 1 Family

4.1 Signaling Pathways and Expression of IL-1 Family and its Receptors in Normal CNS

Members of the IL-1 family include IL-1α, IL-1β, IL-18, and the interleukin IL-1 receptor antagonist (IL-1ra) as well as 6 other proteins with unknown functions (reviewed in Allan et al., 2005). IL-1α is mostly cell-membrane associated and shares little sequence homology with IL-1β. IL-1β and IL-18 are closely related; both possess a similar three-dimensional structure, and their respective precursor forms are inactive until cleaved by the intracellular cysteine protease caspase-1. IL-1α and IL-1β mediate peripheral and central inflammation by binding to the IL-1 type I receptor (IL-1RI) and activating NFkB and stress-activated MAPK signaling cascades. IL-1ra binds to the same receptor and can modulate IL-1 activity. In addition, IL-1α, IL-1β, and IL-1ra bind also to IL-1R2, while IL-18 binds to IL-1R related protein. Recent data suggests that while IL-1α and IL-1β induce identical IL-1 signaling pathways, IL-1β is significantly more potent than IL-1α in stimulating IL-6 release in primary mixed glia (Andre et al., 2005); these differential actions in the CNS raise the possibility that there may be other IL-1 receptor(s) in the brain. Numerous reports have correlated the presence of increased IL-1 in the injured or diseased brain, and its effects on neurons and nonneuronal cells in the CNS, but the importance of IL-1 signaling in normal brain function has only recently been recognized (Basu et al., 2004). Despite their low levels in normal brain, the IL-1 family of proteins has now been shown to exert numerous biological effects in the brain, including induction of acute-phase proteins needed in neuroimmune responses and activation of many inflammatory processes with direct actions on CNS neurons. Evidence that IL-1 is higher in the brain during sleep and that spontaneous sleep can be reduced by IL-1ra implicate this cytokine in sleep physiology. In addition, increased expression of IL-1 after the induction of hippocampal long-term potentiation (LTP) and reversal by IL-1ra establish a connection to synaptic plasticity (reviewed in Allan et al., 2005).

4.2 Role of IL-1 Family in Disease

A number of polymorphisms in the genes that encode IL-1α (IL-1A), IL-1β (IL-1B) and IL-1ra (IL-1RA; IL-1RN) have been implicated in CNS diseases (reviewed in Allan et al., 2005), however many associations are weak or have not been replicated. In contrast, there is substantial evidence that IL-1 and IL-18 are involved in the neuronal injury that occurs in chronic neurodegenerative disorders and the acute damage seen after stroke and brain trauma (Allan et al., 2005). The molecular basis of these cytokine effects remains to be elucidated. For example, in vivo administration of IL-1β (60 μg/kg, i.p.) induced the phosphorylation of ERK1/2 in neurons, astrocytes and microglia in areas at the interface between brain and blood or cerebrospinal fluid: meninges, circumventricular organs, endothelial-like cells of the blood vessels, and in brain nuclei involved in behavioral depression, fever and neuroendocrine activation (paraventricular nucleus of the hypothalamus, supraoptic nucleus, central amygdala and arcuate nucleus). In addition, recent studies support a crucial role for IL-18 in mediating neuroinflammation and neurodegeneration in the CNS under pathological conditions, such as bacterial and viral infection, autoimmune demyelinating disease, and hypoxic-ischemic, hyperoxic and traumatic brain injuries (Felderhoff-Mueser et al., 2005). In psychologically and physically stressed organisms, IL-18 influences pathological and physiological processes by participating in stress-related disruption of host defenses (Sekiyama et al., 2005).

The use of transgenic and gene-knockout mice has provided the opportunity to substantiate the physiological significance of IL-1-family members in the CNS (Table 5.2). Mice deficient in IL-1α and IL-1β show significantly reduced ischemic cell death, whereas deletion of either IL-1α or IL-1β alone does not protect. It has also been demonstrated that IL-1 exacerbates ischemic injury in mice in the absence of IL-1R1, again suggesting the existence of novel IL-1 receptors in the brain. Mice that are deficient in IL-1R1 sleep less, show deficits in synaptic plasticity, and perform worse in a spatial memory task than control animals (reviewed in Allan et al., 2005). Although most of the neurodegenerative effects of IL-1 are thought to be mediated through IL-1β, data from genetic models implicate IL-1α in excitotoxic cell death as well. Furthermore, intrastriatal administration of the excitotoxin AMPA in the rat brain leads to marked increases in IL-1 cytokine expression in the frontoparietal cortex and drastically exacerbates neuronal loss in this region. Enhancement of cell death pathways by IL-1 results in increased limbic seizures (Seripa et al., 2003). IL-1β (Cunningham et al., 1996; Curran et al., 2003) and IL-18 (Curran and O'Connor, 2001) have also been reported to have inhibitory effects on LTP. From human and animal studies, we can conclude that IL-1β and IL-18 participate in fundamental inflammatory processes that increase during the aging process as well as in the febrile response (see below). Lastly, transgenic mouse models with astrocyte-directed overexpression of the IL-1ra have demonstrated an important role of IL-1 signaling in the CNS. Two additional GFAP-hsIL-1ra strains have been generated and characterized further: GILRA2 and GILRA4. These strains show a brain-specific expression of the hsIL-1ra at the mRNA and protein levels in the CSF which are 13- and 28-fold higher respectively, compared to wild-type (Lundkvist et al., 1999).

4.2.1 Febrile Response

Fever is a normal adaptation in response to a pyrogenic stimulus (tissue injury, pathogenic bacteria, etc.) resulting in the generation of cytokines and prostaglandins (Boutin et al., 2003). The body produces a wide array of pyrogenic cytokines such as interleukins (IL-1, IL-6), interferon, and TNF which are sensed by the circumventricular organ system (CVOS) lying at the interface of the blood–brain barrier. When pyrogenic cytokines are detected by the CVOS, prostaglandin synthesis, especially cyclo-oxygenase-dependent prostaglandin E2, is induced as part of the febrile response. Once the hypothalamus receives the signal, autonomic, endocrine, and behavioral processes are activated until the hypothalamic set-point is reset downward as a consequence of a reduction in pyrogen content or antipyretic therapy, with subsequent heat loss. The febrile response elicited by IL-1β (50 ng/mouse i.c.v.) was abolished in hsIL-1ra-overexpressing animals, suggesting that the central IL-1 receptors were occupied by antagonist.

4.2.2 Traumatic Brain Injury and Epilepsy

Increases in IL-1 early after brain injury lead to increases in neuronal excitability through modulation of the balance between excitatory and inhibitory synaptic transmission, induction of neurotoxin production, leukocyte infiltration, activation of microglial cells and astrocytes (reviewed by Allan et al., 2005).

4.2.3 Ischemic Stroke and Excitotoxicity

IL-1 exerts a number of diverse actions in the brain, and it is currently well accepted that it contributes to experimentally induced neurodegeneration. IL-1 is one of the key modulators of the inflammatory response after cerebral stroke, and its activity is critically regulated by its receptor antagonist IL-1ra. Much of this is based on studies using IL-1ra, which inhibits cell death caused by ischemia, brain injury, or excitotoxins (Hailer et al., 2005; Pinteaux et al., 2006).

4.2.4 Inflammation in Aging

Patients with mutations in the NALP3 gene, which controls the activity of caspase-1, readily secrete more IL-1β and IL-18 and suffer from systemic inflammatory diseases (Dinarello, 2006). Patients with defects in this gene have high circulating concentrations of IL-6, serum amyloid, and C-reactive protein, each of which decrease rapidly upon blockade of the IL-1 receptor, which suggests that IL-1β contributes to the elevation of these markers as part of the inflammatory mechanisms associated with aging.

4.2.5 Alzheimer's Disease

IL-1-dependent inflammation may contribute to the pathophysiology of AD since it is expressed at abnormally high levels by glial cells in AD in the vicinity of amyloid plaques and can lead to neuronal injury and activation of p38 MAPK pathways and phosphorylation of tau (Sheng et al., 2001; Li et al., 2003; Cacquevel et al., 2004). IL-1 promoter polymorphisms were among several gene polymorphisms in inflammation-related genes suspected to compose a susceptibility profile for AD risk that also included IL-6, TNF, α-2-macroglobulin (A2M), and α-1-antichymotrypsin (AACT). Although several individual reports of polymorphisms in genes of various IL-1 family ligands (IL-1α -889, IL-1β -511, IL-1β + 3953) suggested associations between specific alleles and AD in that their presence appeared to increase the risk for AD or modified the age-at-onset of AD (Du et al., 2000; Grimaldi et al., 2000; Nicoll et al., 2000; Kolsch et al., 2001; Murphy et al., 2001; Ehl et al., 2003), other studies revealed no significant associations between these polymorphisms and AD (Ki et al., 2001; Green et al., 2002). Therefore, as is the case for other cytokines, further evaluation of the association of IL-1 gene polymorphisms with AD and their role in pathogenesis is needed.

4.2.6 Parkinson's Disease

The CSF of PD patients has been reported to contain high concentrations of IL-1 (Blum-Degen et al., 1995), but the role of IL-1β in this disease is still unclear. Patients with AD and Lewy body dementia showed co-localization of IL-1β -expressing microglia with neurons that were highly immunoreactive for β-amyloid precursor protein (βAPP) and contained both Lewy Bodies and neurofibrillary tangles (Grigoryan et al., 2000), raising the possibility that the clinical and neuropathological overlap between AD and PD may be mediated by IL-1β (Mrak and Griffin 2007). Although numerous in vitro and in vivo studies have demonstrated that IL-1β contributes to the degeneration of neurons in SNpc (reviewed in Allan et al., 2005), a recent report suggested it may also have a neuroprotective role by eliciting GDNF release from astrocytes under acute inflammatory conditions (Saavedra et al., 2007). Consistent with this finding, IL-1β promoter polymorphisms appear to be protective in PD (Nishimura et al., 2001, 2005); additional studies will be needed to confirm these findings.

4.3   IL-1 Family Members as Therapeutic Targets

Because the IL-1 family coordinates both systemic and CNS host defense responses to pathogens and to injury and is at or near the top of the hierarchical cytokine signaling cascade in the CNS that activates neuroinflammation, it may provide an attractive target for therapeutic intervention in diseases where the latter contributes to destruction of vulnerable neuronal populations. For example, IL-1β is the major inducer of central cyclo-oxygenase 2 (COX2) activity and as such has an important role in production of prostaglandin E2 and other prostanoids which sensitize peripheral nociceptive terminals and produce localized pain hypersensitivity. In contrast, inhibiting IL-1 actions (by intracerebroventricular (i.c.v) injection of IL-1ra, neutralizing antibody to IL-1 or caspase-1 inhibitor) significantly reduces ischemic brain damage (Boutin et al., 2001, 2003). Similarly, controlling the caspase-1 activating pathway to suppress IL-18 levels may provide preventative means against stress-related disruption of host defenses.

5 Interleukin 6

5.1 Signaling Pathways and Expression of IL-6 and its Receptors in Normal CNS

Cellular sources of Interleukin -6 (IL-6) are glial cells and neurons; and its receptors are found in the CNS where they signal via the Janus kinase – signal transducer and activator of transcription (JAK-STAT) pathway involving STAT3 (Taga, 1996). Glial production of IL-6 has been studied intensively, but comparatively little is known about the induction of IL-6 in neurons. Interestingly, because neuronal IL-6 expression is induced by excitatory amino acids or membrane depolarization, it may be an activity-dependent physiological neuromodulator of brain functions (Juttler et al., 2002). In fact, cytokines such as IL-1, IL-6, and TNF, previously thought to only have immune-related functions, have been shown to participate in activity-dependent structural synaptic changes in specific neurochemical circuitries in the normal brain (Tonelli and Postolache, 2005). These processes range from the refinement of synaptic connections in sensory systems to learning and memory storage functions of the hippocampus. Therefore, the theme emerges that mechanisms of defense against pathogens can drastically affect brain structure and function by inducing changes in cognition, mood and behavior.

5.2 Role of IL-6 in CNS Disease

High levels of IL-6 along with TNF and IL-1β have been shown to impair cognition (Tonelli and Postolache, 2005). Insight into the role of IL-6 in chronic-progressive neurodegenerative disorders has come from in vivo studies with transgenic mice in which the cytokine ligand/receptor system has been over-expressed, deleted or blocked (Table 5.2). Specifically, astrocyte-targeted overexpression of IL-6 resulted in severe neurologic disease with runting, ataxia, seizures; neurodegeneration, astrocytosis, acute phase proteins and cognitive decline (Campbell et al., 1997). Yet IL-6 ko mice appear to be no different from wild-type littermates in ambulatory, exploratory, and stereotypic activities in home or novel cages, in an open field (OF), in the multicompartment chamber (MCC), or in the elevated plus-maze (EPM). In general, their feeding, exploratory, anxiety- and depression-related behaviors were normal and the plasma corticosterone and basal concentrations of catecholamines, indoleamines and their metabolites in several brain regions were similar to those of wildtype mice (Swiergiel and Dunn, 2006). In contrast, use of IL-6-deficient mice in experimental models of MS suggest that this cytokine may be critical for the activation and differentiation of autoreactive T cells in vivo and that blocking IL-6 function can be an effective means to prevent EAE (Samoilova et al., 1998b). Although autoreactive T cells recognizing self myelin antigens are present in most individuals, autoimmune disease of the central nervous system is a relatively rare medical condition perhaps because development of autoimmune disease requires that autoreactive T cells actually become activated. It has been proposed that activation of T cells requires a minimum of two signals: an antigen-specific signal delivered by MHC-peptide complex and a second signal delivered by costimulatory molecules or cytokines. IL-6-deficient mice were completely resistant to EAE induced by myelin oligodendrocyte glycoprotein (MOG), whereas IL-6-competent control mice developed EAE characterized by focal inflammation and demyelination in the CNS and developed neurological deficits. Furthermore, it was demonstrated that the resistance to EAE in IL-6 ko mice was associated with a failure of MOG-specific T cells to differentiate into either Th1 or Th2 type effector cells in vivo. Polymorphisms in the IL-6 gene have been investigated for links to human disease, including stroke and ischemia (Balding et al., 2004), intracranial hemorrhage (Pawlikowska et al., 2004; Chen et al., 2006), cerebral malaria (Jakobsen et al., 1994), and Parkinson's Disease (Hakansson et al., 2005). As is the case with other cytokine polymorphisms, confirmation by replicative studies and meta-analyses of multiple such association studies will be needed to assess their overall genetic effect.

5.2.1　Cerebral Ischemia

IL-6 expression transiently increases in the acute phase of cerebral ischemia. Studies using an IL-6 receptor monoclonal antibody demonstrated that endogenous IL-6 plays a critical role in preventing damaged neurons from undergoing apoptosis in the acute phase of cerebral ischemia and that its role may be mediated by STAT 3 activation (Yamashita et al., 2005).

5.2.2　Nerve Injury and Regeneration

Sciatic nerve injury has been shown to induce IL-6 (Bolin et al., 1995) and immunostaining for phospho-STAT 3 downtream of IL-6 indicated that activated STAT 3 levels were elevated in the nuclei of dorsal root ganglion (DRG) neurons after a conditioning lesion (Liu and Snider, 2001). Sciatic nerve transection resulted in a time-dependent phosphorylation and activation of STAT3 in DRG neurons (Qiu et al., 2005) and the effect appeared to be specific to peripheral injuries; it did not occur when the dorsal column was crushed. Infusion of the JAK2 kinase inhibitor AG490 to the proximal nerve stump blocked sciatic nerve-induced STAT3 phosphorylation, resulting in reduced neurite outgrowth in vitro and also significantly attenuated dorsal column axonal regeneration in the adult spinal cord. Therefore, STAT3 activation appears to be necessary for repair of DRG neurons and improved axonal regeneration after a conditioning lesion. Consistent with these observations, IL-6, along with leukemia inhibitory factor (LIF), appear to be involved in the conditioning lesioning response of DRG neurons, as illustrated by the observation that this conditioning effect is significantly attenuated in sensory neurons from LIF ko mice (Cafferty et al., 2001) as well as IL-6 ko mice (Cafferty et al., 2004) compared to wild type. Conditional knockout of STAT3 revealed that this pathway impacts survival (but not necessary regeneration) of motor neurons (Schweizer et al., 2002). A neuronal regeneration associated protein that may function downstream of STAT3 is the small proline-rich protein 1 A (SPRR1A) given that it has been identified as a downstream target of gp130 signaling by IL-6 in cardiomyocytes (Pradervand et al., 2004). It is not detectable in uninjured neurons but is induced in DRG neurons in culture and after peripheral axonal damage since it was identified by microarray profiling after sciatic nerve transection (Bonilla et al., 2002). Reduction of SPRR1A function by anti-sense oligonucleotides or SPRR1A antibody in axotomized axons restricted axonal outgrowth. Whether IL-6 also induces SPRR1A in neurons remains to be confirmed, but the latter could well mediate the neurogenic function of the gp130 cytokines.

5.3 IL-6 as a Target for Therapeutic Intervention

Although the complex intracellular signaling cascades inside neurons initiated by ligand-receptor binding are only beginning to be understood, localized therapeutic administration of IL-6 may lead to sufficient STAT3 activation for repair of DRG neurons and improved axonal regeneration after a lesion whereas inhibition of IL-6 may protect from cerebral ischemia in cases where administration of TNF inhibitors would be contraindicated because of the trophic effect of TNF on hippocampal neurons and stroke-induced neurogenesis.

6 Interleukin 10

6.1 Signaling Pathway and Expression in Normal CNS

Interleukin-10 is an immune modulatory cytokine with potent anti-inflammatory and immunosuppressive properties (reviewed in Pestka et al., 2004; Zdanov, 2004). The biologically active form is a homodimer that binds to IL-10 receptors R1 and R2 and signals via kinases Jak1, Tyk2 and the transcription factors Stat3 (Pestka et al., 2004). At least eight cytokines related to IL-10 have recently been described including IL-19 and IL-22 and some share the IL-10R2 receptor

(Pestka et al., 2004; Zdanov, 2004). IL-10 is produced by T cells and monocyte/ macrophages and regulates the function of T cells, B cells, and monocytes among other cells in part by inhibiting cytokine synthesis and expression of costimulatory molecules. In the CNS, activated microglia and astrocytes can produce IL-10 (Wu et al., 2005) and at least in vitro, these cells also express IL-10 receptors (Ledeboer et al., 2002).

6.2   IL-10 Signaling in CNS Disease

The effects of IL-10 in the brain seem to be overwhelmingly beneficial, involving anti-inflammatory and neuroprotective actions. In accord with an effect on immune cells, peripherally administered IL-10 ameliorated pneumococcal meningitis in rats while intrathecal delivery was ineffective (Koedel et al., 1996). Similarly in EAE, systemic administration of IL-10 reduced disease severity in rats (Rott et al., 1994) and mice (Crisi et al., 1995). IL-10 produced from T cells (Bettelli et al., 1998) or MHC class II expressing cells (Cua et al., 1999) in transgenic mice also prevented EAE and IL-10 ko mice showed exacerbated disease (Bettelli et al., 1998; Samoilova et al., 1998a) (Table 5.2). IL-10 producing T cells may also be responsible for mediating protection conferred by intranasal, tolerance inducing, immunization with myelin peptides (O'Neill et al., 2006) but probably not with copaxone/glatiramer acetate-induced tolerogenic T cells (Jee et al., 2006). Thus, copaxone, which is used for the treatment of MS in humans and results in the induction of tolerogenic T cells, was similarly effective in IL-10 ko as in wildtype mice (Jee et al., 2006). Interestingly, intravenous delivery of IL-10 can also exacerbate EAE (Cannella et al., 1996) indicating that the exact timing and pattern of IL-10 expression may be critical for its effects. While the aforementioned effects are consistent with systemic action of IL-10 or IL-10 produced by CNS infiltrating cells, viral delivery of IL-10 into the brain was sufficient to potently inhibit disease progression and prevent relapses (Cua et al., 2001). In contrast, virus-induced IL-10 production in the periphery, which led to significant increases in serum IL-10, was not protective (Cua et al., 2001). Nevertheless, it is likely that local and infiltrating cells are important sources of IL-10 in autoimmune inflammation in the CNS.

Administration of recombinant IL-10 protects rats against traumatic brain injury most likely involving effects on peripheral immune cells (Knoblach and Faden, 1998). Administration of IL-10 is also protective in experimental spinal cord injury and IL-10 ko mice show more damage, at least at early time points in this model (Abraham et al., 2004). Similarly, IL-10 ko mice are more susceptible to ischemia and primary neuronal cell cultures show that IL-10 acts directly on CNS cells, possibly even on neurons (Grilli et al., 2000). Following up on this observation, Bachis et al. (2001) show that IL-10 protects primary neurons exposed to excitotoxins, possibly by inhibiting NF-kB and caspase 3 activity.

Lack of IL-10 in knockout mice also results in increased susceptibility to prion disease and a shorter incubation time (Thackray et al., 2004) (Table 5.2). This is associated with a strong increase in TNF suggesting that IL-10 inhibits inflammation in this paradigm. In conclusion, IL-10 seems to exert its protective effects in part on immune cells infiltrating the brain, but local synthesis by glial cells is likely to contribute as well. Understanding the regulation of IL-10 in the CNS and characterizing its target cells will be necessary to target this pathway therapeutically in CNS disease.

7 Interleukin 12

7.1 Signaling Pathway and Expression in Normal CNS

Interleukin-12 is a key regulator of cellular immune responses (reviewed in Hunter, 2005). The bioactive peptide IL-12p70 (named thereafter IL-12), is a heterodimer made of the p35 subunit encoded by *IL12A* and p40 encoded by *IL12B*. An IL-12p40 homodimer exists as well and can antagonize IL-12p70 function. While IL-12p70 binds to IL-12 receptors RB1 and RB2, the p40 homodimer binds only RB1 (Hunter, 2005). Receptor activation leads to STAT4-dependent signaling involving the kinases Jak2 and Tyk2. Importantly, two new cytokines, IL-23 and IL-27, have recently been added to the IL-12 family: IL-23 is a heterodimer of IL-12p40 and IL-23p19 and uses IL-12-RB1 as one of its receptor components. IL-27 is structurally related to IL-12 but does not share its subunits or receptors (Hunter, 2005). Because some of the earlier work on IL-12 involved manipulation of the p40 subunit, IL-23 levels may have been affected as well and results should be evaluated accordingly. IL-12 is produced by monocyte/macrophages, dendritic cells and B cells and in the CNS, by microglia and astrocytes (Park and Shin, 1996; Stalder et al., 1997). Its receptors are found on T cells where they promote a Th1 type immune response. In addition, IL-12 receptors are present on NK cells. IL-23 is also produced by cells of the myeloid lineage and promotes the development of Th17 type T cells. In the CNS, it is produced by microglia and infiltrating macrophages (Cua et al., 2003).

7.2 IL-12 Signaling in CNS Disease

Given its prominent role in regulating T cell responses in the periphery, IL-12 has been studied mostly in T cell-dependent CNS disease. IL-12p40 is increased in active MS lesions in humans (Windhagen et al., 1995) and CNS expression correlates with disease severity in rodents with EAE (Issazadeh et al., 1995a; Smith et al., 1997; Bright et al., 1998). Using knockout mice it became clear recently that IL-23, not IL-12, is the major pro-inflammatory cytokine in EAE, probably by promoting Th17 T cell subset expansion and survival. Thus, IL-12p40 knockout mice (no IL-12, no IL-23; (Segal et al., 1998; Becher et al., 2002; Cua et al., 2003)) and IL-23p19 knockout mice were resistant to EAE while IL-12p35 knockout mice (no IL-12) had even more severe disease than wildtype animals (Becher et al., 2002; Cua et al., 2003) (Table 5.2). Interestingly, overexpression of IL-12p35 and IL-12-p40 from astrocytes of transgenic mice leads to spontaneous perivascular and parenchymal inflammatory lesions that contain T cells and NK cells (Pagenstecher et al., 2000). These mice are also more susceptible to EAE (Pagenstecher et al., 2000) and simple peripheral administration of mycobacterial extracts lead to EAE-like disease (Lassmann et al., 2001). Whether these mice have also increased IL-23 levels as a result of p40 transgene expression has not been described. Antibodies against p40 are potent inhibitors of EAE in rodents (Leonard et al., 1995; Bright et al., 1998; Segal et al., 1998) and MS-like disease in nonhuman primates (Hart et al., 2005), and it may thus have therapeutic application in human MS as well.

IL-12 expression in the CNS is also increased by peripheral bacterial infection or simply by administration of bacterial toxins (Park and Shin, 1996; Stalder et al., 1997; Lassmann et al., 2001). Of potential relevance for neurodegeneration, IL-12p35 knockout mice had less neuronal injury and microglial activation in an acute model of excitotoxicity (Chen et al., 2004) (Table 5.2). Furthermore, viral delivery of IL-12 to gliomas in mice led to infiltration of Th1 type cytotoxic T cells into the tumor and increased survival of mice (Liu et al., 2002).

References

Abraham KE, McMillen D, Brewer KL (2004) The effects of endogenous interleukin-10 on gray matter damage and the development of pain behaviors following excitotoxic spinal cord injury in the mouse. *Neuroscience* 124:945–952.

Aggarwal BB, Samanta A, Feldmann M (2000a) TNF receptors. In: *Cytokine Reference* (Oppenheim JJ and Feldmann M, ed.), pp. 1620–1632. London: Academic Press.

Aggarwal BB, Samanta A, Feldmann M (2000b) TNFa. In: *Cytokine Reference* (Oppenheim JJ and Feldmann M, ed.), pp. 414–434. London: Academic Press.

Akassoglou K, Probert L, Kontogeorgos G, Kollias G (1997) Astrocyte-specific but not neuron-specific transmembrane TNF triggers inflammation and degeneration in the central nervous system of transgenic mice. *J Immunol* 158:438–445.

Akassoglou K, Bauer J, Kassiotis G, Pasparakis M, Lassmann H, Kollias G, Probert L (1998) Oligodendrocyte apoptosis and primary demyelination induced by local TNF/p55TNF receptor signaling in the central nervous system of transgenic mice: models for multiple sclerosis with primary oligodendrogliopathy. *Am J Pathol* 153:801–813.

Akassoglou K, Bauer J, Kassiotis G, Lassmann H, Kollias G, Probert L (1999) Transgenic models of TNF induced demyelination. *Adv Exp Med Biol* 468:245–259.

Akassoglou K, Douni E, Bauer J, Lassmann H, Kollias G, Probert L (2003) Exclusive tumor necrosis factor (TNF) signaling by the p75TNF receptor triggers inflammatory ischemia in the CNS of transgenic mice. *Proc Natl Acad Sci U S A* 100:709–714.

Albensi BC, Mattson MP (2000) Evidence for the involvement of TNF and NF-kappaB in hippocampal synaptic plasticity. *Synapse* 35:151–159.

Allan SM, Rothwell NJ (2001) Cytokines and acute neurodegeneration. *Nat Rev Neurosci* 2:734–744.

Allan SM, Tyrrell PJ, Rothwell NJ (2005) Interleukin-1 and neuronal injury. *Nat Rev Immunol* 5:629–640.

Aloe L, Fiore M (1997) TNF-alpha expressed in the brain of transgenic mice lowers central tyroxine hydroxylase immunoreactivity and alters grooming behavior. *Neurosci Lett* 238:65–68.

Aloe L, Probert L, Kollias G, Bracci-Laudiero L, Micera A, Mollinari C, Levi-Montalcini R (1993) Level of nerve growth factor and distribution of mast cells in the synovium of tumour necrosis factor transgenic arthritic mice. *Int J Tissue React* 15:139–143.

Aloe L, Fiore M, Probert L, Turrini P, Tirassa P (1999a) Overexpression of tumour necrosis factor alpha in the brain of transgenic mice differentially alters nerve growth factor levels and choline acetyltransferase activity. *Cytokine* 11:45–54.

Aloe L, Properzi F, Probert L, Akassoglou K, Kassiotis G, Micera A, Fiore M (1999b) Learning abilities, NGF and BDNF brain levels in two lines of TNF-alpha transgenic mice, one characterized by neurological disorders, the other phenotypically normal. *Brain Res* 840:125–137.

Alvarez A, Cacabelos R, Sanpedro C, Garcia-Fantini M, Aleixandre M (2007) Serum TNF-alpha levels are increased and correlate negatively with free IGF-I in Alzheimer disease. *Neurobiol Aging* 28(4):533–536.

Andre R, Pinteaux E, Kimber I, Rothwell NJ (2005) Differential actions of IL-1 alpha and IL-1 beta in glial cells share common IL-1 signalling pathways. *Neuroreport* 16:153–157.

Andrews ZB, Zhao H, Frugier T, Meguro R, Grattan DR, Koishi K, McLennan IS (2006) Transforming growth factor beta2 haploinsufficient mice develop age-related nigrostriatal dopamine deficits. *Neurobiol Dis* 21(3):568–575.

Arnett HA, Mason J, Marino M, Suzuki K, Matsushima GK, Ting JP (2001) TNF alpha promotes proliferation of oligodendrocyte progenitors and remyelination. *Nat Neurosci* 4:1116–1122.

Bachis A, Colangelo AM, Vicini S, Doe PP, De Bernardi MA, Brooker G, Mocchetti I (2001) Interleukin-10 prevents glutamate-mediated cerebellar granule cell death by blocking caspase-3-like activity. *J Neurosci* 21:3104–3112.

Baker D, Butler D, Scallon BJ, O'Neill JK, Turk JL, Feldmann M (1994) Control of established experimental allergic encephalomyelitis by inhibition of tumor necrosis factor (TNF) activity within the central nervous system using monoclonal antibodies and TNF receptor-immunoglobulin fusion proteins. *Eur J Immunol* 24:2040–2048.

Balding J, Livingstone WJ, Pittock SJ, Mynett-Johnson L, Ahern T, Hodgson A, Smith OP (2004) The IL-6 G-174C polymorphism may be associated with ischaemic stroke in patients without a history of hypertension. *Ir J Med Sci* 173:200–203.

Balosso S, Ravizza T, Perego C, Peschon J, Campbell IL, De Simoni MG, Vezzani A (2005) Tumor necrosis factor-alpha inhibits seizures in mice via p75 receptors. *Ann Neurol* 57:804–812.

Barcia C, de Pablos V, Bautista-Hernandez V, Sanchez-Bahillo A, Bernal I, Fernandez-Villalba E, Martin J, Banon R, Fernandez-Barreiro A, Herrero MT (2005) Increased plasma levels of TNF-alpha but not of IL1-beta in MPTP-treated monkeys one year after the MPTP administration. *Parkinsonism Relat Disord* 11:435–439.

Basu A, Krady JK, Levison SW (2004) Interleukin-1: a master regulator of neuroinflammation. *J Neurosci Res* 78:151–156.

Beattie EC, Stellwagen D, Morishita W, Bresnahan JC, Ha BK, Von Zastrow M, Beattie MS, Malenka RC (2002) Control of synaptic strength by glial TNFalpha. *Science* 295: 2282–2285.

Becher B, Durell BG, Noelle RJ (2002) Experimental autoimmune encephalitis and inflammation in the absence of interleukin-12. *J Clin Invest* 110:493–497.

Benveniste EN, Benos DJ (1995) TNF-alpha- and IFN-gamma-mediated signal transduction pathways: effects on glial cell gene expression and function. *FASEB J* 9:1577–1584.

Benzing T, Kottgen M, Johnson M, Schermer B, Zentgraf H, Walz G, Kim E (2002) Interaction of 14–3–3 protein with regulator of G protein signaling 7 is dynamically regulated by tumor necrosis factor-alpha. *J Biol Chem* 277:32954–32962.

Bernardino L, Xapelli S, Silva AP, Jakobsen B, Poulsen FR, Oliveira CR, Vezzani A, Malva JO, Zimmer J (2005) Modulator effects of interleukin-1beta and tumor necrosis factor-alpha on

AMPA-induced excitotoxicity in mouse organotypic hippocampal slice cultures. *J Neurosci* 25:6734–6744.

Bertolotto A, Capobianco M, Malucchi S, Manzardo E, Audano L, Bergui M, Bradac GB, Mutani R (1999) Transforming growth factor beta1 (TGFbeta1) mRNA level correlates with magnetic resonance imaging disease activity in multiple sclerosis patients. *Neurosci Lett* 263:21–24.

Bessler H, Djaldetti R, Salman H, Bergman M, Djaldetti M (1999) IL-1 beta, IL-2, IL-6 and TNF-alpha production by peripheral blood mononuclear cells from patients with Parkinson's disease. *Biomed Pharmacother* 53:141–145.

Bettelli E, Das MP, Howard ED, Weiner HL, Sobel RA, Kuchroo VK (1998) IL-10 is critical in the regulation of autoimmune encephalomyelitis as demonstrated by studies of IL-10- and IL-4-deficient and transgenic mice. *J Immunol* 161:3299–3306.

Bettelli E, Carrier Y, Gao W, Korn T, Strom TB, Oukka M, Weiner HL, Kuchroo VK (2006) Reciprocal developmental pathways for the generation of pathogenic effector TH17 and regulatory T cells. *Nature* 441:235–238.

Bierie B, Moses HL (2006) TGF-beta and cancer. *Cytokine Growth Factor Rev* 17:29–40.

Billings LM, Oddo S, Green KN, McGaugh JL, LaFerla FM (2005) Intraneuronal Abeta causes the onset of early Alzheimer's disease-related cognitive deficits in transgenic mice. *Neuron* 45:675–688.

Blasko I, Lederer W, Oberbauer H, Walch T, Kemmler G, Hinterhuber H, Marksteiner J, Humpel C (2006) Measurement of thirteen biological markers in CSF of patients with Alzheimer's disease and other dementias. *Dement Geriatr Cogn Disord* 21:9–15.

Bliss TV, Collingridge GL (1993) A synaptic model of memory: long-term potentiation in the hippocampus. *Nature* 361:31–39.

Blum-Degen D, Muller T, Kuhn W, Gerlach M, Przuntek H, Riederer P (1995) Interleukin-1 beta and interleukin-6 are elevated in the cerebrospinal fluid of Alzheimer's and de novo Parkinson's disease patients. *Neurosci Lett* 202:17–20.

Boche D, Cunningham C, Gauldie J, Perry VH (2003) Transforming growth factor-beta 1-mediated neuroprotection against excitotoxic injury in vivo. *J Cereb Blood Flow Metab* 23:1174–1182.

Bodmer JL, Schneider P, Tschopp J (2002) The molecular architecture of the TNF superfamily. *Trends Biochem Sci* 27:19–26.

Boka G, Anglade P, Wallach D, Javoy-Agid F, Agid Y, Hirsch EC (1994) Immunocytochemical analysis of tumor necrosis factor and its receptors in Parkinson's disease. *Neurosci Lett* 172:151–154.

Bolin LM, Verity AN, Silver JE, Shooter EM, Abrams JS (1995) Interleukin-6 production by Schwann cells and induction in sciatic nerve injury. *J Neurochem* 64:850–858.

Bongartz T, Sutton AJ, Sweeting MJ, Buchan I, Matteson EL, Montori V (2006) Anti-TNF antibody therapy in rheumatoid arthritis and the risk of serious infections and malignancies: systematic review and meta-analysis of rare harmful effects in randomized controlled trials. *JAMA* 295:2275–2285.

Bonilla IE, Tanabe K, Strittmatter SM (2002) Small proline-rich repeat protein 1A is expressed by axotomized neurons and promotes axonal outgrowth. *J Neurosci* 22:1303–1315.

Border WA, Noble NA (1997) TGF-β in kidney fibrosis: a target for gene therapy. Kidney Int 51:1388–1396.

Boulanger LM, Huh GS, Shatz CJ (2001) Neuronal plasticity and cellular immunity: shared molecular mechanisms. *Curr Opin Neurobiol* 11:568–578.

Bourdeau A, Dumont DJ, Letarte M (1999) A murine model of hereditary hemorrhagic telangiectasia. *J Clin Invest* 104:1343–1351.

Bourdeau A, Faughnan ME, McDonald M-L, Paterson AD, Wanless IR, Letarte M (2001) Potential role of modifier genes influencing transforming growth factor-β1 levels in the development of vascular defects in endoglin heterozygous mice with hereditary hemorrhagic telangiectasia. *Am J Pathol* 158:2011–2020.

Boutin H, LeFeuvre RA, Horai R, Asano M, Iwakura Y, Rothwell NJ (2001) Role of IL-1alpha and IL-1beta in ischemic brain damage. *J Neurosci* 21:5528–5534.

Boutin H, Kimber I, Rothwell NJ, Pinteaux E (2003) The expanding interleukin-1 family and its receptors: do alternative IL-1 receptor/signaling pathways exist in the brain? *Mol Neurobiol* 27:239–248.

Breder CD, Tsujimoto M, Terano Y, Scott DW, Saper CB (1993) Distribution and characterization of tumor necrosis factor-alpha-like immunoreactivity in the murine central nervous system. *J Comp Neurol* 337:543–567.

Bright JJ, Musuro BF, Du C, Sriram S (1998) Expression of IL-12 in CNS and lymphoid organs of mice with experimental allergic encephalitis. *J Neuroimmunol* 82:22–30.

Brionne TC, Tesseur I, Masliah E, Wyss-Coray T (2003) Loss of TGF-beta1 leads to increased neuronal cell death and microgliosis in mouse brain. *Neuron* 40:1133–1145.

Bruce AJ, Boling W, Kindy MS, Peschon J, Kraemer PJ, Carpenter MK, Holtsberg FW, Mattson MP (1996) Altered neuronal and microglial responses to excitotoxic and ischemic brain injury in mice lacking TNF receptors. *Nat Med* 2:788–794.

Buckwalter M, Wyss-Coray T (2004) Modelling neuroinflammatory phenotypes in vivo. *J Neuroinflammation* 1:10.

Buckwalter MS, Yamane M, Coleman BS, Ormerod BK, Chin JT, Palmer T, Wyss-Coray T (2006a) Chronically increased transforming growth factor-beta1 strongly inhibits hippocampal neurogenesis in aged mice. *Am J Pathol* 169:154–164.

Buckwalter MS, Coleman BS, Buttini M, Barbour R, Schenk D, Games D, Seubert P, Wyss-Coray T (2006b) Increased T cell recruitment to the CNS after amyloid beta 1–42 immunization in Alzheimer's mice overproducing transforming growth factor-beta 1. *J Neurosci* 26:11437–11441.

Butler MP, O'Connor JJ, Moynagh PN (2004) Dissection of tumor-necrosis factor-alpha inhibition of long-term potentiation (LTP) reveals a p38 mitogen-activated protein kinase-dependent mechanism which maps to early-but not late-phase LTP. *Neuroscience* 124:319–326.

Cacabelos R, Fernandez-Novoa L, Lombardi V, Kubota Y, Takeda M (2005) Molecular genetics of Alzheimer's disease and aging. *Methods Find Exp Clin Pharmacol* 27 Suppl A:1–573.

Cacquevel M, Lebeurrier N, Cheenne S, Vivien D (2004) Cytokines in neuroinflammation and Alzheimer's disease. *Curr Drug Targets* 5:529–534.

Cafferty WB, Gardiner NJ, Gavazzi I, Powell J, McMahon SB, Heath JK, Munson J, Cohen J, Thompson SW (2001) Leukemia inhibitory factor determines the growth status of injured adult sensory neurons. *J Neurosci* 21:7161–7170.

Cafferty WB, Gardiner NJ, Das P, Qiu J, McMahon SB, Thompson SW (2004) Conditioning injury-induced spinal axon regeneration fails in interleukin-6 knock-out mice. *J Neurosci* 24:4432–4443.

Callahan JF, Burgess JL, Fornwald JA, Gaster LM, Harling JD, Harrington FP, Heer J, Kwon C, Lehr R, Mathur A, Olson BA, Weinstock J, Laping NJ (2002) Identification of novel inhibitors of the transforming growth factor β1 (TGF-β1) type 1 receptor (ALK5). *J Med Chem* 45:999–1001.

Campbell IL, Abraham CR, Masliah E, Kemper P, Inglis JD, Oldstone MBA, Mucke L (1993) Neurologic disease induced in transgenic mice by cerebral overexpression of interleukin 6. *Proc Natl Acad Sci U S A* 90:10061–10065.

Campbell IL, Stalder AK, Chiang CS, Bellinger R, Heyser CJ, Steffensen S, Masliah E, Powell HC, Gold LH, Henriksen SJ, Siggins GR (1997) Transgenic models to assess the pathogenic actions of cytokines in the central nervous system. *Mol Psychiatry* 2:125–129.

Cannella B, Gao YL, Brosnan C, Raine CS (1996) IL-10 fails to abrogate experimental autoimmune encephalomyelitis. *J Neurosci Res* 45:735–746.

Carvey PM, Chen EY, Lipton JW, Tong CW, Chang QA, Ling ZD (2005) Intra-parenchymal injection of tumor necrosis factor-alpha and interleukin 1-beta produces dopamine neuron loss in the rat. *J Neural Transm* 112:601–612.

Chen LZ, Hochwald GM, Huang C, Dakin G, Tao H, Cheng C, Simmons WJ, Dranoff G, Thorbecke GJ (1998) Gene therapy in allergic encephalomyelitis using myelin basic protein-specific T cells engineered to express latent transforming growth factor-beta1. *Proc Natl Acad Sci U S A* 95:12516–12521.

Chen H, Zhang SM, Hernan MA, Schwarzschild MA, Willett WC, Colditz GA, Speizer FE, Ascherio A (2003) Nonsteroidal anti-inflammatory drugs and the risk of Parkinson disease. *Arch Neurol* 60:1059–1064.

Chen Z, Duan RS, Q HC, Wu Q, Mix E, Winblad B, Ljunggren HG, Zhu J (2004) IL-12p35 deficiency alleviates kainic acid-induced hippocampal neurodegeneration in C57BL/6 mice. *Neurobiol Dis* 17:171–178.

Chen Y, Pawlikowska L, Yao JS, Shen F, Zhai W, Achrol AS, Lawton MT, Kwok PY, Yang GY, Young WL (2006) Interleukin-6 involvement in brain arteriovenous malformations. *Ann Neurol* 59:72–80.

Chin J, Angers A, Cleary LJ, Eskin A, Byrne JH (2002) Transforming growth factor β1 alters synapsin distribution and modulates synaptic depression in *Aplysia*. *J Neurosci* 22:1–6.

Choi DW (1988) Glutamate neurotoxicity and diseases of the nervous system. *Neuron* 1:623–634.

Chun J (2001) Selected comparison of immune and nervous system development. *Adv Immunol* 77:297–322.

Chung CY, Seo H, Sonntag KC, Brooks A, Lin L, Isacson O (2005) Cell type-specific gene expression of midbrain dopaminergic neurons reveals molecules involved in their vulnerability and protection. *Hum Mol Genet* 14:1709–1725.

Cleren C, Calingasan NY, Chen J, Beal MF (2005) Celastrol protects against MPTP- and 3-nitropropionic acid-induced neurotoxicity. *J Neurochem* 94:995–1004.

Collins JS, Perry RT, Watson B, Jr., Harrell LE, Acton RT, Blacker D, Albert MS, Tanzi RE, Bassett SS, McInnis MG, Campbell RD, Go RC (2000) Association of a haplotype for tumor necrosis factor in siblings with late-onset Alzheimer disease: the NIMH Alzheimer Disease Genetics Initiative. *Am J Med Genet* 96:823–830.

Corti A, Ghezzi P (2004) Tumor Necrosis Factor: Methods and Protocols, First Edition. Totowa, NJ: Humana Press.

Crews L, Wyss-Coray T, Masliah E (2004) Insights into the pathogenesis of hydrocephalus from transgenic and experimental animal models. *Brain Pathol* 14:312–316.

Crisi GM, Santambrogio L, Hochwald GM, Smith SR, Carlino JA, Thorbecke GJ (1995) Staphylococcal enterotoxin B and tumor-necrosis factor-alpha-induced relapses of experimental allergic encephalomyelitis: protection by transforming growth factor-beta and interleukin-10. *Eur J Immunol* 25:3035–3040.

Cua DJ, Groux H, Hinton DR, Stohlman SA, Coffman RL (1999) Transgenic interleukin 10 prevents induction of experimental autoimmune encephalomyelitis. *J Exp Med* 189:1005–1010.

Cua DJ, Hutchins B, LaFace DM, Stohlman SA, Coffman RL (2001) Central nervous system expression of IL-10 inhibits autoimmune encephalomyelitis. *J Immunol* 166:602–608.

Cua DJ, Sherlock J, Chen Y, Murphy CA, Joyce B, Seymour B, Lucian L, To W, Kwan S, Churakova T, Zurawski S, Wiekowski M, Lira SA, Gorman D, Kastelein RA, Sedgwick JD (2003) Interleukin-23 rather than interleukin-12 is the critical cytokine for autoimmune inflammation of the brain. *Nature* 421:744–748.

Cunningham AJ, Murray CA, O'Neill LA, Lynch MA, O'Connor JJ (1996) Interleukin-1 beta (IL-1 beta) and tumour necrosis factor (TNF) inhibit long-term potentiation in the rat dentate gyrus in vitro. *Neurosci Lett* 203:17–20.

Curran B, O'Connor JJ (2001) The pro-inflammatory cytokine interleukin-18 impairs long-term potentiation and NMDA receptor-mediated transmission in the rat hippocampus in vitro. *Neuroscience* 108:83–90.

Curran BP, Murray HJ, O'Connor JJ (2003) A role for c-Jun N-terminal kinase in the inhibition of long-term potentiation by interleukin-1beta and long-term depression in the rat dentate gyrus in vitro. *Neuroscience* 118:347–357.

De Groot CJ, Montagne L, Barten AD, Sminia P, Van Der Valk P (1999) Expression of transforming growth factor (TGF)-beta1, -beta2, and -beta3 isoforms and TGF-beta type I and type II receptors in multiple sclerosis lesions and human adult astrocyte cultures. *J Neuropathol Exp Neurol* 58:174–187.

de Sousa Lopes SM, Carvalho RL, van den Driesche S, Goumans MJ, ten Dijke P, Mummery CL (2003) Distribution of phosphorylated Smad2 identifies target tissues of TGF beta ligands in mouse development. *Gene Expr Patterns* 3:355–360.

Deboer T, Fontana A, Tobler I (2002) Tumor necrosis factor (TNF) ligand and TNF receptor deficiency affects sleep and the sleep EEG. *J Neurophysiol* 88:839–846.

Dennler S, Goumans M-J, Dijke Pt (2002) Transforming growth factor β signal transduction. *J Leukoc Biol* 71:731–740.

Derynck R, Zhang YE (2003) Smad-dependent and Smad-independent pathways in TGF-beta family signalling. *Nature* 425:577–584.

Dickson MC, Martin JS, Cousins FM, Kulkarni AB, Karlsson S, Akhurst RJ (1995) Defective haematopoiesis and vasculogenesis in transforming growth factor-β1 knock out mice. *Development* 121:1845–1854.

Dinarello CA (2006) Interleukin 1 and interleukin 18 as mediators of inflammation and the aging process. *Am J Clin Nutr* 83:447S–455S.

Dopp JM, Mackenzie-Graham A, Otero GC, Merrill JE (1997) Differential expression, cytokine modulation, and specific functions of type-1 and type-2 tumor necrosis factor receptors in rat glia. *J Neuroimmunol* 75:104–112.

Du Y, Dodel RC, Eastwood BJ, Bales KR, Gao F, Lohmuller F, Muller U, Kurz A, Zimmer R, Evans RM, Hake A, Gasser T, Oertel WH, Griffin WS, Paul SM, Farlow MR (2000) Association of an interleukin 1 alpha polymorphism with Alzheimer's disease. *Neurology* 55:480–483.

Ehl C, Kolsch H, Ptok U, Jessen F, Schmitz S, Frahnert C, Schlosser R, Rao ML, Maier W, Heun R (2003) Association of an interleukin-1beta gene polymorphism at position -511 with Alzheimer's disease. *Int J Mol Med* 11:235–238.

Fan L, Young PR, Barone FC, Feuerstein GZ, Smith DH, McIntosh TK (1996) Experimental brain injury induces differential expression of tumor necrosis factor-alpha mRNA in the CNS. *Brain Res Mol Brain Res* 36:287–291.

Felderhoff-Mueser U, Schmidt OI, Oberholzer A, Buhrer C, Stahel PF (2005) IL-18: a key player in neuroinflammation and neurodegeneration? *Trends Neurosci* 28:487–493.

Ferger B, Leng A, Mura A, Hengerer B, Feldon J (2004) Genetic ablation of tumor necrosis factor-alpha (TNF-alpha) and pharmacological inhibition of TNF-synthesis attenuates MPTP toxicity in mouse striatum. *J Neurochem* 89:822–833.

Fernandez LA, Sanz-Rodriguez F, Blanco FJ, Bernabeu C, Botella LM (2006) Hereditary hemorrhagic telangiectasia, a vascular dysplasia affecting the TGF-beta signaling pathway. *Clin Med Res* 4:66–78.

Feuerstein GZ, Liu T, Barone FC (1994) Cytokines, inflammation, and brain injury: role of tumor necrosis factor-alpha. *Cerebrovasc Brain Metab Rev* 6:341–360.

Fillit H, Ding WH, Buee L, Kalman J, Altstiel L, Lawlor B, Wolf-Klein G (1991) Elevated circulating tumor necrosis factor levels in Alzheimer's disease. Neurosci Lett 129:318–320.

Finch CE, Laping NJ, Morgan TE, Nichols NR, Pasinetti GM (1993) TGF-β1 is an organizer of responses to neurodegeneration. J Cell Biochem 53:314–322.

Fiore M, Probert L, Kollias G, Akassoglou K, Alleva E, Aloe L (1996) Neurobehavioral alterations in developing transgenic mice expressing TNF-alpha in the brain. *Brain Behav Immun* 10:126–138.

Flanders KC, Burmester JK (2003) Medical applications of transforming growth factor-beta. *Clin Med Res* 1:13–20.

Flanders KC, Ren RF, Lippa CF (1998) Transforming growth factor–βs in neurodegenerative disease. *Prog Neurobiol* 54:71–85.

Flanders KC, Kim ES, Roberts AB (2001) Immunohistochemical expression of Smads 1–6 in the 15-day gestation mouse embryo: signaling by BMPs and TGF-betas. *Dev Dyn* 220:141–154.

Gaertner RF, Wyss-Coray T, Von Euw D, Lesne S, Vivien D, Lacombe P (2005) Reduced brain tissue perfusion in TGF-beta 1 transgenic mice showing Alzheimer's disease-like cerebrovascular abnormalities. *Neurobiol Dis* 19:38–46.

Galbreath E, Kim S-J, Park K, Brenner M, Messing A (1995) Overexpression of TGF-β1 in the central nervous system of transgenic mice results in hydrocephalus. *J Neuropathol Exp Neurol* 54:339–349.

Gao HM, Hong JS, Zhang W, Liu B (2002a) Distinct role for microglia in rotenone-induced degeneration of dopaminergic neurons. *J Neurosci* 22:782–790.

Gao HM, Jiang J, Wilson B, Zhang W, Hong JS, Liu B (2002b) Microglial activation-mediated delayed and progressive degeneration of rat nigral dopaminergic neurons: relevance to Parkinson's disease. *J Neurochem* 81:1285–1297.

Gao HM, Liu B, Zhang W, Hong JS (2003) Novel anti-inflammatory therapy for Parkinson's disease. *Trends Pharmacol Sci* 24:395–401.

Gayle DA, Ling Z, Tong C, Landers T, Lipton JW, Carvey PM (2002) Lipopolysaccharide (LPS)-induced dopamine cell loss in culture: roles of tumor necrosis factor-alpha, interleukin-1beta, and nitric oxide. *Brain Res Dev Brain Res* 133:27–35.

Glaser KB, Li J, Aakre ME, Morgan DW, Sheppard G, Stewart KD, Pollock J, Lee P, O'Connor CZ, Anderson SN, Mussatto DJ, Wegner CW, Moses HL (2002) Transforming growth factor beta mimetics: discovery of 7-[4-(4-cyanophenyl)phenoxy]-heptanohydroxamic acid, a biaryl hydroxamate inhibitor of histone deacetylase. *Mol Cancer Ther* 1:759–768.

Golan H, Levav T, Mendelsohn A, Huleihel M (2004) Involvement of tumor necrosis factor alpha in hippocampal development and function. *Cereb Cortex* 14:97–105.

Goodman JC, Robertson CS, Grossman RG, Narayan RK (1990) Elevation of tumor necrosis factor in head injury. *J Neuroimmunol* 30:213–217.

Goumans MJ, Valdimarsdottir G, Itoh S, Rosendahl A, Sideras P, ten Dijke P (2002) Balancing the activation state of the endothelium via two distinct TGF-β type I receptors. *EMBO J* 21:1743–1753.

Goumans MJ, Valdimarsdottir G, Itoh S, Lebrin F, Larsson J, Mummery C, Karlsson S, ten Dijke P (2003) Activin receptor-like kinase (ALK)1 is an antagonistic mediator of lateral TGFbeta/ALK5 signaling. *Mol Cell* 12:817–828.

Grammas P, Ovase R (2002) Cerebrovascular transforming growth factor-β contributes to inflammation in the Alzheimer's disease brain. *Am J Pathol* 160:1583–1587.

Green EK, Harris JM, Lemmon H, Lambert JC, Chartier-Harlin MC, St Clair D, Mann DM, Iwatsubo T, Lendon CL (2002) Are interleukin-1 gene polymorphisms risk factors or disease modifiers in AD? *Neurology* 58:1566–1568.

Grell M, Wajant H, Zimmermann G, Scheurich P (1998) The type 1 receptor (CD120a) is the high-affinity receptor for soluble tumor necrosis factor. *Proc Natl Acad Sci U S A* 95:570–575.

Grigoryan GA, Gray JA, Rashid T, Chadwick A, Hodges H (2000) Conditionally immortal neuroepithelial stem cell grafts restore spatial learning in rats with lesions at the source of cholinergic forebrain projections cholinergic forebrain projections Conditionally immortal neuroepithelial stem cell grafts restore spatial learning in rats with lesions at the source of cholinergic forebrain projections. *Restor Neurol Neurosci* 17:1.

Grilli M, Barbieri I, Basudev H, Brusa R, Casati C, Lozza G, Ongini E (2000) Interleukin-10 modulates neuronal threshold of vulnerability to ischaemic damage. *Eur J Neurosci* 12:2265–2272.

Grimaldi LM, Casadei VM, Ferri C, Veglia F, Licastro F, Annoni G, Biunno I, De Bellis G, Sorbi S, Mariani C, Canal N, Griffin WS, Franceschi M (2000) Association of early-onset Alzheimer's disease with an interleukin-1alpha gene polymorphism. *Ann Neurol* 47:361–365.

Gross CE, Bednar MM, Howard DB, Sporn MB (1993) Transforming growth factor-β 1 reduces infarct size after experimental cerebral ischemia in a rabbit model. *Stroke* 24:558–562.

Guo Z, Iyun T, Fu W, Zhang P, Mattson MP (2004) Bone marrow transplantation reveals roles for brain macrophage/microglia TNF signaling and nitric oxide production in excitotoxic neuronal death. *Neuromolecular Med* 5:219–234.

Hailer NP, Vogt C, Korf HW, Dehghani F (2005) Interleukin-1beta exacerbates and interleukin-1 receptor antagonist attenuates neuronal injury and microglial activation after excitotoxic damage in organotypic hippocampal slice cultures. *Eur J Neurosci* 21:2347–2360.

Hakansson A, Westberg L, Nilsson S, Buervenich S, Carmine A, Holmberg B, Sydow O, Olson L, Johnels B, Eriksson E, Nissbrandt H (2005) Interaction of polymorphisms in the genes encoding interleukin-6 and estrogen receptor beta on the susceptibility to Parkinson's disease. *Am J Med Genet B Neuropsychiatr Genet* 133:88–92.

Haridas V, Darnay BG, Natarajan K, Heller R, Aggarwal BB (1998) Overexpression of the p80 TNF receptor leads to TNF-dependent apoptosis, nuclear factor-kappa B activation, and c-Jun kinase activation. *J Immunol* 160:3152–3162.

Harrington JF, Messier AA, Levine A, Szmydynger-Chodobska J, Chodobski A (2005) Shedding of tumor necrosis factor type 1 receptor after experimental spinal cord injury. *J Neurotrauma* 22:919–928.

Hasegawa Y, Inagaki T, Sawada M, Suzumura A (2000) Impaired cytokine production by peripheral blood mononuclear cells and monocytes/macrophages in Parkinson's disease. *Acta Neurol Scand* 101:159–164.

Heldmann U, Thored P, Claasen JH, Arvidsson A, Kokaia Z, Lindvall O (2005) TNF-alpha antibody infusion impairs survival of stroke-generated neuroblasts in adult rat brain. *Exp Neurol* 196:204–208.

Henrich-Noack P, Prehn JHM, Krieglstein J (1996) TGF-β1 protects hippocampal neurons against degeneration caused by transient global ischemia. Dose-response relationship and potential neuroprotective mechanisms. *Stroke* 27:1609–1615.

Hermann GE, Rogers RC, Bresnahan JC, Beattie MS (2001) Tumor necrosis factor-alpha induces cFOS and strongly potentiates glutamate-mediated cell death in the rat spinal cord. *Neurobiol Dis* 8:590–599.

Hirsch EC, Hunot S, Damier P, Faucheux B (1998) Glial cells and inflammation in Parkinson's disease: a role in neurodegeneration? *Ann Neurol* 44:S115–120.

Hjelmeland MD, Hjelmeland AB, Sathornsumetee S, Reese ED, Herbstreith MH, Laping NJ, Friedman HS, Bigner DD, Wang XF, Rich JN (2004) SB-431542, a small molecule transforming growth factor-beta-receptor antagonist, inhibits human glioma cell line proliferation and motility. *Mol Cancer Ther* 3:737–745.

Hofman FM, Hinton DR, Johnson K, Merrill JE (1989) Tumor necrosis factor identified in multiple sclerosis brain. *J Exp Med* 170:607–612.

Huang S, Flanders KC, Roberts AB (2000) Characterization of the mouse Smad1 gene and its expression pattern in adult mouse tissues. *Gene* 258:43–53.

Hunot S, Dugas N, Faucheux B, Hartmann A, Tardieu M, Debre P, Agid Y, Dugas B, Hirsch EC (1999) FcepsilonRII/CD23 is expressed in Parkinson's disease and induces, in vitro, production of nitric oxide and tumor necrosis factor-alpha in glial cells. *J Neurosci* 19:3440–3447.

Hunter CA (2005) New IL-12-family members: IL-23 and IL-27, cytokines with divergent functions. *Nat Rev Immunol* 5:521–531.

Idriss HT, Naismith JH (2000) TNF alpha and the TNF receptor superfamily: structure-function relationship(s). *Microsc Res Tech* 50:184–195.

Isacson O (2002) Models of repair mechanisms for future treatment modalities of Parkinson's disease. *Brain Res Bull* 57:839–846.

Issazadeh S, Ljungdahl A, Hojeberg B, Mustafa M, Olsson T (1995a) Cytokine production in the central nervous system of Lewis rats with experimental autoimmune encephalomyelitis: dynamics of mRNA expression for interleukin-10, interleukin-12, cytolysin, tumor necrosis factor alpha and tumor necrosis factor beta. *J Neuroimmunol* 61:205–212.

Issazadeh S, Mustafa M, Ljungdahl A, Hojeberg B, Dagerlind A, Elde R, Olsson T (1995b) Interferon gamma, interleukin 4 and transforming growth factor beta in experimental autoimmune encephalomyelitis in Lewis rats: dynamics of cellular mRNA expression in the central nervous system and lymphoid cells. *J Neurosci* Res 40:579–590.

Itoh S, Itoh F, Goumans M-J, ten Dijke P (2000) Signaling of transforming growth factor-β family members through Smad proteins. *Eur J Biochem* 267:6954–6967.

Jakobsen PH, McKay V, Morris-Jones SD, McGuire W, van Hensbroek MB, Meisner S, Bendtzen K, Schousboe I, Bygbjerg IC, Greenwood BM (1994) Increased concentrations of interleukin-6 and interleukin-1 receptor antagonist and decreased concentrations of beta-2-glycoprotein I in Gambian children with cerebral malaria. *Infect Immun* 62:4374–4379.

Janelsins MC, Mastrangelo MA, Oddo S, LaFerla FM, Federoff HJ, Bowers WJ (2005) Early correlation of microglial activation with enhanced tumor necrosis factor-alpha and monocyte chemoattractant protein-1 expression specifically within the entorhinal cortex of triple transgenic Alzheimer's disease mice. *J Neuroinflammation* 2:23.

Janssens K, Vanhoenacker F, Bonduelle M, Verbruggen L, Van Maldergem L, Ralston S, Guanabens N, Migone N, Wientroub S, Divizia MT, Bergmann C, Bennett C, Simsek S, Melancon S, Cundy T, Van Hul W (2006) Camurati–Engelmann disease: review of the clinical, radiological, and molecular data of 24 families and implications for diagnosis and treatment. *J Med Genet* 43:1–11.

Jee Y, Liu R, Bai XF, Campagnolo DI, Shi FD, Vollmer TL (2006) Do Th2 cells mediate the effects of glatiramer acetate in experimental autoimmune encephalomyelitis? *Int Immunol* 18:537–544.

John GR, Lee SC, Brosnan CF (2003) Cytokines: powerful regulators of glial cell activation. *Neuroscientist* 9:10–22.

Johns LD, Sriram S (1993) Experimental allergic encephalomyelitis: neutralizing antibody to TGF beta 1 enhances the clinical severity of the disease. *J Neuroimmunol* 47:1–7.

Johns LD, Flanders KC, Ranges GE, Sriram S (1991) Successful treatment of experimental allergic encephalomyelitis with transforming growth factor. *J Immunol* 147:1792.

Johnson DW, Berg JN, Baldwin MA, Gallione CJ, Marondel I, Yoon SJ, Stenzel TT, Speer M, Pericak-Vance MA, Diamond A, Guttmacher AE, Jackson CE, Attisano L, Kucherlapati R, Porteous ME, Marchuk DA (1996) Mutations in the activin receptor-like kinase 1 gene in hereditary haemorrhagic telangiectasia type 2. *Nat Genet* 13:189–195.

Juttler E, Tarabin V, Schwaninger M (2002) Interleukin-6 (IL-6): a possible neuromodulator induced by neuronal activity. *Neuroscientist* 8:268–275.

Kalinovsky A, Scheiffele P (2004) Transcriptional control of synaptic differentiation by retrograde signals. *Curr Opin Neurobiol* 14:272–279.

Karahan ZC, Deda G, Sipahi T, Elhan AH, Akar N (2005) TNF-alpha -308G/A and IL-6 -174 G/C polymorphisms in the Turkish pediatric stroke patients. *Thromb Res* 115:393–398.

Kassiotis G, Pasparakis M, Kollias G, Probert L (1999) TNF accelerates the onset but does not alter the incidence and severity of myelin basic protein-induced experimental autoimmune encephalomyelitis. *Eur J Immunol* 29:774–780.

Kelliher MA, Grimm S, Ishida Y, Kuo F, Stanger BZ, Leder P (1998) The death domain kinase RIP mediates the TNF-induced NF-kappaB signal. *Immunity* 8:297–303.

Ki CS, Na DL, Kim HJ, Kim JW (2001) Alpha-1 antichymotrypsin and alpha-2 macroglobulin gene polymorphisms are not associated with Korean late-onset Alzheimer's disease. *Neurosci Lett* 302:69–72.

Kim S-J, Angel P, Lafyatis R, Hattori K, Kim KY, Sporn MB, Karin M, Roberts AB (1990) Autoinduction of transforming growth factor β1 is mediated by the AP-1 complex. *Mol Cell Biol* 10:1492–1497.

Kim DK, Kim J, Park HJ (2004) Design, synthesis, and biological evaluation of novel 2-pyridinyl-[1,2,4]triazoles as inhibitors of transforming growth factor beta1 type 1 receptor. *Bioorg Med Chem* 12:2013–2020.

Kinouchi K, Brown G, Pasternak G, Donner DB (1991) Identification and characterization of receptors for tumor necrosis factor-alpha in the brain. *Biochem Biophys Res Commun* 181:1532–1538.

Kitazawa K, Tada T (1994) Elevation of transforming growth factor-beta 1 level in cerebrospinal fluid of patients with communicating hydrocephalus after subarachnoid hemorrhage. *Stroke* 25:1400–1404.

Kitazawa M, Oddo S, Yamasaki TR, Green KN, LaFerla FM (2005) Lipopolysaccharide-induced inflammation exacerbates tau pathology by a cyclin-dependent kinase 5-mediated pathway in a transgenic model of Alzheimer's disease. *J Neurosci* 25:8843–8853.

Knight JC, Udalova I, Hill AV, Greenwood BM, Peshu N, Marsh K, Kwiatkowski D (1999) A polymorphism that affects OCT-1 binding to the TNF promoter region is associated with severe malaria. *Nat Genet* 22:145–150.

Knoblach SM, Faden AI (1998) Interleukin-10 improves outcome and alters proinflammatory cytokine expression after experimental traumatic brain injury. *Exp Neurol* 153:143–151.

Knoblach SM, Fan L, Faden AI (1999) Early neuronal expression of tumor necrosis factor-alpha after experimental brain injury contributes to neurological impairment. *J Neuroimmunol* 95:115–125.

Koedel U, Bernatowicz A, Frei K, Fontana A, Pfister HW (1996) Systemically (but not intrathecally) administered IL-10 attenuates pathophysiologic alterations in experimental pneumococcal meningitis. *J Immunol* 157:5185–5191.

Kollias G (2005) TNF pathophysiology in murine models of chronic inflammation and autoimmunity. *Semin Arthritis Rheum* 34:3–6.

Kolsch H, Ptok U, Bagli M, Papassotiropoulos A, Schmitz S, Barkow K, Kockler M, Rao ML, Maier W, Heun R (2001) Gene polymorphisms of interleukin-1alpha influence the course of Alzheimer's disease. *Ann Neurol* 49:818–819.

Konig HG, Kogel D, Rami A, Prehn JH (2005) TGF-{beta}1 activates two distinct type I receptors in neurons: implications for neuronal NF-{kappa}B signaling. *J Cell Biol* 168:1077–1086.

Korner H, Lemckert FA, Chaudhri G, Etteldorf S, Sedgwick JD (1997) Tumor necrosis factor blockade in actively induced experimental autoimmune encephalomyelitis prevents clinical disease despite activated T cell infiltration to the central nervous system. *Eur J Immunol* 27:1973–1981.

Krieglstein K, Rufer M, Suter-Crazzolara C, Unsicker K (1995) Neural functions of the transforming growth factors beta. *Int J Dev Neurosci* 13:301–315.

Kulkarni AB, Huh C-H, Becker D, Geiser A, Lyght M, Flanders KC, Roberts AB, Sporn M, Ward JM, Karlson S (1993) Transforming growth factor-β1 null mutation in mice causes excessive inflammatory response and early death. *Proc Natl Acad Sci U S A* 90:770–774.

Kumar NM, Sigurdson SL, Sheppard D, Lwebuga-Mukasa JS (1995) Differential modulation of integrin receptors and extracellular matrix laminin by transforming growth factor-beta 1 in rat alveolar epithelial cells. *Exp Cell Res* 221:385–394.

Kuruvilla AP, Shah R, Hochwald GM, Liggitt HD, Palladino MA, Thorbecke GJ (1991) Protective effect of transforming growth factor beta 1 on experimental autoimmune diseases in mice. *Proc Natl Acad Sci U S A* 88:2918–2921.

Langrish CL, Chen Y, Blumenschein WM, Mattson J, Basham B, Sedgwick JD, McClanahan T, Kastelein RA, Cua DJ (2005) IL-23 drives a pathogenic T cell population that induces autoimmune inflammation. *J Exp Med* 201:233–240.

Lassmann S, Kincaid C, Asensio VC, Campbell IL (2001) Induction of type 1 immune pathology in the brain following immunization without central nervous system autoantigen in transgenic mice with astrocyte-targeted expression of IL-12. *J Immunol* 167:5485–5493.

Ledeboer A, Breve JJ, Wierinckx A, van der Jagt S, Bristow AF, Leysen JE, Tilders FJ, Van Dam AM (2002) Expression and regulation of interleukin-10 and interleukin-10 receptor in rat astroglial and microglial cells. *Eur J Neurosci* 16:1175–1185.

Leng A, Mura A, Feldon J, Ferger B (2005) Tumor necrosis factor-alpha receptor ablation in a chronic MPTP mouse model of Parkinson's disease. *Neurosci Lett* 375:107–111.

Leon LR (2002) Invited review: cytokine regulation of fever: studies using gene knockout mice. *J Appl Physiol* 92:2648–2655.

Leonard JP, Waldburger KE, Goldman SJ (1995) Prevention of experimental autoimmune encephalomyelitis by antibodies against interleukin 12. *J Exp Med* 181:381–386.

Leonoudakis D, Braithwaite SP, Beattie MS, Beattie EC (2004) TNFalpha-induced AMPA-receptor trafficking in CNS neurons; relevance to excitotoxicity? *Neuron Glia Biol* 1:263–273.

Letterio JJ (2000) Murine models define the role of TGF-β as a master regulator of immune cell function. *Cytokine Growth Factor Rev* 11:81–87.

Li DY, Sorensen LK, Brooke BS, Urness LD, Davis EC, Taylor DG, Boak BB, Wendel DP (1999) Defective angiogenesis in mice lacking endoglin. *Science* 284:1534–1537.

Li Y, Liu L, Barger SW, Griffin WS (2003) Interleukin-1 mediates pathological effects of microglia on tau phosphorylation and on synaptophysin synthesis in cortical neurons through a p38-MAPK pathway. *J Neurosci* 23:1605–1611.

Li MO, Wan YY, Sanjabi S, Robertson AK, Flavell RA (2006) Transforming growth factor-beta regulation of immune responses. *Annu Rev Immunol* 24:99–146.

Lieberman AP, Pitha PM, Shin HS, Shin ML (1989) Production of tumor necrosis factor and other cytokines by astrocytes stimulated with lipopolysaccharide or a neurotropic virus. *Proc Natl Acad Sci U S A* 86:6348–6352.

Lin S, Wei X, Xu Y, Yan C, Dodel R, Zhang Y, Liu J, Klaunig JE, Farlow M, Du Y (2003) Minocycline blocks 6-hydroxydopamine-induced neurotoxicity and free radical production in rat cerebellar granule neurons. *Life Sci* 72:1635–1641.

Lin AH, Luo J, Mondshein LH, Ten Dijke P, Vivien D, Contag CH, Wyss-Coray T (2005) Global analysis of Smad2/3-dependent TGF-{beta} signaling in living mice reveals prominent tissue-specific responses to injury. *J Immunol* 175:547–554.

Ling ZD, Potter ED, Lipton JW, Carvey PM (1998) Differentiation of mesencephalic progenitor cells into dopaminergic neurons by cytokines. *Exp Neurol* 149:411–423.

Ling ZD, Robie HC, Tong CW, Carvey PM (1999) Both the antioxidant and D3 agonist actions of pramipexole mediate its neuroprotective actions in mesencephalic cultures. *J Pharmacol Exp Ther* 289:202–210.

Ling Z, Gayle DA, Ma SY, Lipton JW, Tong CW, Hong JS, Carvey PM (2002) In utero bacterial endotoxin exposure causes loss of tyrosine hydroxylase neurons in the postnatal rat midbrain. *Mov Disord* 17:116–124.

Link J, Sîderstrîm M, Olsson T, Hîjeberg B, Ljungdahl è, Link H (1994) Increased transforming growth factor-β, interleukin-4, and interferon-gamma in multiple sclerosis. *Ann Neurol* 36:379–386.

Lippa CF, Flanders KC, Kim ES, Croul S (1998) TGF-beta receptors-I and -II immunoexpression in Alzheimer's disease: a comparison with aging and progressive supranuclear palsy. *Neurobiol Aging* 19:527–533.

Liu ZG (2005) Molecular mechanism of TNF signaling and beyond. *Cell Res* 15:24–27.

Liu RY, Snider WD (2001) Different signaling pathways mediate regenerative versus developmental sensory axon growth. *J Neurosci* 21:RC164.

Liu T, Clark RK, McDonnell PC, Young PR, White RF, Barone FC, Feuerstein GZ (1994) Tumor necrosis factor-alpha expression in ischemic neurons. *Stroke* 25:1481–1488.

Liu B, Du L, Hong JS (2000) Naloxone protects rat dopaminergic neurons against inflammatory damage through inhibition of microglia activation and superoxide generation. *J Pharmacol Exp Ther* 293:607–617.

Liu Y, Ehtesham M, Samoto K, Wheeler CJ, Thompson RC, Villarreal LP, Black KL, Yu JS (2002) In situ adenoviral interleukin 12 gene transfer confers potent and long-lasting cytotoxic immunity in glioma. *Cancer Gene Ther* 9:9–15.

Locksley RM, Killeen N, Lenardo MJ (2001) The TNF and TNF receptor superfamilies: integrating mammalian biology. *Cell* 104:487–501.

Loeys BL, Chen J, Neptune ER, Judge DP, Podowski M, Holm T, Meyers J, Leitch CC, Katsanis N, Sharifi N, Xu FL, Myers LA, Spevak PJ, Cameron DE, De Backer J, Hellemans J, Chen Y, Davis EC, Webb CL, Kress W, Coucke P, Rifkin DB, De Paepe AM, Dietz HC (2005) A syndrome of altered cardiovascular, craniofacial, neurocognitive and skeletal development caused by mutations in TGFBR1 or TGFBR2. *Nat Genet* 37:275–281.

Luckenbill-Edds L (1997) Laminin and the mechanism of neuronal outgrowth. *Brain Res Rev* 23:1–27.

Luethviksson BR, Gunnlaugsdottir B (2003) Transforming growth factor-beta as a regulator of site-specific T-cell inflammatory response. *Scand J Immunol* 58:129–138.

Lund S, Porzgen P, Mortensen AL, Hasseldam H, Bozyczko-Coyne D, Morath S, Hartung T, Bianchi M, Ghezzi P, Bsibsi M, Dijkstra S, Leist M (2005) Inhibition of microglial inflammation by the MLK inhibitor CEP-1347. *J Neurochem* 92:1439–1451.

Lundkvist J, Sundgren-Andersson AK, Tingsborg S, Ostlund P, Engfors C, Alheim K, Bartfai T, Iverfeldt K, Schultzberg M (1999) Acute-phase responses in transgenic mice with CNS over-expression of IL-1 receptor antagonist. *Am J Physiol* 276:R644–651.

Luo J, Lin AH, Masliah E, Wyss-Coray T (2006) Bioluminescence imaging of Smad signaling in living mice shows correlation with excitotoxic neurodegeneration. *Proc Natl Acad Sci U S A* 103:18326–18331.

Luterman JD, Haroutunian V, Yemul S, Ho L, Purohit D, Aisen PS (2000) Cytokine gene expression as a function of the clinical progression of Alzheimer disease dementia. *Arch Neurol* 2000:1153–1160.

MacEwan DJ (2002) TNF receptor subtype signalling: differences and cellular consequences. *Cell Signal* 14:477–492.

Madri JA, Bell L, Merwin JR (1992) Modulation of vascular cell behavior by transforming growth factors β. *Mol Reprod Dev* 32:121–126.

Mangan PR, Harrington LE, O'Quinn DB, Helms WS, Bullard DC, Elson CO, Hatton RD, Wahl SM, Schoeb TR, Weaver CT (2006) Transforming growth factor-beta induces development of the T(H)17 lineage. *Nature* 441:231–234.

Marchetti B, Abbracchio MP (2005) To be or not to be (inflamed) – is that the question in anti-inflammatory drug therapy of neurodegenerative disorders? *Trends Pharmacol Sci* 26:517–525

Marchetti L, Klein M, Schlett K, Pfizenmaier K, Eisel UL (2004) Tumor necrosis factor (TNF)-mediated neuroprotection against glutamate-induced excitotoxicity is enhanced by *N*-methyl-D-aspartate receptor activation. Essential role of a TNF receptor 2-mediated phosphatidylinositol 3-kinase-dependent NF-kappa B pathway. *J Biol Chem* 279:32869–32881.

Marklund N, Keck C, Hoover R, Soltesz K, Millard M, LeBold D, Spangler Z, Banning A, Benson J, McIntosh TK (2005) Administration of monoclonal antibodies neutralizing the inflammatory mediators tumor necrosis factor alpha and interleukin -6 does not attenuate acute behavioral deficits following experimental traumatic brain injury in the rat. *Restor Neurol Neurosci* 23:31–42.

Massagué J, Blain SW, Lo RS (2000) TGF-β signaling in growth control, cancer, and heritable disorders. *Cell* 103:295–309.

McAllister KA, Grogg KM, Johnson DW, Gallione CJ, Baldwin MA, Jackson CE, Helmbold EA, Markel DS, McKinnon WC, Murrell J (1994) Endoglin, a TGF-beta binding protein of endothelial cells, is the gene for hereditary haemorrhagic telangiectasia type 1. *Nat Genet* 8:345–351.

McCoy MK, Martinez TN, Ruhn KA, Szymkowski DE, Smith CG, Botterman BR, Tansey KE, Tansey MG (2006) Blocking soluble tumor necrosis factor signaling with dominant-negative TNF inhibitor attenuates loss of dopaminergic neurons in models of Parkinsons Disease. *J Neurosci* 26:9365–9375.

McCusker SM, Curran MD, Dynan KB, McCullagh CD, Urquhart DD, Middleton D, Patterson CC, McIlroy SP, Passmore AP (2001) Association between polymorphism in regulatory region of gene encoding tumour necrosis factor alpha and risk of Alzheimer's disease and vascular dementia: a case-control study. *Lancet* 357:436–439.

McGeer PL, McGeer EG (2001) Polymorphisms in inflammatory genes and the risk of Alzheimer disease. *Arch Neurol* 58:1790–1792.

McGuire W, Hill AV, Allsopp CE, Greenwood BM, Kwiatkowski D (1994) Variation in the TNF-alpha promoter region associated with susceptibility to cerebral malaria. *Nature* 371:508–510.

McGuire W, Knight JC, Hill AV, Allsopp CE, Greenwood BM, Kwiatkowski D (1999) Severe malarial anemia and cerebral malaria are associated with different tumor necrosis factor promoter alleles. *J Infect Dis* 179:287–290.

McGuire SO, Ling ZD, Lipton JW, Sortwell CE, Collier TJ, Carvey PM (2001) Tumor necrosis factor alpha is toxic to embryonic mesencephalic dopamine neurons. *Exp Neurol* 169:219–230.

Mehlhorn G, Hollborn M, Schliebs R (2000) Induction of cytokines in glial cells surrounding cortical beta-amyloid plaques in transgenic Tg2576 mice with Alzheimer pathology. *Int J Dev Neurosci* 18:423–431.

Merrill JE (1991) Effects of interleukin-1 and tumor necrosis factor-alpha on astrocytes, microglia, oligodendrocytes, and glial precursors in vitro. *Dev Neurosci* 13:130–137.

Mogi M, Harada M, Riederer P, Narabayashi H, Fujita K, Nagatsu T (1994) Tumor necrosis factor-alpha (TNF-alpha) increases both in the brain and in the cerebrospinal fluid from parkinsonian patients. *Neurosci Lett* 165:208–210.

Mogi M, Togari A, Tanaka K, Ogawa N, Ichinose H, Nagatsu T (1999) Increase in level of tumor necrosis factor (TNF)-alpha in 6-hydroxydopamine-lesioned striatum in rats without influence of systemic L-DOPA on the TNF-alpha induction. *Neurosci Lett* 268:101–104.

Mogi M, Togari A, Kondo T, Mizuno Y, Komure O, Kuno S, Ichinose H, Nagatsu T (2000) Caspase activities and tumor necrosis factor receptor R1 (p55) level are elevated in the substantia nigra from parkinsonian brain. *J Neural Transm* 107:335–341.

Moon JA, Kim HT, Cho IS, Sheen YY, Kim DK (2006) IN-1130, a novel transforming growth factor-beta type I receptor kinase (ALK5) inhibitor, suppresses renal fibrosis in obstructive nephropathy. *Kidney Int* 70:1234–1243.

Morganti-Kossman MC, Lenzlinger PM, Hans V, Stahel P, Csuka E, Ammann E, Stocker R, Trentz O, Kossmann T (1997) Production of cytokines following brain injury: beneficial and deleterious for the damaged tissue. *Mol Psychiatry* 2:133–136.

Mrak, RE, Griffin WS (2007) Common inflammatory mechanisms in Lewy body disease and Alzheimer disease. *J Neuropathol Exp Neurol* 66(8):683–6

Munch G, Apelt J, Rosemarie Kientsch E, Stahl P, Luth HJ, Schliebs R (2003) Advanced glycation endproducts and pro-inflammatory cytokines in transgenic Tg2576 mice with amyloid plaque pathology. *J Neurochem* 86:283–289.

Murphy GM, Jr., Claassen JD, DeVoss JJ, Pascoe N, Taylor J, Tinklenberg JR, Yesavage JA (2001) Rate of cognitive decline in AD is accelerated by the interleukin-1 alpha -889 *1 allele. *Neurology* 56:1595–1597.

Nagatsu T, Sawada M (2005) Inflammatory process in Parkinson's disease: role for cytokines. *Curr Pharm Des* 11:999–1016.

Nagatsu T, Mogi M, Ichinose H, Togari A (2000a) Changes in cytokines and neurotrophins in Parkinson's disease. *J Neural Transm* Suppl:277–290.

Nagatsu T, Mogi M, Ichinose H, Togari A (2000b) Cytokines in Parkinson's disease. *J Neural Transm* Suppl:143–151.

Nicoletti F, DiMarco R, Patti F, Reggio E, Nicoletti A, Zaccone P, Stivala F, Meroni PL, Reggio A (1998) Blood levels of transforming growth factor-beta 1 (TGF-beta1) are elevated in both relapsing remitting and chronic progressive multiple sclerosis (MS) patients and are further augmented by treatment with interferon-beta 1b (IFN-beta1b). *Clin Exp Immunol* 113:96–99.

Nicoll JA, Mrak RE, Graham DI, Stewart J, Wilcock G, MacGowan S, Esiri MM, Murray LS, Dewar D, Love S, Moss T, Griffin WS (2000) Association of interleukin-1 gene polymorphisms with Alzheimer's disease. *Ann Neurol* 47:365–368.

Nishimura M, Mizuta I, Mizuta E, Yamasaki S, Ohta M, Kaji R, Kuno S (2001) Tumor necrosis factor gene polymorphisms in patients with sporadic Parkinson's disease. *Neurosci Lett* 311:1–4.

Nishimura M, Kuno S, Kaji R, Yasuno K, Kawakami H (2005) Glutathione-S-transferase-1 and interleukin-1beta gene polymorphisms in Japanese patients with Parkinson's disease. *Mov Disord* 20:901–902.

O'Neill EJ, Day MJ, Wraith DC (2006) IL-10 is essential for disease protection following intranasal peptide administration in the C57BL/6 model of EAE. *J Neuroimmunol* 178:1–8.

Oh SP, Seki T, Goss KA, Imamura T, Yi Y, Donahue PK, Li L, Miyazono K, Dijke P, Kim S, Li E (2000) Activin receptor-like kinase 1 modulates transforming growth factor-β1 signaling in the regulation of angiogensis. *Proc Natl Acad Sci U S A* 97:2626–2631.

Paganelli R, Di Iorio A, Patricelli L, Ripani F, Sparvieri E, Faricelli R, Iarlori C, Porreca E, Di Gioacchino M, Abate G (2002) Proinflammatory cytokines in sera of elderly patients with dementia: levels in vascular injury are higher than those of mild-moderate Alzheimer's disease patients. *Exp Gerontol* 37:257–263.

Pagenstecher A, Lassmann S, Carson MJ, Kincaid CL, Stalder AK, Campbell IL (2000) Astrocyte-targeted expression of IL-12 induces active cellular immune responses in the central nervous system and modulates experimental allergic encephalomyelitis. *J Immunol* 164:4481–4492.

Pang L, Ye W, Che X-M, Roessler BJ, Betz AL, Yang G-Y (2001) Reduction of inflammatory response in the mouse brain with adenoviral-mediated transforming growth factor-β1 expression. *Stroke* 32:544–552.

Park JH, Shin SH (1996) Induction of IL-12 gene expression in the brain in septic shock. *Biochem Biophys Res Commun* 224:391–396.

Park H, Li Z, Yang XO, Chang SH, Nurieva R, Wang YH, Wang Y, Hood L, Zhu Z, Tian Q, Dong C (2005) A distinct lineage of CD4 T cells regulates tissue inflammation by producing interleukin 17. *Nat Immunol* 6:1133–1141.

Pasparakis M, Alexopoulou L, Episkopou V, Kollias G (1996) Immune and inflammatory responses in TNF alpha-deficient mice: a critical requirement for TNF alpha in the formation of primary B cell follicles, follicular dendritic cell networks and germinal centers, and in the maturation of the humoral immune response. *J Exp Med* 184:1397–1411.

Pavlov VA, Tracey KJ (2005) The cholinergic anti-inflammatory pathway. Brain Behav Immun 19:493–499.

Pavlov VA, Ochani M, Gallowitsch-Puerta M, Ochani K, Huston JM, Czura CJ, Al-Abed Y, Tracey KJ (2006) Central muscarinic cholinergic regulation of the systemic inflammatory response during endotoxemia. *Proc Natl Acad Sci U S A* 103:5219–5223.

Pawlikowska L, Tran MN, Achrol AS, McCulloch CE, Ha C, Lind DL, Hashimoto T, Zaroff J, Lawton MT, Marchuk DA, Kwok PY, Young WL (2004) Polymorphisms in genes involved in inflammatory and angiogenic pathways and the risk of hemorrhagic presentation of brain arteriovenous malformations. *Stroke* 35:2294–2300.

Pepper MS (1997) Transforming growth factor-β: Vasculogenesis, angiogenesis, and vessel wall integrity. *Cytokine Growth Factor Rev* 8:21–43.

Peress NS, Perillo E (1995) Differential expression of TGF-beta 1, 2 and 3 isotypes in Alzheimer's disease: a comparative immunohistochemical study with cerebral infarction, aged human and mouse control brains. *J Neuropathol Exp Neurol* 54:802–811.

Perry RT, Collins JS, Harrell LE, Acton RT, Go RC (2001a) Investigation of association of 13 polymorphisms in eight genes in southeastern African American Alzheimer disease patients as compared to age-matched controls. *Am J Med Genet* 105:332–342.

Perry RT, Collins JS, Wiener H, Acton R, Go RC (2001b) The role of TNF and its receptors in Alzheimer's disease. *Neurobiol Aging* 22:873–883.

Pestka S, Krause CD, Sarkar D, Walter MR, Shi Y, Fisher PB (2004) Interleukin-10 and related cytokines and receptors. *Annu Rev Immunol* 22:929–979.

Pickering M, Cumiskey D, O'Connor JJ (2005) Actions of TNF-alpha on glutamatergic synaptic transmission in the central nervous system. *Exp Physiol* 90:663–670.

Pimentel-Muinos FX, Seed B (1999) Regulated commitment of TNF receptor signaling: a molecular switch for death or activation. *Immunity* 11:783–793.

Pinteaux E, Rothwell NJ, Boutin H (2006) Neuroprotective actions of endogenous interleukin-1 receptor antagonist (IL-1ra) are mediated by glia. *Glia* 53:551–556.

Pittet JF, Griffiths MJ, Geiser T, Kaminski N, Dalton SL, Huang X, Brown LA, Gotwals PJ, Koteliansky VE, Matthay MA, Sheppard D (2001) TGF-beta is a critical mediator of acute lung injury. *J Clin Invest* 107:1537–1544.

Pradervand S, Yasukawa H, Muller OG, Kjekshus H, Nakamura T, St Amand TR, Yajima T, Matsumura K, Duplain H, Iwatate M, Woodard S, Pedrazzini T, Ross J, Firsov D, Rossier BC, Hoshijima M, Chien KR (2004) Small proline-rich protein 1A is a gp130 pathway- and stress-inducible cardioprotective protein. *EMBO J* 23:4517–4525.

Prehn JH, Bindokas VP, Marcuccilli CJ, Krajewski S, Reed JC, Miller RJ (1994) Regulation of neuronal Bcl2 protein expression and calcium homeostasis by transforming growth factor type beta confers wide-ranging protection on rat hippocampal neurons. *Proc Natl Acad Sci U S A* 91:12599–12603.

Probert L, Akassoglou K (2001) Glial expression of tumor necrosis factor in transgenic animals: how do these models reflect the "normal situation"? *Glia* 36:212–219.

Probert L, Akassoglou K, Pasparakis M, Kontogeorgos G, Kollias G (1995) Spontaneous inflammatory demyelinating disease in transgenic mice showing central nervous system-specific expression of tumor necrosis factor alpha. *Proc Natl Acad Sci U S A* 92:11294–11298.

Probert L, Eugster HP, Akassoglou K, Bauer J, Frei K, Lassmann H, Fontana A (2000) TNFR1 signalling is critical for the development of demyelination and the limitation of T-cell responses during immune-mediated CNS disease. *Brain* 123 (Pt 10):2005–2019.

Qiu J, Cafferty WB, McMahon SB, Thompson SW (2005) Conditioning injury-induced spinal axon regeneration requires signal transducer and activator of transcription 3 activation. *J Neurosci* 25:1645–1653.

Quintana A, Giralt M, Rojas S, Penkowa M, Campbell IL, Hidalgo J, Molinero A (2005) Differential role of tumor necrosis factor receptors in mouse brain inflammatory responses in cryolesion brain injury. *J Neurosci Res* 82:701–716.

Racke MK, Dhib-Jalbut S, Cannella B, Albert PS, Raine CS, McFarlin DE (1991) Prevention and treatment of chronic relapsing experimental allergic encephalomyelitis by transforming growth factor-β_1. *J Immunol* 146:3012–3017.

Raine CS, Bonetti B, Cannella B (1998) Multiple sclerosis: expression of molecules of the tumor necrosis factor ligand and receptor families in relationship to the demyelinated plaque. *Rev Neurol (Paris)* 154:577–585.

Rieckmann P, Albrecht M, Kitze B, Weber T, Tumani H, Broocks A, Luer W, Helwig A, Poser S (1995) Tumor necrosis factor-alpha messenger RNA expression in patients with relapsing-remitting multiple sclerosis is associated with disease activity. *Ann Neurol* 37:82–88.

Roberts AB, Sporn MB (1996) Transforming growth factor-β. In: *The Molecular and Cellular Biology of Wound Repair*, Second Edition (Clark RAF, ed.), pp. 275–308. New York, NY: Plenum Press.

Rott O, Fleischer B, Cash E (1994) Interleukin-10 prevents experimental allergic encephalomyelitis in rats. *Eur J Immunol* 24:1434–1440.

Rousselet E, Callebert J, Parain K, Joubert C, Hunot S, Hartmann A, Jacque C, Perez-Diaz F, Cohen-Salmon C, Launay JM, Hirsch EC (2002) Role of TNF-alpha receptors in mice intoxicated with the parkinsonian toxin MPTP. *Exp Neurol* 177:183–192.

Ruddle NH, Bergman CM, McGrath KM, Lingenheld EG, Grunnet ML, Padula SJ, Clark RB (1990) An antibody to lymphotoxin and tumor necrosis factor prevents transfer of experimental allergic encephalomyelitis. *J Exp Med* 172:1193–1200.

Ruuls SR, Hoek RM, Ngo VN, McNeil T, Lucian LA, Janatpour MJ, Korner H, Scheerens H, Hessel EM, Cyster JG, McEvoy LM, Sedgwick JD (2001) Membrane-bound TNF supports secondary lymphoid organ structure but is subservient to secreted TNF in driving autoimmune inflammation. *Immunity* 15:533–543.

Saavedra A, Baltazar G, Duarte EP (2007) Interleukin-1beta mediates GDNF up-regulation upon dopaminergic injury in ventral midbrain cell cultures. *Neurobiol Dis* 25(1):92–104.

Sairam K, Saravanan KS, Banerjee R, Mohanakumar KP (2003) Non-steroidal anti-inflammatory drug sodium salicylate, but not diclofenac or celecoxib, protects against 1-methyl-4-phenyl pyridinium-induced dopaminergic neurotoxicity in rats. *Brain Res* 966:245–252.

Samoilova EB, Horton JL, Chen Y (1998a) Acceleration of experimental autoimmune encephalomyelitis in interleukin-10-deficient mice: roles of interleukin-10 in disease progression and recovery. *Cell Immunol* 188:118–124.

Samoilova EB, Horton JL, Hilliard B, Liu TS, Chen Y (1998b) IL-6-deficient mice are resistant to experimental autoimmune encephalomyelitis: roles of IL-6 in the activation and differentiation of autoreactive T cells. *J Immunol* 161:6480–6486.

Sanyal S, Kim SM, Ramaswami M (2004) Retrograde regulation in the CNS; neuron-specific interpretations of TGF-beta signaling. *Neuron* 41:845–848.

Satomi J, Mount RJ, Toporsian M, Paterson AD, Wallace MC, Harrison RV, Letarte M (2003) Cerebral vascular abnormalities in a murine model of hereditary hemorrhagic telangiectasia. *Stroke* 34:783–789.

Scherbel U, Raghupathi R, Nakamura M, Saatman KE, Trojanowski JQ, Neugebauer E, Marino MW, McIntosh TK (1999) Differential acute and chronic responses of tumor necrosis factor-deficient mice to experimental brain injury. *Proc Natl Acad Sci U S A* 96:8721–8726.

Schweizer U, Gunnersen J, Karch C, Wiese S, Holtmann B, Takeda K, Akira S, Sendtner M (2002) Conditional gene ablation of Stat3 reveals differential signaling requirements for survival of motoneurons during development and after nerve injury in the adult. *J Cell Biol* 156:287–297.

Sedgwick JD, Riminton DS, Cyster JG, Korner H (2000) Tumor necrosis factor: a master-regulator of leukocyte movement. *Immunol Today* 21:110–113.

Segal BM, Dwyer BK, Shevach EM (1998) An interleukin (IL)-10/IL-12 immunoregulatory circuit controls susceptibility to autoimmune disease. *J Exp Med* 187:537–546.

Sekiyama A, Ueda H, Kashiwamura S, Nishida K, Kawai K, Teshima-kondo S, Rokutan K, Okamura H (2005) IL-18; a cytokine translates a stress into medical science. *J Med Invest* 52 Suppl:236–239.

Selmaj KW, Farooq M, Norton WT, Raine CS, Brosnan CF (1990) Proliferation of astrocytes in vitro in response to cytokines. A primary role for tumor necrosis factor. *J Immunol* 144:129–135.

Selmaj K, Raine CS, Cannella B, Brosnan CF (1991) Identification of lymphotoxin and tumor necrosis factor in multiple sclerosis lesions. *J Clin Invest* 87:949–954.

Selmaj K, Papierz W, Glabinski A, Kohno T (1995) Prevention of chronic relapsing experimental autoimmune encephalomyelitis by soluble tumor necrosis factor receptor I. *J Neuroimmunol* 56:135–141.

Seripa D, Dobrina A, Margaglione M, Matera MG, Gravina C, Vecile E, Fazio VM (2003) Relevance of interleukin-1 receptor antagonist intron-2 polymorphism in ischemic stroke. *Cerebrovasc Dis* 15:276–281.

Sharief MK, Hentges R (1991) Association between tumor necrosis factor-alpha and disease progression in patients with multiple sclerosis. *N Engl J Med* 325:467–472.

Shen HM, Pervaiz S (2006) TNF receptor superfamily-induced cell death: redox-dependent execution. *FASEB J* 20:1589–1598.

Sheng JG, Jones RA, Zhou XQ, McGinness JM, Van Eldik LJ, Mrak RE, Griffin WS (2001) Interleukin-1 promotion of MAPK-p38 overexpression in experimental animals and in Alzheimer's disease: potential significance for tau protein phosphorylation. *Neurochem Int* 39:341–348.

Shi Y, Massague J (2003) Mechanisms of TGF-beta signaling from cell membrane to the nucleus. *Cell* 113:685–700.

Sly LM, Krzesicki RF, Brashler JR, Buhl AE, McKinley DD, Carter DB, Chin JE (2001) Endogenous brain cytokine mRNA and inflammatory responses to lipopolysaccharide are elevated in the Tg2576 transgenic mouse model of Alzheimer's disease. *Brain Res Bull* 56:581–588.

Smith T, Hewson AK, Kingsley CI, Leonard JP, Cuzner ML (1997) Interleukin-12 induces relapse in experimental allergic encephalomyelitis in the Lewis rat. *Am J Pathol* 150:1909–1917.

Sriram K, Matheson JM, Benkovic SA, Miller DB, Luster MI, O'Callaghan JP (2002) Mice deficient in TNF receptors are protected against dopaminergic neurotoxicity: implications for Parkinson's disease. *FASEB J* 16:1474–1476.

Sriram K, Miller DB, O'Callaghan JP (2006a) Minocycline attenuates microglial activation but fails to mitigate striatal dopaminergic neurotoxicity: role of tumor necrosis factor-alpha. *J Neurochem* 96:706–718.

Sriram K, Matheson JM, Benkovic SA, Miller DB, Luster MI, O'Callaghan JP (2006b) Deficiency of TNF receptors suppresses microglial activation and alters the susceptibility of brain regions to MPTP-induced neurotoxicity: role of TNF-alpha. *FASEB J* 20:670–682.

Stalder AK, Pagenstecher A, Yu NC, Kincaid C, Chiang CS, Hobbs MV, Bloom FE, Campbell IL (1997) Lipopolysaccharide-induced IL-12 expression in the central nervous system and cultured astrocytes and microglia. *J Immunol* 159:1344–1351.

Steed PM, Tansey MG, Zalevsky J, Zhukovsky EA, Desjarlais JR, Szymkowski DE, Abbott C, Carmichael D, Chan C, Cherry L, Cheung P, Chirino AJ, Chung HH, Doberstein SK, Eivazi A, Filikov AV, Gao SX, Hubert RS, Hwang M, Hyun L, Kashi S, Kim A, Kim E, Kung J, Martinez SP, Muchhal US, Nguyen DH, O'Brien C, O'Keefe D, Singer K, Vafa O, Vielmetter J, Yoder SC, Dahiyat BI (2003) Inactivation of TNF signaling by rationally designed dominant-negative TNF variants. *Science* 301:1895–1898.

Sukal SA, Nadiminti L, Granstein RD (2006) Etanercept and demyelinating disease in a patient with psoriasis. *J Am Acad Dermatol* 54:160–164.

Suthanthiran M, Li B, Song JO, Ding R, Sharma VK, Schwartz JE, August P (2000) Transforming growth factor–β_1 hyperexpression in African-American hypertensives: a novel mediator of hypertension and/or target organ damage. *Proc Natl Acad Sci U S A* 97:3479–3484.

Sweeney ST, Davis GW (2002) Unrestricted synaptic growth in spinster-a late endosomal protein implicated in TGF-beta-mediated synaptic growth regulation. *Neuron* 36:403–416.

Swiergiel AH, Dunn AJ (2006) Feeding, exploratory, anxiety- and depression-related behaviors are not altered in interleukin-6-deficient male mice. *Behav Brain Res* 171:94–108.

Szymkowski DE (2005) Creating the next generation of protein therapeutics through rational drug design. *Curr Opin Drug Discov Devel* 8:590–600.

Hart BA, Brok HP, Remarque E, Benson J, Treacy G, Amor S, Hintzen RQ, Laman JD, Bauer J, Blezer EL (2005) Suppression of ongoing disease in a nonhuman primate model of multiple sclerosis by a human-anti-human IL-12p40 antibody. *J Immunol* 175:4761–4768.

Tada T, Kanaji M, Kobayashi S (1994) Induction of communicating hydrocephalus in mice by intrathecal injection of human recombinant transforming growth factor-beta 1. *J Neuroimmunol* 50:153–158.

Taga T (1996) Gp130, a shared signal transducing receptor component for hematopoietic and neuropoietic cytokines. *J Neurochem* 67:1–10.

Takizawa T, Tada T, Kitazawa K, Tanaka Y, Hongo K, Kameko M, Uemura KI (2001) Inflammatory cytokine cascade released by leukocytes in cerebrospinal fluid after subarachnoid hemorrhage. *Neurol Res* 23:724–730.

Tarkowski E, Issa R, Sjogren M, Wallin A, Blennow K, Tarkowski A, Kumar P (2002) Increased intrathecal levels of the angiogenic factors VEGF and TGF-beta in Alzheimer's disease and vascular dementia. *Neurobiol Aging* 23:237–243.

Tartaglia LA, Rothe M, Hu YF, Goeddel DV (1993) Tumor necrosis factor's cytotoxic activity is signaled by the p55 TNF receptor. *Cell* 73:213–216.

Taupin V, Renno T, Bourbonniere L, Peterson AC, Rodriguez M, Owens T (1997) Increased severity of experimental autoimmune encephalomyelitis, chronic macrophage/microglial reactivity, and demyelination in transgenic mice producing tumor necrosis factor-alpha in the central nervous system. *Eur J Immunol* 27:905–913.

ten Dijke P, Hill CS (2004) New insights into TGF-beta-Smad signalling. *Trends Biochem Sci* 29:265–273.

Tesseur I, Zou K, Berber E, Zhang H, Wyss-Coray T (2006a) Highly sensitive and specific bioassay for measuring bioactive TGF-beta. *BMC Cell Biol* 7:15.

Tesseur I, Zou K, Esposito L, Bard F, Berber E, Can JV, Lin AH, Crews L, Tremblay P, Mathews P, Mucke L, Masliah E, Wyss-Coray T (2006b) Deficiency in neuronal TGF-beta signaling promotes neurodegeneration and Alzheimer's pathology. *J Clin Invest* 116:3060–3069.

Thackray AM, McKenzie AN, Klein MA, Lauder A, Bujdoso R (2004) Accelerated prion disease in the absence of interleukin-10. *J Virol* 78:13697–13707.

Tian J, Shi J, Bailey K, Mann DM (2003) Negative association between amyloid plaques and cerebral amyloid angiopathy in Alzheimer's disease. *Neurosci Lett* 352:137–140.

Tobinick E, Gross H, Weinberger A, Cohen H (2006) TNF-alpha modulation for treatment of Alzheimer's disease: a 6-month pilot study. *MedGenMed* 8:25.

Tonelli LH, Postolache TT (2005) Tumor necrosis factor alpha, interleukin-1 beta, interleukin-6 and major histocompatibility complex molecules in the normal brain and after peripheral immune challenge. *Neurol Res* 27:679–684.

Tsukazaki T, Chiang TA, Davison AF, Attisano L, Wrana JL (1998) SARA, a FYVE domain protein that recruits Smad2 to the TGFbeta receptor. *Cell* 95:779–791.

Ueberham U, Ueberham E, Bruckner MK, Seeger G, Gartner U, Gruschka H, Gebhardt R, Arendt T (2005) Inducible neuronal expression of transgenic TGF-beta1 in vivo: dissection of short-term and long-term effects. *Eur J Neurosci* 22:50–64.

Uhl M, Aulwurm S, Wischhusen J, Weiler M, Ma JY, Almirez R, Mangadu R, Liu YW, Platten M, Herrlinger U, Murphy A, Wong DH, Wick W, Higgins LS, Weller M (2004) SD-208, a novel transforming growth factor beta receptor I kinase inhibitor, inhibits growth and invasiveness and enhances immunogenicity of murine and human glioma cells in vitro and in vivo. *Cancer Res* 64:7954–7961.

Unsicker K, Krieglstein K (2000) Co-activation of TGF-β and cytokine signaling pathways are required for neurotropic functions. *Cytokine Growth Factor Rev* 11:97–102.

Unsicker K, Krieglstein K (2002) TGF-betas and their roles in the regulation of neuron survival. *Adv Exp Med Biol* 513:353–374.

Unsicker K, Flanders KC, Cissel DS, Lafyatis R, Sporn MB (1991) Transforming growth factor beta isoforms in the adult rat central and peripheral nervous system. *Neuroscience* 44:613–625.

van der Wal EA, Gómez-Pinilla F, Cotman CW (1993) Transforming growth factor-β1 is in plaques in Alzheimer and Down pathologies. *Neuroreport* 4:69–72.

Veldhoen M, Hocking RJ, Flavell RA, Stockinger B (2006a) Signals mediated by transforming growth factor-beta initiate autoimmune encephalomyelitis, but chronic inflammation is needed to sustain disease. *Nat Immunol* 7:1151–1156.

Veldhoen M, Hocking RJ, Atkins CJ, Locksley RM, Stockinger B (2006b) TGFbeta in the context of an inflammatory cytokine milieu supports de novo differentiation of IL-17-producing T cells. *Immunity* 24:179–189.

Venstrom KA, Reichardt LF (1993) Extracellular matrix. 2: Role of extracellular matrix molecules and their receptors in the nervous system. *FASEB J* 7:996–1003.

Vitarbo EA, Chatzipanteli K, Kinoshita K, Truettner JS, Alonso OF, Dietrich WD (2004) Tumor necrosis factor alpha expression and protein levels after fluid percussion injury in rats: the effect of injury severity and brain temperature. *Neurosurgery* 55:416–424; discussion 424–415.

Wajant H, Pfizenmaier K, Scheurich P (2003) Tumor necrosis factor signaling. *Cell Death Differ* 10:45–65.

Weinstein M, Yang X, Deng C-X (2000) Functions of mammalian *Smad* genes as revealed by targeted gene disruption in mice. *Cytokine Growth Factor Rev* 11:49–58.

Wiendl H, Hohlfeld R (2002) Therapeutic approaches in multiple sclerosis: lessons from failed and interrupted treatment trials. *BioDrugs* 16:183–200.

Wilson AG, de Vries N, Pociot F, di Giovine FS, van der Putte LB, Duff GW (1993) An allelic polymorphism within the human tumor necrosis factor alpha promoter region is strongly associated with HLA A1, B8, and DR3 alleles. *J Exp Med* 177:557–560.

Windhagen A, Newcombe J, Dangond F, Strand C, Woodroofe MN, Cuzner ML, Hafler DA (1995) Expression of costimulatory molecules B7–1 (CD80), B7–2 (CD86), and interleukin 12 cytokine in multiple sclerosis lesions. *J Exp Med* 182:1985–1996.

Wu Z, Zhang J, Nakanishi H (2005) Leptomeningeal cells activate microglia and astrocytes to induce IL-10 production by releasing pro-inflammatory cytokines during systemic inflammation. *J Neuroimmunol* 167:90–98.

Wyss-Coray T (2006) Inflammation in Alzheimer disease: driving force, bystander or beneficial response? *Nat Med* 12:1005–1015.

Wyss-Coray T, Mucke L (2002) Inflammation in neurodegenerative disease – a double-edged sword. *Neuron* 35:419–432.

Wyss-Coray T, Feng L, Masliah E, Ruppe MD, Lee HS, Toggas SM, Rockenstein EM, Mucke L (1995) Increased central nervous system production of extracellular matrix components and development of hydrocephalus in transgenic mice overexpressing transforming growth factor-β1. *Am J Pathol* 147:53–67.

Wyss-Coray T, Borrow P, Brooker MJ, Mucke L (1997a) Astroglial overproduction of TGF-β1 enhances inflammatory central nervous system disease in transgenic mice. *J Neuroimmunol* 77:45–50.

Wyss-Coray T, Masliah E, Mallory M, McConlogue L, Johnson-Wood K, Lin C, Mucke L (1997b) Amyloidogenic role of cytokine TGF-β1 in transgenic mice and Alzheimer's disease. *Nature* 389:603–606.

Wyss-Coray T, Lin C, Sanan D, Mucke L, Masliah E (2000) Chronic overproduction of TGF-β1 in astrocytes promotes Alzheimer's disease-like microvascular degeneration in transgenic mice. *Am J Pathol* 156:139–150.

Wyss-Coray T, Lin C, Yan F, Yu G, Rohde M, McConlogue L, Masliah E, Mucke L (2001) TGF-β1 promotes microglial amyloid-β clearance and reduces plaque burden in transgenic mice. *Nat Med* 7:612–618.

Wyss-Coray T, Yan F, Lin AH, Lambris JD, Alexander JJ, Quigg RJ, Masliah E (2002) Prominent neurodegeneration and increased plaque formation in complement-inhibited Alzheimer's mice. *Proc Natl Acad Sci U S A* 99:10837–10842.

Yamashita T, Sawamoto K, Suzuki S, Suzuki N, Adachi K, Kawase T, Mihara M, Ohsugi Y, Abe K, Okano H (2005) Blockade of interleukin-6 signaling aggravates ischemic cerebral damage in mice: possible involvement of Stat3 activation in the protection of neurons. *J Neurochem* 94:459–468.

Yang L, Lindholm K, Konishi Y, Li R, Shen Y (2002) Target depletion of distinct tumor necrosis factor receptor subtypes reveals hippocampal neuron death and survival through different signal transduction pathways. *J Neurosci* 22:3025–3032.

Yang YC, Piek E, Zavadil J, Liang D, Xie D, Heyer J, Pavlidis P, Kucherlapati R, Roberts AB, Bottinger EP (2003) Hierarchical model of gene regulation by transforming growth factor beta. *Proc Natl Acad Sci U S A* 100:10269–10274.

Youdim MB, Buccafusco JJ (2005) Multi-functional drugs for various CNS targets in the treatment of neurodegenerative disorders. *Trends Pharmacol Sci* 26:27–35.

Zdanov A (2004) Structural features of the interleukin-10 family of cytokines. Curr Pharm Des 10:3873–3884.

Zhang F, Endo S, Cleary LJ, Eskin A, Byrne JH (1997) Role of transforming growth factor-β in long-term synaptic facilitation in *Aplysia*. *Science* 275:1318.

Zheng X, Wang J, Haerry TE, Wu AY, Martin J, O'Connor MB, Lee CH, Lee T (2003) TGF-beta signaling activates steroid hormone receptor expression during neuronal remodeling in the Drosophila brain. *Cell* 112:303–315.

Zhu Y, Yang G-Y, Ahlemeyer B, Pang L, Che X-M, Culmsee C, Klumpp S, Krieglstein J (2002) Transforming growth factor-β1 increases bad phosphorylation and protects neurons against damage. *J Neurosci* 22:3898–3909.

Zou JY, Crews FT (2005) TNF alpha potentiates glutamate neurotoxicity by inhibiting glutamate uptake in organotypic brain slice cultures: neuroprotection by NF kappa B inhibition. *Brain Res* 1034:11–24.

6
Arachidonic Acid Metabolites: Function in Neurotoxicity and Inflammation in the Central Nervous System

K. Andreasson and T.J. Montine

1 Preface

Arachidonic acid (AA) is liberated from membrane phospholipids by the action of phospholipases upon stimulation by a variety of extracellular stimuli including neurotransmitters, growth factors, and cytokines (Farooqui et al., 2006). As illustrated in Fig. 6.1, three enzyme systems act on AA to generate prostaglandin H_2 via the cyclooxygenase enzymes (COX-1 and COX-2), leukotrienes via the LOX pathways, and epoxyeicosatrienoic acids via the epoxygenase cytochrome P450 pathway. These lipid products are signaling messengers and regulate a wide variety of physiologic functions including cerebral blood flow and synaptic function under basal conditions. Increases in expression and activity of enzymes in each of these three pathways have been documented in a wide spectrum of neurological diseases, including acute insults such as cerebral ischemia and trauma as well as chronic neurodegenerative diseases such as Alzheimer's disease, Parkinson's disease, and amyotrophic lateral sclerosis. Of the three enzymatic pathways, the cyclooxygenase (COX) pathway has been most studied, and a significant body of evidence links activity of this pathway and downstream prostanoid products to a broad range of neurological diseases.

In addition to these eicosanoids generated by enzyme-catalyzed oxidation of free AA, similar chemistry occurs with free radical attack on AA still esterified to lipid. Given the indiscriminant nature of free radical reactions, multiple products can be formed from AA including reactive aldehydes like 4-hydroxy-2-nonenal that modify cellular nucleophiles in protein, nucleic acid, and lipid, as well as chemically stable products, such as the isoprostanes, that can be used as quantitative in vivo markers of free radical damage and may have receptor activating properties.

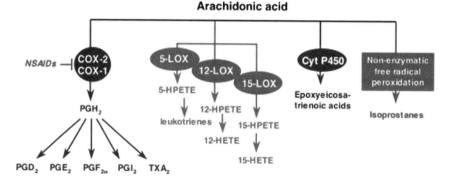

Fig. 6.1 Arachidonic acid is metabolized by three enzyme systems: (1) the cyclooxygenases COX-1 and COX-2 to yield PGH$_2$, which is converted by specific synthases to the five prostanoids, (2) the lipoxygenases 5-LOX, 12-LOX, and 15-LOX to yield leukotrienes and HETEs, and (3) the cytochrome P450 epoxygenases to yield epoxyeicosanotrienoic acids. AA is also non-enzymatically oxidized by free radicals to form a large family of isoprostanes (*See also color plates*).

2   COX-1 and COX-2 in the Central Nervous System

2.1   Metabolism of AA and Production of Prostaglandins

The cyclooxygenases COX-1 and COX-2 catalyze the first committed step in prostanoid and thromboxane synthesis (reviewed in Breyer et al., 2001). The cyclooxygenases are heme-containing bis-oxygenases and convert AA to PGH$_2$, which then serves as a substrate for a family of prostanoid and thromboxane synthases that generate PGE$_2$, PGD$_2$, PGF$_{2\alpha}$, PGI$_2$ (prostacyclin), and TXA$_2$ (thromboxane). These prostaglandins bind to several classes of G protein-coupled receptors (GPRCs), some of which have more than one isoform. These GPRCs are designated EP (for E-prostanoid receptor), FP, DP, IP, and TP, respectively. Further complexity arises in that there can be more than one receptor for each prostaglandin: there are four EP receptors (EP1, EP2, EP3, and EP4) and two DP receptors (DP1 and DP2, aka CRTH2). In brain, these prostaglandin receptors have distinct cellular and anatomical distributions, are differentially regulated in response to exogenous stimuli, and have different downstream signaling cascades. The receptor subtypes and expression patterns in brain are summarized in Table 6.1. Prostaglandin receptor subtypes are distinguished not only by which prostaglandin they bind, but also by the specific signal transduction pathway that is activated upon ligand binding. Activation of PG receptors triggers intracellular signals that lead to modifications in production of cAMP and/or phosphoinositol (PI) turnover. Within one family of receptors, subtypes may have opposing effects on cAMP levels, and thus have functionally different downstream effects. For example, EP2, EP4, DP1 and IP1 activation result

Table 6.1 Summary of prostaglandin receptor classes, anatomic distribution in brain, signaling properties, and phenotype of null mutant mice

PG	Receptor subtype	Distribution in brain	Signaling	Null mutant phenotype
$PGF_{2\alpha}$ FP		Neurons; hippocampal synaptosomes	Increase IP_3, Ca^{2+}	Failure of parturition (Sugimoto et al., 1997)
PGD_2 DP1		Meninges > thalamus, hypothalamus (HTH), brainstem > cortex, hippocampus	Increase cAMP	Sleep phenotype; resistant to immune cell infiltration in asthma (Matsuoka et al., 2000; Mizoguchi et al., 2001)
	DP2 (CRTH2)	Neurons, astrocytes: cortex, hippocampus, thalamus	Increase IP_3, Ca^{2+}	Decreased skin inflammation (Satoh et al., 2006)
PGE_2 EP1		Neurons: hypothalamus/ thalamus > cortex, hippocampus and striatum	Increase IP_3, Ca^{2+}	Disrupted algesia, blood pressure regulation (Stock et al., 2001); decreased stroke volume (Kawano et al., 2006)
	EP2	Neurons, astrocytes: cortex, striatum hippocampus, thalamus; inducible with LPS	Increase cAMP	Abortive expansion of the ovarian cumulus and impaired fertility (Hizaki et al., 1999); decreased inflammation (Liang et al., 2005b; Montine and Milatovic et al., 2002)
	EP3	Neurons >> astrocytes: cortex, striatum, hippocampus, thalamus/HTH	Decrease cAMP	Failure to mount febrile response (Ushikubi et al., 1998)
	EP4	Neurons: HTH, thalamus > striatum cortex; induced in cortex, hippocampus after LPS	Increase cAMP	Patent ductus arteriosus (Segi et al., 1998); impaired bone resorption (Miyaura et al., 2000)
PGI_2 IP		Neurons: cortex, hippocampus, striatum; very high in n. solitary tract, dorsal horn, spinal trigeminal nucleus	Increase cAMP	Enhanced thrombosis, decreased vascular permeability and inflammatory pain response (Murata et al., 1997); decreased susceptibility to renovascular hypertension (Fujino et al., 2004)
TXA_2 TP		Oligodendrocytes, astrocytes, neurons: white matter tracts, hippocampus, cortex	Increase IP_3, Ca^{2+}	Prolonged bleeding time (Thomas et al., 1998)

in increased cAMP levels. However, EP1, TP and FP receptors signal via stimulation of phospholipase C to produce IP3 and di-acyl-glycerol and EP3 and DP2 signal via inhibition of adenylyl cyclase through the inhibitory guanine nucleotide-binding regulator protein (G_i). Genetic knockouts for all the PG receptors have been generated, and phenotypes of these null mutants are listed in Table 6.1.

2.2   COX-1 and COX-2: Localization and Physiologic Function

COX-1 and COX-2 are expressed in brain, albeit with significantly different cellular and anatomic distributions. COX-1 is constitutively expressed at low levels in all cell types. COX-2 however is expressed at moderately robust levels basally in specific neuronal populations including dentate gyrus, hippocampal pyramidal neurons, efferent layers of cerebral cortex, pyriform cortex, amygdala, and in discrete subpopulations in hypothalamus (Breder et al., 1995; Yamagata et al., 1993). Within neurons, COX-2 immunoreactivity is localized to post-synaptic sites in dendritic spines, the sites of NMDA receptor-mediated neurotransmission (Kaufmann et al., 1996). COX-2 expression is upregulated in neurons following N-methyl-D-aspartate (NMDA) receptor-dependent excitatory synaptic activity in paradigms of learning and memory. The fact that COX-2 is specifically localized to post-synaptic structures, and that its expression is increased upon excitatory synaptic activity points to a role for COX-2 in modulating synaptic transmission under physiologic conditions. Electrophysiologic and in vivo modeling in rodent models of learning and memory (Rall et al., 2003; Shaw et al., 2003; Teather et al., 2002) suggest that COX-2, and in particular PGE_2, participate in synaptic plasticity (Chen and Bazan, 2005; Chen et al., 2002; Sang et al., 2005). In the Morris Water Maze, which tests hippocampal-dependent learning and memory, COX-2 inhibitors given either before or after training inhibit both acquisition and consolidation of spatial memory (Rall et al., 2003; Teather et al., 2002). Mechanisms by which COX-2 and its downstream prostaglandins might function in synaptic remodeling include regulation of downstream growth factors, particularly BDNF (Shaw et al., 2003) and neurovascular coupling in which synaptically active neurons promote local hyperemia (Niwa et al., 2000).

2.3   Prostaglandins and Prostaglandin Receptors: Localization and Physiologic Function

2.3.1 $PGF_{2\alpha}$

Binding studies show that FP receptors are enriched in brain synaptosomes and $PGF2_{\alpha}$ has been immunolocalized in vivo to neurons (Linton et al., 1995). Levels of $PGF2_{\alpha}$ are induced in models of excitotoxicity (Nakagomi et al., 1990; Ogawa et al., 1987; Seregi et al., 1985) and following catecholamine-mediated neurotransmission in hippocampal mossy fibers (Separovic and Dorman, 1993). Lerea et al. (1997) have further demonstrated that $PGF2_{\alpha}$ couples NMDA-receptor activation to induction of the immediate early gene *c-fos* in hippocampal neurons, underscoring the importance of PG signaling in the spectrum of NMDA-receptor mediated events ranging from neuroplastic responses to cellular injury. FP receptor activation results in increased PI turnover and elevation of intracellular Ca^{2+}

(Kitanaka et al., 1993), and promotes neuronal death in an in vivo model of spinal cord injury (Liu et al., 2001).

2.3.2 PGD$_2$

PGD$_2$ immunolocalizes to all regions of the brain, notably the hippocampus and cerebral cortex (Watanabe et al., 1985). The DP1 receptor localizes to the leptomeninges (Liang et al., 2005b), and highly to diencephalon, brainstem, and to a lesser extent telencephalon. The DP2 receptor is also highly expressed in brain (Liang et al., 2005b), and is localized to neurons of hippocampus and cerebral cortex by in situ hybridization.

2.3.3 PGE$_2$

The effects of PGE$_2$ are complex because PGE$_2$ can bind to four distinct EP receptors that have distinct second messenger systems. The EP2 and EP3 receptors are widely expressed in neurons of cerebral cortex, hippocampus, thalamus and hypothalamus (Ek et al., 2000; McCullough et al., 2004; Sugimoto et al., 1994). EP4 and EP1 receptors are most abundant in hypothalamus (Batshake et al., 1995) and present at lower levels in cortex, hippocampus and striatum. Importantly, EP2 and EP4 receptor expression is inducible after stimulation with the endotoxin lipopolysaccharide (LPS) or IL1β (Zhang and Rivest, 1999).

2.3.4 PGI$_2$

PGI$_2$ binds to the IP receptor which is present at moderate levels in cortex, hippocampus, striatum and thalamus, with highest concentration in sensory integration areas such as the nucleus of the solitary tract and dorsal horn (Matsumura et al., 1995). In addition, specific binding of a PGI$_2$ analogue to a second putative receptor has been described, suggesting the existence of a second IP receptor that is localized to telencephalic structures (Watanabe et al., 1999).

2.3.5 TXA$_2$

TP receptors are present in endothelium where they promote vasoconstriction. They are also expressed in neurons of hippocampus and can promote glutamate (Hsu and Kan, 1996) and noradrenaline release (Nishihara et al., 1999) and inhibit voltage sensitive Ca^{2+} channels (Hsu et al., 1996). In addition, TP receptors are present on astrocytes (Nakahata et al., 1992), oligodendrocytes and myelin tracts (Blackman et al., 1998), and modulate phospholipase C beta activity with activation of inositol triphosphate signaling. The complex effects of TXA$_2$ will likely depend on the cross activation of TP receptors on multiple cell types.

2.4 COX-2 and Prostaglandins: Pathologic Function

2.4.1 Excitotoxicity and Ischemia

COX-2 neuronal expression is tightly regulated by excitatory synaptic activity via the NMDA receptor (Yamagata et al., 1993). This coupling of neuronal COX-2 expression via the NMDA receptor however is deleterious in paradigms of excitotoxicity, where there is significant release of glutamate and excessive signaling via the NMDA receptor resulting in death of the neuron. Multiple models of acute glutamate toxicity have now been associated with increased COX-2 expression and activity including seizures (Yamagata et al., 1993), excitotoxin injection (Adams et al., 1996), spreading depression (Miettinen et al., 1997) and cerebral ischemia. In ischemic paradigms, COX-2 expression is rapidly induced early on in neurons in response to massive glutamate release, and later in non-neuronal cells, including microglia, astrocytes and infiltrating leukocytes that participate in the post-ischemic inflammatory reaction. In models of cerebral ischemia, pharmacologic inhibition of COX-2 enzymatic activity and genetic deletion of COX-2 produce a reduction in the infarct size (Nakayama et al., 1998; Nogawa et al., 1997). Similarly in a different model of acute neurological injury, a rodent model of spinal cord injury, in which initial injury is mediated by glutamate toxicity and lipid peroxidation, inhibition of COX-2 activity results in improved functional outcome (Resnick et al., 1998).

The mechanism by which COX-2 activity contributes to neuronal injury has not been determined but may involve (1) production of reactive oxygen species (ROS) as a byproduct of the peroxidase activity of the COX enzyme (Hanna et al., 1997; Kukreja et al., 1986), (2) production of PGs and activation of selected PG receptors that impact negatively on neuronal viability, or (3) promotion of PGE_2-mediated glutamate release from astrocytes (Bezzi et al., 1998). With regard to the first mechanism, the extent of production of ROS by COX activity and its effects on neuronal viability are not fully defined. ROS species are formed during the metabolism of arachidonic acid as a byproduct of the peroxidase activity of the enzyme. The peroxidase moiety requires a reducing co-substrate to donate a single electron to the peroxidase intermediate, and a co-substrate free radical is then produced for each reaction. Thus, increased COX-2 activity may increase the overall production of ROS in the setting of ischemia, adding to that generated by mitochondrial respiration and nitric oxide metabolism. Older studies have linked NSAID with a reduction of ROS following hypoxia (Hall et al., 1993; Patel et al., 1993).

More recent efforts have focused on what specific prostaglandin receptors promote COX-2 neurotoxicity in cerebral ischemia. Genetic deletion of the EP1 receptor or pharmacological inhibition of EP1 signaling rescues tissue in a model of transient focal ischemia (Kawano et al., 2006). In vitro examination of EP1 receptor function in cultured neurons demonstrated that the EP1 receptor enhanced the Ca^{2+} dysregulation induced by NMDA receptor activation by impairing Na^+-Ca^{2+} exchange (Kawano et al., 2006). Conversely, other studies have demonstrated a paradoxical neuroprotection in models of glutamate toxicity and ischemia, particularly with respect to PG receptors that are positively coupled to cAMP production. In primary hippocampal neurons or organotypic hippocampal and spinal cord slices subjected to glutamate toxicity, PGE_2 signaling via the EP2 receptor promotes a paradoxical neuroprotection (Bilak et al., 2004; Liu et al., 2005; McCullough et al., 2004), consistent with previous studies in primary neurons (Akaike et al., 1994; Cazevielle et al., 1994). The neuroprotective effect of the EP2 receptor has been confirmed in vivo in models of focal ischemia (Liu et al., 2005; McCullough et al., 2004). PGD_2, signaling via its DP1 receptor similarly promotes neuroprotection against glutamate toxicity in dispersed hippocampal neurons, organotypic hippocampal slices, and spinal cord slices (Liang et al., 2005a). For both EP2 and DP1 mediated neuroprotection, inactivation of protein kinase A (PKA) abolishes neuroprotection, indicating that neuroprotection in this paradigm is dependent on cAMP signaling (Bilak et al., 2004; Liang et al., 2005b; McCullough et al., 2004). More recently, the EP4 receptor, which like the EP2 and DP1 receptors is coupled to $G_{\alpha s}$, has also been shown to be neuroprotective in glutamate toxicity (Ahmad et al., 2005).

2.4.2 Chronic Neuroinflammatory Diseases

Evidence is accumulating suggesting a substantial role for COX-2 activity and prostaglandin signaling in mediating secondary neuronal injury in chronic degenerative neurological diseases characterized by a predominantly inflammatory pathology.

Alzheimer's disease. Alzheimer's disease (AD) is a common cause of age-dependent dementia, affecting nearly 1 in 3 adults over the age of 80. Hallmark pathological features include deposition of A-beta peptide into amyloid plaques in brain parenchyma, neurofibrillary tangles with extensive hippocampal and cortical neuronal loss, and glial activation. A central role of inflammation is postulated in the development and progression of AD (Akiyama et al., 2000). In support of this concept is the fact that in normal aging populations, chronic use of non-steroidal anti-inflammatory drugs (NSAIDs), which block COX activity, is associated with a significantly lower risk of developing AD (in t' Veld et al., 2001; McGeer et al., 1996; Stewart et al., 1997). Conversely, the administration of NSAIDs or COX-2 inhibitors to patients already symptomatic with AD has little benefit (Aisen et al., 2003). These studies indicate that in aging populations, NSAIDs may be effective in the prevention and not the therapy of AD, and that inflammation is an early and potentially reversible pre-clinical event.

Modeling the effect of NSAIDs in vivo using transgenic mouse models of mutant human alleles of amyloid precursor protein (APP) and presenilin-1 (PS1) has demonstrated that NSAIDs can reduce amyloid deposition in association with a reversal of behavioral deficits (Jantzen et al., 2002; Lim et al., 2000, 2001; Yan et al., 2003). The mechanism by which NSAIDs promote this effect has been attributed to anti-inflammatory effects through inhibition of COX and production of pro-inflammatory prostaglandins. Recent in vitro studies have demonstrated that high concentrations of some but not all NSAIDs can alter γ-secretase activity and reduce the ratio of Aβ 42–38 in cultured cells (Weggen et al., 2001, 2003a, b), although whether this mechanism is responsible for the preventive effects of NSAIDs in AD is not clear (Eriksen et al., 2003; Lanz et al., 2005). Pathologic evidence of significantly decreased levels of microglial activation has been described in patients taking NSAIDs (Mackenzie and Munoz, 1998). In transgenic models of Familial AD, NSAIDs reduce Aβ accumulation and markers of microglial activation including CD11b (Mac1) and CD45 (Yan et al., 2003).

Parkinson's disease. Parkinson's disease (PD) is another age-related neurodegenerative disease involving selective degeneration of the dopaminergic neurons of the substantia nigra. The predominant symptoms include motor symptoms of tremor, rigidity, and bradykinesia, although additional higher order cognitive and psychiatric symptoms frequently develop later in the disease. As in AD, a major role for inflammation is now suspected in PD as evidenced by significant microglial activation with complement activation (reviewed in McGeer and McGeer, 2004) evident in autopsied brains of patients with PD. Significantly, COX-2 and downstream prostaglandin signaling are believed to play a role in disease progression in rodent models of PD. In these models, inhibition of COX-2, either using genetic COX-2 knockout strategies or pharmacological COX-2 inhibitors protect against development of PD-like symptoms (Feng et al., 2002; Teismann et al., 2003).

Amyotrophic lateral sclerosis. Amyotrophic lateral sclerosis (ALS) is a devastating neurodegenerative disease, characterized by the progressive loss of motor neurons in ventral spinal cord and motor cortex. Pathologic findings in ALS suggest an inflammatory component to this disease involving activated microglia, and astrocytes (McGeer and McGeer, 2002), and increased COX-2 expression (Maihofner et al., 2003; Yasojima et al., 2001; Yiangou et al., 2006) and CSF PGE_2 levels (Almer et al., 2001, 2002), but see (Cudkowicz et al., 2006). An extensively characterized familial form of ALS is caused by mutations in the copper/zinc superoxide dismutase (SOD1) gene (Bruijn et al., 1998; Gurney et al., 1994; Wong et al., 1995). Transgenic mice that express mutant forms of SOD1 recapitulate the human disease, and develop paralysis associated with loss of motor neurons and premature death. Levels of COX-2 mRNA and PGE_2 are increased in mutant SOD mice (Almer et al., 2001; Klivenyi et al., 2004) and administration of COX-2 inhibitors beginning before onset of hindlimb paralysis reduces levels of CSF PGE_2 and extends survival (Drachman et al., 2002; Klivenyi et al., 2004; Pompl et al., 2003). These observations have suggested that COX-2, via its downstream prostaglandin products, can promote motor neuron injury in this model.

2.4.3 Prostaglandin Receptor Signaling and Inflammatory Injury

Brain inflammation underlies a large number of neurodegenerative diseases as well as normal aging. The prostaglandin products of COX-1 and COX-2 enzymatic activity are important mediators of the inflammatory response. The identification of which prostaglandin receptors promote COX-2 toxicity is currently an active area of investigation. Because of the number of prostaglandin receptor signaling systems and the complexity of cellular distributions, inducibility, and downstream signaling cascades, this undertaking is likely to be quite complex. However, recent studies using genetic mouse models are beginning to uncover important neurotoxic and pro-inflammatory functions of selected prostaglandin receptors. Significant headway has been made in examining the function of the PGE_2 EP2 receptor in the LPS model of innate immunity and in a transgenic model of Familial AD and amyloid deposition.

PGE_2 and the innate immune response to LPS. A first line of defense in the CNS against injury or infection is the innate immune response, in which activated microglia produce cytotoxic compounds such as ROS and nitric oxide and phagocytose the offending agents. Injection of the endotoxin LPS constitutes a primary model of innate immunity, and has been used to study mechanisms of immune activation, including microglial activation and production of ROS, nitric oxide, cytokines and chemokines, as well as phagocytosis. LPS is a potent and specific stimulus of innate immune response that binds the co-receptors CD14 receptor and TLR4; this is followed by activation of mitogen-activated protein kinases and NFkappaB which regulate transcription of pro-inflammatory genes such as COX-2 and iNOS that function in microglial activation. LPS stimulation also leads to activation of microglial NADPH oxidase, a major source of superoxide in inflammation. LPS-induced inflammatory changes can enhance cognitive decline in aging rodents (Hauss-Wegrzyniak et al., 2002); ROS-induced injury leads to abnormal aging, and impairments in learning-impaired aged rats are associated with higher levels of oxidative damage as compared to normal aged controls (Nicolle et al., 2001).

The EP2 receptor functions prominently in the elaboration of oxidative damage in innate immunity triggered by intraventricular (ICV) injection of LPS (Milatovic et al., 2004, 2005; Montine et al., 2002a). In wild type (wt) mice, ICV LPS activates glial innate immunity, leading to transient elevation of cerebral F2-IsoPs (discussed below), a stable measure of oxidative damage (Montine et al., 2002a). We found that this LPS-induced cerebral oxidative damage is abolished in EP2 null mice (Montine et al., 2002b). This protective effect is not associated with changes in cerebral eicosanoid production, but is partially related to reduced induction of nitric oxide synthase (NOS) activity. We further observed that mice lacking EP2 are completely protected from CD14-dependent synaptodendritic degeneration (Milatovic et al., 2004, 2005). Following ICV LPS injection, hippocampal CA1 pyramidal neurons undergo a delayed, reversible decrease in dendrite length and spine density without neuron death that reaches its nadir at approximately 24 h and recovers to near basal levels by 72 h; no change in these endpoints is seen in EP2 null mice under the same conditions.

PGE_2 signaling in the APPSwe-PS1ΔE9 model of Familial AD/amyloidosis. Mouse transgenic models of Familial Alzheimer's disease (FAD) consist of overexpression of mutant APP and PS1 proteins. These mice develop age-dependent amyloid accumulation in concert with significant microglial activation, production of inflammatory cytokines and increased ROS. In aging APPSwe-PS1ΔE9 mice,

deletion of the EP2 receptor results in marked decreases in lipid peroxidation, with significant reductions in neuronal lipid peroxidation (Liang et al., 2005a). The effect of EP2 deletion on lipid peroxidation in this model parallels what is seen in the simpler LPS model. In spite of the significant complexity of the amyloid model, the CD14-dependent innate immune response to LPS nevertheless is relevant to the immune response to accumulating Aβ peptide because microglial activation and elaboration of inflammatory mediators are in part CD14-dependent (Fassbender et al., 2004; Milatovic et al., 2004). In aged APPSwe-PS1ΔE9 mice, deletion of the EP2 receptor is also associated with lower levels total Aβ 40 and 42 peptides and accumulated amyloid deposits (Liang et al., 2005a). This finding raises the question of whether the lower oxidative damage in the EP2–/– background results in a delayed and less severe accumulation of Aβ, or conversely, whether transgenic Aβ production fails to elicit a vigorous oxidative response, which in turn results in a decreased accumulation of Aβ over time. The temporal and causal relationship of inflammatory oxidative stress and Aβ peptide accumulation is of central relevance to the pathogenesis of AD, where oxidative stress may precede and possibly trigger Aβ deposition (Pratico et al., 2001).

The hypothesis that the EP2 receptor functions in the innate immune response to Aβ peptides is supported by evidence in non-neuronal systems. Expression of the EP2 receptor is upregulated in models of innate immunity in peripheral macrophages and antigen presenting cells (Harizi et al., 2003; Hubbard et al., 2001) where the EP2 receptor regulates expression of inflammatory mediators, including TNF-α (Akaogi et al., 2004; Fennekohl et al., 2002; Vassiliou et al., 2003), IL-6 (Akaogi et al., 2004; Treffkorn et al., 2004), MCP-1 (Largo et al., 2004), ICAM (Noguchi et al., 2001), and iNOS (Minghetti et al., 1997). Separate studies point to additional roles of EP2 receptor in modulating macrophage migration (Baratelli et al., 2004), and inhibiting phagocytosis of bacterial components by lung alveolar macrophages (Aronoff et al., 2004). Regarding its function in phagocytosis, an intriguing recent study demonstrates that EP2–/– microglia exhibit a strong phagocytic response to deposited Aβ not observed in wild type microglia (Shie et al., 2005). These studies are consistent with others that demonstrate that PGE$_2$ inhibits phagocytosis by macrophages by a process that is dependent on increased cAMP levels (Aronoff et al., 2004; Borda et al., 1998; Canning et al., 1991; Hutchison and Myers, 1987). Conversely, NSAIDs potentiate phagocytosis by macrophages in many studies (Bjornson et al., 1988; Gilmour et al., 1993; Gurer et al., 2002; Laegreid et al., 1989). Thus the anti-amyloidogenic properties of NSAIDs in transgenic mutant APP models (Jantzen et al., 2002; Lim et al., 2000; Yan et al., 2003) might be explained by indirect interruption of PGE$_2$ signaling via the EP2 receptor with COX inhibition, resulting in reduced oxidation and increased phagocytosis and clearance of Aβ. Similarly, in APPSwe-PS1ΔE9 mice, deletion of the EP2 receptor may decrease oxidative stress and increase clearance of Aβ, resulting in the final picture of decreased Aβ load. Taken together, these studies suggest a possible dual role for the EP2 receptor in promoting oxidative damage and inhibiting phagocytosis of Aβ. In AD and in models of Familial AD, an imbalance develops with aging between the accumulation and clearance of Aβ. This might be occuring because the microglial clearance of Aβ does not keep up with the fibrillization and accumulation of Aβ, which are themselves accelerated by the increased oxidative stress elicited by increasing levels of Aβ. An important question in the in vivo development of pathology is whether defective phagocytosis leads to further inflammation, oxidative damage and secondary neurotoxicity, and whether EP2 receptor activation promotes this.

3   LOX Metabolism of AA and Bioactivity of HPETEs

3.1   Products of LOXs and Their Bioactivity

LOXs are a family of cytosolic enzymes that catalyze the oxygenation of polyunsaturated fatty acids (PUFAs) to form lipid hydroperoxides; for AA, these are called hydroperoxyeicosatetraenoic acids (HPETEs) (Campbell and Halushka, 2001). Different LOXs vary in the placement of the hydroperoxy group and are so named. Thus 5-LOX catalyzes the formation of 5-HPETEs, 12-LOXs the formation of 12-HPETEs, and so on. HPETEs are unstable intermediates, like PGH_2, and are converted to potent autocrine and paracrine factors, their corresponding hydroxyl (HETEs), either non-enzymatically or by peroxidases. Products of 5-LOX may be further metabolized to leukotriene (LT) A_4, which can be subsequently hydrolyzed to LTB_4 or LTC_4. LTC_4 is metabolized via the mercapturic acid pathway to the cysteintyl-LT's (CysLTs): LTD_4, LTE_4, and LTF_4. Elements in this arm of LT biosynthesis are well known potent components of local inflammatory response and indeed comprise the "slow reacting substance of anaphylaxis". The CysLTs activate two classes of receptors, $CysLT_1$ and $CysLT_2$, that are targets for antagonists being evaluated for efficacy in inflammatory diseases such as asthma (Norel and Brink, 2004). In addition to LTs, LOX can catalyze the formation of lipoxins that function in the resolution of inflammatory responses and act via the ALX receptor (Norel and Brink, 2004).

Unlike the COX pathway for which there is epidemiologic and mechanistic data in support of a pathogenic role in neurodegenerative diseases, we are unaware of any epidemiologic, pharmacologic, or animal model data that yet point to a significant contribution of LOX pathway metabolites in these diseases; however, inhibitors of LT formation or receptor activation have not been in use for very long, so it is too early to discount this pathway. Nevertheless, 5-LOX is expressed in neurons, including the hippocampus, and its expression may increase with age (Sugaya et al., 2000). One autopsy-based study has associated increased activity in 12/15-LOX pathways with late-stage AD (Pratico et al., 2004).

4   The Epoxygenase System

In addition to the COX and LOX pathways described above, AA and other PUFA's are substrates for enzyme-catalyzed oxidation by cytochome P450's, the epoxygenase pathway. While the biological role for these eicosanoid products is less clear than products of COX or LOX, there is growing evidence that at least some products in this pathway play important roles in vascular physiology (Fisslthaler et al., 1999). Along this line, it has been proposed that glial generation of oxygenated lipids by this pathway functionally contributes to neurovascular coupling (Lovick et al., 2005; Metea and Newman, 2006). We are unaware of reports connecting this pathway to specific elements in the pathogenesis of neurodegenerative diseases.

5 Non Enzymatic Metabolites for AA

5.1 Free Radical Damage to AA

A central hypothesis for the pathogenesis of several neurodegenerative diseases is that increased free radical damage contributes to the initiation and progression of disease (Montine et al., 2002b). It is critical to note that unlike enzyme-catalyzed reactions described above, free radical damage is an indiscriminate process that will simultaneously modify multiple targets including nucleic acid, protein, and lipids (Montine et al., 2003). PUFAs including AA are among the most vulnerable targets for free radical damage, a process termed lipid peroxidation (Porter et al., 1995). This complex process directly damages membranes and generates a number of oxygenated products that can be classified as either chemically reactive or stable products (Esterbauer and Ramos, 1996).

5.1.1 Reactive Products of Lipid Peroxidation and Their Bioactivity

Recently, considerable progress has been made in understanding the potential contribution of chemically reactive products of lipid peroxidation to neurodegeneration. The presumed mechanism of action of all of these electrophilic products is adduction of nucleophilic groups in protein or nucleic acid. For example, adduction of a critical amino acid residue in an enzyme or transporter may lead to its dysfunction (Esterbauer and Ramos, 1996; Keller et al., 1997; Mark et al., 1997). However, interpreting experiments that investigate the contribution of reactive products of lipid peroxidation to disease pathogenesis is limited by their lack of biochemical specificity.

Despite this limitation to understanding the precise biochemical mechanisms of action, many studies in a variety of model systems and autopsy-derived tissue have implicated reactive products of lipid peroxidation in the pathogenesis of AD (Ando et al., 1998; Lovell et al., 1997; Markesbery and Lovell, 1998; McGrath et al., 2001; Montine et al., 1997, 1998; Sayre et al., 1997). One class of chemically reactive products of lipid peroxidation that has been studied in great detail is diffusible low molecular weight aldehydes. By far, the most extensively studied of these are 4-hydroxy-2-nonenal (HNE), generated by peroxidation of ω-6 PUFAs like AA

(Esterbauer et al., 1991). While the pathophysiologic consequences of overproduction of HNE have been highlighted in numerous studies, it is noteworthy that these reactive aldehydes also are generated at low levels in all cells and appear to have a role in normal physiologic signaling (Forman and Dickinson, 2004). Indeed, several highly polymorphic enzyme systems have evolved apparently to metabolize specifically these lipid peroxidation products and thereby terminate their signaling or detoxify them (Picklo et al., 2002); two of these have been tentatively associated with an increased risk of AD (Kamino et al., 2000; Li et al., 2003). Recently, another class of chemically reactive lipid peroxidation products has been identified: γ-ketoaldehyde isoketals (IsoKs) derived from AA (Bernoud-Hubac et al., 2001). These γ-ketoaldehydes are much more reactive with cellular nucleophiles than HNE and, unlike the structurally similar COX-derived levuglandins, IsoKs remain esterified to phospholipids. In light of the ability of LGs to significantly accelerate oligomerization of Aβ peptides in vitro (Boutaud et al., 2002), IsoKs are now being explored for related mechanisms of neurotoxicity (Davies et al., 2002).

5.1.2 Stable Products of Lipid Peroxidation and Their Bioactivity

In the early 1990's, Morrow and colleagues demonstrated that free radical-mediated damage to AA followed by oxygen insertion and cyclization generated products that were isomeric to PG products of COX. These newly discovered compounds were termed isoprostanes (IsoPs) (Morrow et al., 1990). There are three important differences between PGs and IsoPs. First, IsoPs are a large class of molecules consisting of 64 enantiomers contained within four regioisomeric families. Second, IsoPs are formed in situ while esterified to phospholipids and may be subsequently released by hydrolysis. Third, while some IsoPs do activate G protein-coupled receptors, the extent of their receptor-mediated activity remains unclear. What is clear is that predicting receptor activity based on similarity to isomeric PGs is limited. Since the discovery of potent renal vasoconstrictor activity for 15-F_{2t}-IsoP, there has been an explosion of interest in the PG receptor-mediated activity of IsoPs, especially effects of 15-F_{2t}-IsoP in vasculature, kidney, lungs, and platelets (Morrow and Roberts, 1997). Much of the receptor-mediated activity of 15-F_{2t}-IsoP occurs via TP (Audoly et al., 2000). The contribution of IsoP-mediated receptor activation to neurodegenerative diseases is not known.

5.1.3 Quantitative In vivo Biomarkers of Lipid Peroxidation

A final aspect to consider for lipid oxidation products is their use as quantitative biomarkers of free radical-mediated damage in vivo. Every lipid peroxidation product discussed above has been used as a measure of oxidative damage. Since we have reviewed this topic recently, we will not go into great detail here, however, an important point must be kept in mind (Montine et al., 2002b). The goal of a biomarker is to quantitatively reflect changes in free radical mediated damage. Interpretation of biomarkers that are chemically reactive or that are extensively metabolized has inherent limitations. Consider HNE: although several robust methods exist for its quantification, how do you interpret a change in its concentration? Was it due to a change in lipid peroxidation, a change in the concentration or availability of intracellular nucleophiles like glutathione, a change in the rate of its metabolism by one of several highly efficient enzymes, or some combination of these?

Exquisitely sensitive assays have been developed for IsoPs and, because of the chemical stability and relatively limited metabolism of F_2-IsoPs in situ, they have emerged as a leading quantitative biomarker of lipid peroxidation in vivo (Morrow and Roberts, 1994). F_2-IsoPs can be measured in tissue samples where this product of lipid peroxidation remains esterified to phospholipids. Several groups have used measurements of hydrolyzed F2-IsoPs in body fluids in an attempt to quantify the magnitude of oxidative damage in vivo. In AD, there is broad agreement that CSF F_2-IsoPs are increased in patients with mild dementia and even in individuals with prodromal dementia (Montine et al., 1999a, b, c, 2001; Pratico et al., 2000, 2002). Similar to the attempts to identify peripheral biomarkers of other neurodegenerative diseases, attempts to use plasma or urine F_2-IsoPs in AD patients have not yielded reproducible results across centers using a variety of techniques (Bohnstedt et al., 2003; Feillet-Coudray et al., 1999; Montine et al., 1999b, 2000, 2001; Pratico et al., 2000, 2002; Tuppo et al., 2001). This is perhaps not surprising given the small amounts of brain-derived F_2-IsoPs relative to peripheral organ-derived F_2-IsoPs and the many systemic, dietary, and environmental factors that modulate peripheral F_2-IsoPs independent of disease.

References

Adams, J., Collaco-Moraes, Y., and de Belleroche, J. (1996). Cyclooxygenase-2 induction in cerebral cortex: an intracellular response to synaptic excitation. *J Neurochem, 66*(1), 6–13.

Ahmad, A. S., Ahmad, M., de Brum-Fernandes, A. J., and Dore, S. (2005). Prostaglandin EP4 receptor agonist protects against acute neurotoxicity. *Brain Res, 1066*(1–2), 71–77.

Aisen, P. S., Schafer, K. A., Grundman, M., Pfeiffer, E., Sano, M., Davis, K. L., et al. (2003). Effects of rofecoxib or naproxen vs placebo on Alzheimer disease progression: a randomized controlled trial. *JAMA, 289*(21), 2819–2826.

Akaike, A., Kaneko, S., Tamura, Y., Nakata, N., Shiomi, H., Ushikubi, F., and Narumiya, S. (1994). Prostaglandin E2 protects cultured cortical neurons against N-methyl-D-aspartate receptor-mediated glutamate cytotoxicity. *Brain Res, 663*, 237–244.

Akaogi, J., Yamada, H., Kuroda, Y., Nacionales, D. C., Reeves, W. H., and Satoh, M. (2004). Prostaglandin E2 receptors EP2 and EP4 are up-regulated in peritoneal macrophages and joints of pristane-treated mice and modulate TNF-alpha and IL-6 production. *J Leukoc Biol, 76*(1), 227–236.

Akiyama, H., Barger, S., Barnum, S., Bradt, B., Bauer, J., Cole, G. M., et al. (2000). Inflammation and Alzheimer's disease. *Neurobiol Aging, 21*(3), 383–421.

Almer, G., Guegan, C., Teismann, P., Naini, A., Rosoklija, G., Hays, A. P., et al. (2001). Increased expression of the pro-inflammatory enzyme cyclooxygenase-2 in amyotrophic lateral sclerosis. *Ann Neurol, 49*(2), 176–185.

Almer, G., Teismann, P., Stevic, Z., Halaschek-Wiener, J., Deecke, L., Kostic, V., and Przedborski, S. (2002). Increased levels of the pro-inflammatory prostaglandin PGE2 in CSF from ALS patients. *Ann Neurol, 58*, 1277–1279.

Ando, Y., Brannstrom, T., Uchida, K., Nyhlin, N., Nasman, B., Suhr, O., et al. (1998). Histochemical detection of 4-hydroxynonenal protein in Alzheimer amyloid. *J Neurol Sci, 156*(2), 172–176.

Aronoff, D. M., Canetti, C., and Peters-Golden, M. (2004). Prostaglandin E2 inhibits alveolar macrophage phagocytosis through an E-prostanoid 2 receptor-mediated increase in intracellular cyclic AMP. *J Immunol, 173*(1), 559–565.

Audoly, L. P., Rocca, B., Fabre, J. E., Koller, B. H., Thomas, D., Loeb, A. L., et al. (2000). Cardiovascular responses to the isoprostanes iPF(2alpha)-III and iPE(2)-III are mediated via the thromboxane A(2) receptor in vivo. *Circulation, 101*(24), 2833–2840.

Baratelli, F. E., Heuze-Vourc'h, N., Krysan, K., Dohadwala, M., Riedl, K., Sharma, S., et al. (2004). Prostaglandin E2-dependent enhancement of tissue inhibitors of metalloproteinases-1 production limits dendritic cell migration through extracellular matrix. *J Immunol, 173*(9), 5458–5466.

Batshake, B., Nilsson, C., and Sundelin, J. (1995). Molecular characterization of the mouse prostanoid EP1 receptor gene. *Eur J Biochem, 231*(3), 809–814.

Bernoud-Hubac, N., Davies, S. S., Boutaud, O., Montine, T. J., and Roberts, L. J., 2nd. (2001). Formation of highly reactive gamma-ketoaldehydes (neuroketals) as products of the neuroprostane pathway. *J Biol Chem, 276*(33), 30964–30970.

Bezzi, P., Carmignoto, G., Pasti, L., Vesce, S., Rossi, D., Rizzini, B. L., Pozzan, T., and Volterra, A. (1998). Prostaglandins stimulate calcium dependent glutamate release in astrocytes. *Nature, 391*, 281–285.

Bilak, M., Wu, L., Wang, Q., Haughey, N., Conant, K., St Hillaire, C., et al. (2004). PGE2 receptors rescue motor neurons in a model of amyotrophic lateral sclerosis. *Ann Neurol, 56*(2), 240–248.

Bjornson, A. B., Knippenberg, R. W., and Bjornson, H. S. (1988). Nonsteroidal anti-inflammatory drugs correct the bactericidal defect of polymorphonuclear leukocytes in a guinea pig model of thermal injury. *J Infect Dis, 157*(5), 959–967.

Blackman, S., Dawson, G., Antonakis, K., and Le Breton, G. (1998). The identification and characterization of oligodendrocyte thromboxane A2 receptors. *J Biol Chem, 273*, 475–483.

Bohnstedt, K. C., Karlberg, B., Wahlund, L. O., Jonhagen, M. E., Basun, H., and Schmidt, S. (2003). Determination of isoprostanes in urine samples from Alzheimer patients using porous graphitic carbon liquid chromatography-tandem mass spectrometry. *J Chromatogr B Analyt Technol Biomed Life Sci, 796*(1), 11–19.

Borda, E. S., Tenenbaum, A., Sales, M. E., Rumi, L., and Sterin-Borda, L. (1998). Role of arachidonic acid metabolites in the action of a beta adrenergic agonist on human monocyte phagocytosis. *Prostaglandins Leukot Essent Fatty Acids, 58*(2), 85–90.

Boutaud, O., Ou, J. J., Chaurand, P., Caprioli, R. M., Montine, T. J., and Oates, J. A. (2002). Prostaglandin H2 (PGH2) accelerates formation of amyloid beta1-42 oligomers. *J Neurochem, 82*(4), 1003–1006.

Breder, C. D., Dewitt, D., and Kraig, R. P. (1995). Characterization of inducible cyclooxygenase in rat brain. *J Comp Neurol, 355*(2), 296–315.

Breyer, R. M., Bagdassarian, C. K., Myers, S. A., and Breyer, M. D. (2001). Prostanoid receptors: subtypes and signaling. *Annu Rev Pharmacol Toxicol, 41*, 661–690.

Bruijn, L. I., Houseweart, M. K., Kato, S., Anderson, K. L., Anderson, S. D., Ohama, E., et al. (1998). Aggregation and motor neuron toxicity of an ALS-linked SOD1 mutant independent from wild-type SOD1. *Science, 281*(5384), 1851–1854.

Campbell, W. B. and Halushka, P. V. (2001). Lipid-derived autocoids. In J. G. Hardman and L. E. Limbird (Eds.), *Goodman and Gilman's the Pharmacological Basis of Therapeutics* (9th ed., pp. 601–616). New York, NY: McGraw-Hill.

Canning, B. J., Hmieleski, R. R., Spannhake, E. W., and Jakab, G. J. (1991). Ozone reduces murine alveolar and peritoneal macrophage phagocytosis: the role of prostanoids. *Am J Physiol, 261*(4 Pt 1), L277–282.

Cazevielle, C., Muller, A., Meynier, F., Dutrait, N., and Bonne, C. (1994). Protection by prostaglandins from glutamate toxicity in cortical neurons. *Neurochem Int, 24*, 156–159.

Chen, C. and Bazan, N. G. (2005). Endogenous PGE2 regulates membrane excitability and synaptic transmission in hippocampal CA1 pyramidal neurons. *J Neurophysiol, 93*(2), 929–941.

Chen, C., Magee, J. C., and Bazan, N. G. (2002). Cyclooxygenase-2 regulates prostaglandin E2 signaling in hippocampal long-term synaptic plasticity. *J Neurophysiol, 87*(6), 2851–2857.

Cudkowicz, M. E., Shefner, J. M., Schoenfeld, D. A., Zhang, H., Andreasson, K. I., Rothstein, J. D., et al. (2006). Trial of celecoxib in amyotrophic lateral sclerosis. *Ann Neurol, 60*(1), 22–31.

Davies, S. S., Amarnath, V., Montine, K. S., Bernoud-Hubac, N., Boutaud, O., Montine, T. J., et al. (2002). Effects of reactive gamma-ketoaldehydes formed by the isoprostane pathway (isoketals) and cyclooxygenase pathway (levuglandins) on proteasome function. *FASEB J, 16*(7), 715–717.

Drachman, D. B., Frank, K., Dykes-Hoberg, M., Teismann, P., Almer, G., Przedborski, S., et al. (2002). Cyclooxygenase 2 inhibition protects motor neurons and prolongs survival in a transgenic mouse model of ALS. *Ann Neurol, 52*(6), 771–778.

Ek, M., Arias, C., Sawchenko, P., and Ericsson-Dahlstrand, A. (2000). Distribution of the EP3 prostaglandin E(2) receptor subtype in the rat brain: relationship to sites of interleukin-1-induced cellular responsiveness. *J Comp Neurol, 428*(1), 5–20.

Eriksen, J. L., Sagi, S. A., Smith, T. E., Weggen, S., Das, P., McLendon, D. C., et al. (2003). NSAIDs and enantiomers of flurbiprofen target gamma-secretase and lower Abeta 42 in vivo. *J Clin Invest, 112*(3), 440–449.

Esterbauer, H. and Ramos, P. (1996). Chemistry and pathophysiology of oxidation of LDL. *Rev Physiol Biochem Pharmacol, 127*, 31–64.

Esterbauer, H., Schaur, R. J., and Zollner, H. (1991). Chemistry and biochemistry of 4-hydroxynonenal, malonaldehyde and related aldehydes. *Free Radic Biol Med, 11*(1), 81–128.

Farooqui, A. A., Ong, W. Y., and Horrocks, L. A. (2006). Inhibitors of brain phospholipase A2 activity: their neuropharmacological effects and therapeutic importance for the treatment of neurologic disorders. *Pharmacol Rev, 58*(3), 591–620.

Fassbender, K., Walter, S., Kuhl, S., Landmann, R., Ishii, K., Bertsch, T., et al. (2004). The LPS receptor (CD14) links innate immunity with Alzheimer's disease. *FASEB J, 18*(1), 203–205.

Feillet-Coudray, C., Tourtauchaux, R., Niculescu, M., Rock, E., Tauveron, I., Alexandre-Gouabau, M. C., et al. (1999). Plasma levels of 8-epiPGF2alpha, an in vivo marker of oxidative stress, are not affected by aging or Alzheimer's disease. *Free Radic Biol Med, 27*(3–4), 463–469.

Feng, Z. H., Wang, T. G., Li, D. D., Fung, P., Wilson, B. C., Liu, B., Ali, S. F, Langenbach, R., and Hong, J. S. (2002). Cyclooxygenase-2-deficient mice are resistant to 1-methyl-4-phenyl1, 2, 3, 6-tetrahydropyridine-induced damage of dopaminergic neurons in the substantia nigra. *Neurosci Lett, 329*, 354–358.

Fennekohl, A., Sugimoto, Y., Segi, E., Maruyama, T., Ichikawa, A., and Puschel, G. P. (2002). Contribution of the two Gs-coupled PGE2-receptors EP2-receptor and EP4-receptor to the inhibition by PGE2 of the LPS-induced TNFalpha-formation in Kupffer cells from EP2-or EP4-receptor-deficient mice. Pivotal role for the EP4-receptor in wild type Kupffer cells. *J Hepatol, 36*(3), 328–334.

Fisslthaler, B., Popp, R., Kiss, L., Potente, M., Harder, D. R., Fleming, I., et al. (1999). Cytochrome P450 2C is an EDHF synthase in coronary arteries. *Nature, 401*(6752), 493–497.

Forman, H. J. and Dickinson, D. A. (2004). Introduction to serial reviews on 4-hydroxy-2-nonenal as a signaling molecule. *Free Radic Biol Med, 37*(5), 594–596.

Fujino, T., Nakagawa, N., Yuhki, K., Hara, A., Yamada, T., Takayama, K., et al. (2004). Decreased susceptibility to renovascular hypertension in mice lacking the prostaglandin I2 receptor IP. *J Clin Invest, 114*(6), 805–812.

Gilmour, M. I., Park, P., Doerfler, D., and Selgrade, M. K. (1993). Factors that influence the suppression of pulmonary antibacterial defenses in mice exposed to ozone. *Exp Lung Res, 19*(3), 299–314.

Gurer, U. S., Palanduz, A., Gurbuz, B., Yildirmak, Y., Cevikbas, A., and Kayaalp, N. (2002). Effect of antipyretics on polymorphonuclear leukocyte functions in children. *Int Immunopharmacol, 2*(11), 1599–1602.

Gurney, M. E., Pu, H., Chiu, A. Y., Dal Canto, M. C., Polchow, C. Y., Alexander, D. D., et al. (1994). Motor neuron degeneration in mice that express a human Cu,Zn superoxide dismutase mutation. *Science, 264*(5166), 1772–1775.

Hall, E. D., Andrus, P. K., Althaus, J. S., and VonVoigtlander, P. F. (1993). Hydroxyl radical production and lipid peroxidation parallels selective post-ischemic vulnerability in gerbil brain. *J Neurosci Res, 34*, 107–112.

Hanna, N., Peri, K., Abran, D., Hardy, P., Doke, A., Lachapelle, P., Roy, M-S., Orquin, J., Varma, D., and Chemtob, S. (1997). Light induces peroxidation in retina by activating prostaglandin G/H synthase. *Free Radic Biol Med, 23*, 885–897.

Harizi, H., Grosset, C., and Gualde, N. (2003). Prostaglandin E2 modulates dendritic cell function via EP2 and EP4 receptor subtypes. *J Leukoc Biol, 73*(6), 756–763.

Hauss-Wegrzyniak, B., Lynch, M. A., Vraniak, P. D., and Wenk, G. L. (2002). Chronic brain inflammation results in cell loss in the entorhinal cortex and impaired LTP in perforant path-granule cell synapses. *Exp Neurol, 176*(2), 336–341.

Hizaki, H., Segi, E., Sugimoto, Y., Hirose, M., Saji, T., Ushikubi, F., et al. (1999). Abortive expansion of the cumulus and impaired fertility in mice lacking the prostaglandin E receptor subtype EP(2). *Proc Natl Acad Sci U S A, 96*(18), 10501–10506.

Hsu, K. and Kan, W. (1996). Thromboxane A2 agonist modulation of excitatory synaptic transmission in the rat hippocampal slice. *Br J Pharmacol, 118*, 2220–2227.

Hsu, K., Huang, C., Kan, W., and Gean, P. (1996). TXA2 agonists inhibit high-voltage-activated calcium channels in rat hippocampal CA1 neurons. *Am J Physiol, 271*, C1269–1277.

Hubbard, N. E., Lee, S., Lim, D., and Erickson, K. L. (2001). Differential mRNA expression of prostaglandin receptor subtypes in macrophage activation. *Prostaglandins Leukot Essent Fatty Acids, 65*(5–6), 287–294.

Hutchison, D. L. and Myers, R. L. (1987). Prostaglandin-mediated suppression of macrophage phagocytosis of Listeria monocytogenes. *Cell Immunol, 110*(1), 68–76.

in t' Veld, B. A., Ruitenberg, A., Hofman, A., Launer, L. J., van Duijn, C. M., Stijnen, T., et al. (2001). Nonsteroidal antiinflammatory drugs and the risk of Alzheimer's disease. *N Engl J Med, 345*(21), 1515–1521.

Jantzen, P. T., Connor, K. E., DiCarlo, G., Wenk, G., Wallace, J., Rojiani, A., Coppola, D., Morgan, D., and Gordon, M. (2002). Microglial activation and β-amyloid deposit reduction caused by a nitric oxide-releasing nonsteroidal anti-inflammatory drug in amyloid precursor protein plus presenilin-1 transgenic mice. *J Neurosci, 22*, 2246–2254.

Kamino, K., Nagasaka, K., Imagawa, M., Yamamoto, H., Yoneda, H., Ueki, A., et al. (2000). Deficiency in mitochondrial aldehyde dehydrogenase increases the risk for late-onset Alzheimer's disease in the Japanese population. *Biochem Biophys Res Commun, 273*(1), 192–196.

Kaufmann, W. E., Worley, P. F., Pegg, J., Bremer, M., and Isakson, P. (1996). COX-2, a synaptically induced enzyme, is expressed by excitatory neurons at postsynaptic sites in rat cerebral cortex. *Proc Natl Acad Sci U S A, 93*(6), 2317–2321.

Kawano, T., Anrather, J., Zhou, P., Park, L., Wang, G., Frys, K. A., et al. (2006). Prostaglandin E2 EP1 receptors: downstream effectors of COX-2 neurotoxicity. *Nat Med, 12*(2), 225–229.

Keller, J. N., Mark, R. J., Bruce, A. J., Blanc, E., Rothstein, J. D., Uchida, K., et al. (1997). 4-Hydroxynonenal, an aldehydic product of membrane lipid peroxidation, impairs glutamate transport and mitochondrial function in synaptosomes. *Neuroscience, 80*(3), 685–696.

Kitanaka, J., Ishibashi, T., and Baba, A. (1993). Phloretin as an antagonist of prostaglandin F2 alpha receptor in cultured rat astrocytes. *J Neurochem, 60*(2), 704–708.

Klivenyi, P., Kiaei, M., Gardian, G., Calingasan, N. Y., and Beal, M. F. (2004). Additive neuro-protective effects of creatine and cyclooxygenase 2 inhibitors in a transgenic mouse model of amyotrophic lateral sclerosis. *J Neurochem, 88*(3), 576–582.

Kukreja, R., Kontos, H., Hess, H., and Ellis, E. (1986). PGH synthase and lipoxygenase generate superoxide in the presence of NADH or NADPH. *Circ Res, 59*, 612–619.

Laegreid, W. W., Liggitt, H. D., Silflow, R. M., Evermann, J. R., Taylor, S. M., and Leid, R. W. (1989). Reversal of virus-induced alveolar macrophage bactericidal dysfunction by cyclooxy-genase inhibition in vitro. *J Leukoc Biol, 45*(4), 293–300.

Lanz, T. A., Fici, G. J., and Merchant, K. M. (2005). Lack of Specific Amyloid-{beta}(1–42) Suppression by Nonsteroidal anti-inflammatory drugs in young, plaque-free Tg2576 mice and in guinea pig neuronal cultures. *J Pharmacol Exp Ther, 312*(1), 399–406.

Largo, R., Diez-Ortego, I., Sanchez-Pernaute, O., Lopez-Armada, M. J., Alvarez-Soria, M. A., Egido, J., et al. (2004). EP2/EP4 signalling inhibits monocyte chemoattractant protein-1 pro-duction induced by interleukin 1beta in synovial fibroblasts. *Ann Rheum Dis, 63*(10), 1197–1204.

Lerea, L. S., Carlson, N. G., Simonato, M., Morrow, J., Roberts, J. L., and McNamara, J. O. (1997). Prostaglandin F 2alpha is required for NMDA receptor-mediated induction of c-fos mRNA in dentate gyrus neurons. *J Neurosci, 17*, 117–124.

Li, Y. J., Oliveira, S. A., Xu, P., Martin, E. R., Stenger, J. E., Scherzer, C. R., et al. (2003). Glutathione S-transferase omega-1 modifies age-at-onset of Alzheimer disease and Parkinson disease. *Hum Mol Genet, 12*(24), 3259–3267.

Liang, X., Wang, Q., Hand, T., Wu, L., Breyer, R. M., Montine, T. J., et al. (2005a). Deletion of the prostaglandin E2 EP2 receptor reduces oxidative damage and amyloid burden in a model of Alzheimer's disease. *J Neurosci, 25*(44), 10180–10187.

Liang, X., Wu, L., Hand, T., and Andreasson, K. (2005b). Prostaglandin D2 mediates neuronal protection via the DP1 receptor. *J Neurochem, 92*(3), 477–486.

Lim, G. P., Yang, F., Chu, T., Chen, P., Beech, W., Teter, B., et al. (2000). Ibuprofen suppresses plaque pathology and inflammation in a mouse model for Alzheimer's disease. *J Neurosci, 20*(15), 5709–5714.

Lim, G. P., Yang, F., Chu, T., Gahtan, E., Ubeda, O., Beech, W., et al. (2001). Ibuprofen effects on Alzheimer pathology and open field activity in APPsw transgenic mice. *Neurobiol Aging, 22*(6), 983–991.

Linton, M. F., Atkinson, J. B., and Fazio, S. (1995). Prevention of atherosclerosis in apolipopro-tein E-deficient mice by bone marrow transplantation. *Science, 267*(5200), 1034–1037.

Liu, D., Li, L. and Augustus, L. (2001). Prostaglandin release by spinal cord injury mediates pro-duction of hydroxyl radical, malondialdehyde and cell death: a site of the neuroprotective action of methylprednisolone. *J Neurochem, 77*(4), 1036–1047.

Liu, D., Wu, L., Breyer, R., Mattson, M. P., and Andreasson, K. (2005). Neuroprotection by the PGE2 EP2 receptor in permanent focal cerebral ischemia. *Ann Neurol, 57*(5), 758–761.

Lovell, M. A., Ehmann, W. D., Mattson, M. P., and Markesbery, W. R. (1997). Elevated 4-hydrox-ynonenal in ventricular fluid in Alzheimer's disease. *Neurobiol Aging, 18*(5), 457–461.

Lovick, T. A., Brown, L. A., and Key, B. J. (2005). Neuronal activity-related coupling in cortical arterioles: involvement of astrocyte-derived factors. *Exp Physiol, 90*(1), 131–140.

Mackenzie, I. R. and Munoz, D. G. (1998). Nonsteroidal anti-inflammatory drug use and Alzheimer-type pathology in aging. *Neurology, 50*(4), 986–990.

Maihofner, C., Probst-Cousin, S., Bergmann, M., Neuhuber, W., Neundorfer, B., and Heuss, D. (2003). Expression and localization of cyclooxygenase-1 and -2 in human sporadic amyo-trophic lateral sclerosis. *Eur J Neurosci, 18*(6), 1527–1534.

Mark, R. J., Lovell, M. A., Markesbery, W. R., Uchida, K., and Mattson, M. P. (1997). A role for 4-hydroxynonenal, an aldehydic product of lipid peroxidation, in disruption of ion home-ostasis and neuronal death induced by amyloid beta-peptide. *J Neurochem, 68*(1), 255–264.

Markesbery, W. R. and Lovell, M. A. (1998). Four-hydroxynonenal, a product of lipid peroxida-tion, is increased in the brain in Alzheimer's disease. *Neurobiol Aging, 19*(1), 33–36.

Matsumura, K., Watanabe, Y., Onoe, H., and Watanabe, Y. (1995). Prostacyclin receptor in the brain and central terminals of the primary sensory neurons: an autoradiographic study using a stable prostacyclin analogue [3H]iloprost. *Neuroscience, 65*, 493–503.

Matsuoka, T., Hirata, M., Tanaka, H., Takahashi, Y., Murata, T., Kabashima, K., et al. (2000). Prostaglandin D2 as a mediator of allergic asthma. *Science, 287*(5460), 2013–2017.

McCullough, L., Wu, L., Haughey, N., Liang, X., Hand, T., Wang, Q., et al. (2004). Neuroprotective function of the PGE2 EP2 receptor in cerebral ischemia. *J Neurosci, 24*(1), 257–268.

McGeer, P. L. and McGeer, E. G. (2002). Inflammatory processes in amyotrophic lateral sclerosis. *Muscle Nerve, 26*, 459–470.

McGeer, P. L. and McGeer, E. G. (2004). Inflammation and neurodegeneration in Parkinson's disease. *Parkinsonism Relat Disord, 10 Suppl 1*, S3–S7.

McGeer, P. L., Schulzer, M., and McGeer, E. G. (1996). Arthritis and anti-inflammatory agents as possible protective factors for Alzheimer's disease: a review of 17 epidemiologic studies. *Neurology, 47*(2), 425–432.

McGrath, L. T., McGleenon, B. M., Brennan, S., McColl, D., Mc, I. S., and Passmore, A. P. (2001). Increased oxidative stress in Alzheimer's disease as assessed with 4-hydroxynonenal but not malondialdehyde. *QJM, 94*(9), 485–490.

Metea, M. R. and Newman, E. A. (2006). Glial cells dilate and constrict blood vessels: a mechanism of neurovascular coupling. *J Neurosci, 26*(11), 2862–2870.

Miettinen, S., Fusco, F. R., Yrjanheikki, J., Keinanen, R., Hirvonen, T., Roivainen, R., et al. (1997). Spreading depression and focal brain ischemia induce cyclooxygenase-2 in cortical neurons through N-methyl-D-aspartic acid-receptors and phospholipase A2. *Proc Natl Acad Sci U S A, 94*(12), 6500–6505.

Milatovic, D., Zaja-Milatovic, S., Montine, K. S., Shie, F. S., and Montine, T. J. (2004). Neuronal oxidative damage and dendritic degeneration following activation of CD14-dependent innate immune response in vivo. *J Neuroinflammation, 1*(1), 20.

Milatovic, D., Zaja-Milatovic, S., Montine, K. S., Nivison, M., and Montine, T. J. (2005). CD14-dependent innate immunity-mediated neuronal damage in vivo is suppressed by NSAIDs and ablation of a prostaglandin E2 receptor, EP2. *CNS Agents Med Chem, 5*(2), 151–156.

Minghetti, L., Nicolini, A., Polazzi, E., Creminon, C., Maclouf, J., and Levi, G. (1997). Prostaglandin E2 downregulates inducible nitric oxide synthase expression in microglia by increasing cAMP levels. *Adv Exp Med Biol, 433*, 181–184.

Miyaura, C., Inada, M., Suzawa, T., Sugimoto, Y., Ushikubi, F., Ichikawa, A., et al. (2000). Impaired bone resorption to prostaglandin E2 in prostaglandin E receptor EP4-knockout mice. *J Biol Chem, 275*(26), 19819–19823.

Mizoguchi, A., Eguchi, N., Kimura, K., Kiyohara, Y., Qu, W. M., Huang, Z. L., et al. (2001). Dominant localization of prostaglandin D receptors on arachnoid trabecular cells in mouse basal forebrain and their involvement in the regulation of non-rapid eye movement sleep. *Proc Natl Acad Sci U S A, 98*(20), 11674–11679.

Montine, K. S., Olson, S. J., Amarnath, V., Whetsell, W. O., Jr., Graham, D. G., and Montine, T. J. (1997). Immunohistochemical detection of 4-hydroxy-2-nonenal adducts in Alzheimer's disease is associated with inheritance of APOE4. *Am J Pathol, 150*(2), 437–443.

Montine, K. S., Reich, E., Neely, M. D., Sidell, K. R., Olson, S. J., Markesbery, W. R., et al. (1998). Distribution of reducible 4-hydroxynonenal adduct immunoreactivity in Alzheimer disease is associated with APOE genotype. *J Neuropathol Exp Neurol, 57*(5), 415–425.

Montine, T. J., Beal, M. F., Cudkowicz, M. E., O'Donnell, H., Margolin, R. A., McFarland, L., et al. (1999a). Increased CSF F2-isoprostane concentration in probable AD. *Neurology, 52*(3), 562–565.

Montine, T. J., Beal, M. F., Robertson, D., Cudkowicz, M. E., Biaggioni, I., O'Donnell, H., et al. (1999b). Cerebrospinal fluid F2-isoprostanes are elevated in Huntington's disease. *Neurology, 52*(5), 1104–1105.

Montine, T. J., Sidell, K. R., Crews, B. C., Markesbery, W. R., Marnett, L. J., Roberts, L. J., 2nd, et al. (1999c). Elevated CSF prostaglandin E2 levels in patients with probable AD. *Neurology, 53*(7), 1495–1498.

Montine, T. J., Shinobu, L., Montine, K. S., Roberts, L. J., 2nd, Kowall, N. W., Beal, M. F., et al. (2000). No difference in plasma or urinary F2-isoprostanes among patients with Huntington's disease or Alzheimer's disease and controls. *Ann Neurol, 48*(6), 950.

Montine, T. J., Kaye, J. A., Montine, K. S., McFarland, L., Morrow, J. D., and Quinn, J. F. (2001). Cerebrospinal fluid abeta42, tau, and f2-isoprostane concentrations in patients with Alzheimer disease, other dementias, and in age-matched controls. *Arch Pathol Lab Med, 125*(4), 510–512.

Montine, T. J., Milatovic, D., Gupta, R. C., Valyi-Nagy, T., Morrow, J. D., and Breyer, R. M. (2002a). Neuronal oxidative damage from activated innate immunity is EP2 receptor-dependent. *J Neurochem, 83*(2), 463–470.

Montine, T. J., Neely, M. D., Quinn, J. F., Beal, M. F., Markesbery, W. R., Roberts, L. J., et al. (2002b). Lipid peroxidation in aging brain and Alzheimer's disease. *Free Radic Biol Med, 33*(5), 620–626.

Montine, K. S., Quinn, J. F., and Montine, T. J. (2003). Membrane lipid peroxidation. In M. P. Mattson (Ed.), *Membrane Lipid Signaling in Aging and Age-Related Disease* (pp. 11–26). Amsterdam: Elsevier.

Morrow, J. D., Hill, K. E., Burk, R. F., Nammour, T. M., Badr, K. F., and Roberts, L. J., 2nd. (1990). A series of prostaglandin F2-like compounds are produced in vivo in humans by a non-cyclooxygenase, free radical-catalyzed mechanism. *Proc Natl Acad Sci U S A, 87*(23), 9383–9387.

Morrow, J. D. and Roberts, L. J. (1997). The isoprostanes: unique bioactive products of lipid peroxidation. *Prog Lipid Res, 36*(1), 1–21.

Morrow, J. D. and Roberts, L. J., 2nd. (1994). Mass spectrometry of prostanoids: F2-isoprostanes produced by non-cyclooxygenase free radical-catalyzed mechanism. *Methods Enzymol, 233*, 163–174.

Murata, T., Ushikubi, F., Matsuoka, T., hirata, M, Yamazaki, A., Sugimoto, Y., Ichikawa, A., Aze, Y., Tanaka, T, Yoshida, N., Ueno, A., Oh-ishi, S., and Narumiya, S. (1997). Altered pain perception and inflammatory response in mice lacking prostacyclin receptor. *Nature, 388*, 678–682.

Nakagomi, T., Sasaki, T., Ogawa, H., Noguchi, M., Saito, I., and Takakura, K. (1990). Immunohistochemical localization of prostaglandin F2 alpha in reperfused gerbil brain. *Neurol Med Chir (Tokyo), 30*(4), 223–228.

Nakahata, N., Ishimoto, H., Kurita, M., Ohmori, K., Takahashi, A., and Nakanishi, H. (1992). The presence of thromboxane A2 receptors in cultured astrocytes from rabbit brain. *Brain Res, 583*, 100–104.

Nakayama, M., Uchimura, K., Zhu, R. L., Nagayama, T., Rose, M. E., Stetler, R. A., et al. (1998). Cyclooxygenase-2 inhibition prevents delayed death of CA1 hippocampal neurons following global ischemia. *Proc Natl Acad Sci U S A, 95*(18), 10954–10959.

Nicolle, M. M., Gonzalez, J., Sugaya, K., Baskerville, K. A., Bryan, D., Lund, K., et al. (2001). Signatures of hippocampal oxidative stress in aged spatial learning-impaired rodents. *Neuroscience, 107*(3), 415–431.

Nishihara, M., Yokotani, K., Inoue, S., and Osumi, Y. (1999). U-46619, a selective Thromboxane A2 mimetic, inhibits the release of endogenous noradrenaline from the rat hippocampus in vitro. *Jpn J Pharmacol, 82*, 226–231.

Niwa, K., Araki, E., Morham, S. G., Ross, M. E., and Iadecola, C. (2000). Cyclooxygenase-2 contributes to functional hyperemia in whisker-barrel cortex. *J Neurosci, 20*(2), 763–770.

Nogawa, S., Zhang, F., Ross, M. E., and Iadecola, C. (1997). Cyclo-oxygenase-2 gene expression in neurons contributes to ischemic brain damage. *J Neurosci, 17*(8), 2746–2755.

Noguchi, K., Iwasaki, K., Shitashige, M., Umeda, M., Izumi, Y., Murota, S., et al. (2001). Downregulation of lipopolysaccharide-induced intercellular adhesion molecule-1 expression via EP2/EP4 receptors by prostaglandin E2 in human fibroblasts. *Inflammation, 25*(2), 75–81.

Norel, X. and Brink, C. (2004). The quest for new cysteinyl-leukotriene and lipoxin receptors: recent clues. *Pharmacol Ther, 103*(1), 81–94.

Ogawa, H., Sasaki, T., Kassell, N. F., Nakagomi, T., Lehman, R. M., and Hongo, K. (1987). Immunohistochemical demonstration of increase in prostaglandin F2-alpha after recirculation in global ischemic rat brains. *Acta Neuropathol, 75*(1), 62–68.

Patel, P., Drummond, J., Sano, T., Cole, D., Kalkman, C., and Yaksh, T. (1993). Effect of ibuprofen on regional eicosanoid production and neuronal injury after forebrain ischemia in rats. *Brain Res, 614*, 315–324.

Picklo, M. J., Montine, T. J., Amarnath, V., and Neely, M. D. (2002). Carbonyl toxicology and Alzheimer's disease. *Toxicol Appl Pharmacol, 184*(3), 187–197.

Pompl, P. N., Ho, L., Bianchi, M., McManus, T., Qin, W., and Pasinetti, G. M. (2003). A therapeutic role for cyclooxygenase-2 inhibitors in a transgenic mouse model of amyotrophic lateral sclerosis. *FASEB J, 17*(6), 725–727.

Porter, N. A., Caldwell, S. E., and Mills, K. A. (1995). Mechanisms of free radical oxidation of unsaturated lipids. *Lipids, 30*(4), 277–290.

Pratico, D., Clark, C. M., Lee, V. M., Trojanowski, J. Q., Rokach, J., and FitzGerald, G. A. (2000). Increased 8,12-iso-iPF2alpha-VI in Alzheimer's disease: correlation of a noninvasive index of lipid peroxidation with disease severity. *Ann Neurol, 48*(5), 809–812.

Pratico, D., Uryu, K., Leight, S., Trojanoswki, J. Q., and Lee, V. M. (2001). Increased lipid peroxidation precedes amyloid plaque formation in an animal model of Alzheimer amyloidosis. *J Neurosci, 21*(12), 4183–4187.

Pratico, D., Clark, C. M., Liun, F., Rokach, J., Lee, V. Y., and Trojanowski, J. Q. (2002). Increase of brain oxidative stress in mild cognitive impairment: a possible predictor of Alzheimer disease. *Arch Neurol, 59*(6), 972–976.

Pratico, D., Zhukareva, V., Yao, Y., Uryu, K., Funk, C. D., Lawson, J. A., et al. (2004). 12/15-lipoxygenase is increased in Alzheimer's disease: possible involvement in brain oxidative stress. *Am J Pathol, 164*(5), 1655–1662.

Rall, J. M., Mach, S. A., and Dash, P. K. (2003). Intrahippocampal infusion of a cyclooxygenase-2 inhibitor attenuates memory acquisition in rats. *Brain Res, 968*(2), 273–276.

Resnick, D. K., Graham, S. H., Dixon, C. E., and Marion, D. W. (1998). Role of cyclooxygenase 2 in acute spinal cord injury. *J Neurotrauma, 15*(12), 1005–1013.

Sang, N., Zhang, J., Marcheselli, V., Bazan, N. G., and Chen, C. (2005). Postsynaptically synthesized prostaglandin E2 (PGE2) modulates hippocampal synaptic transmission via a presynaptic PGE2 EP2 receptor. *J Neurosci, 25*(43), 9858–9870.

Satoh, T., Moroi, R., Aritake, K., Urade, Y., Kanai, Y., Sumi, K., et al. (2006). Prostaglandin D2 plays an essential role in chronic allergic inflammation of the skin via CRTH2 receptor. *J Immunol, 177*(4), 2621–2629.

Sayre, L. M., Zelasko, D. A., Harris, P. L., Perry, G., Salomon, R. G., and Smith, M. A. (1997). 4-Hydroxynonenal-derived advanced lipid peroxidation end products are increased in Alzheimer's disease. *J Neurochem, 68*(5), 2092–2097.

Segi, E., Sugimoto, Y., Yamasaki, A., Aze, Y., Oida, H., Nishimura, T., et al. (1998). Patent ductus arteriosus and neonatal death in prostaglandin receptor EP4-deficient mice. *Biochem Biophys Res Commun, 246*(1), 7–12.

Separovic, D., and Dorman, R. V. (1993). Prostaglandin F2 alpha synthesis in the hippocampal mossy fiber synaptosomal preparation: I. Dependence in arachidonic acid, phospholipase A2, calcium availability and membrane depolarization. *Prostaglandins Leukot Essent Fatty Acids, 48*(2), 127–137.

Seregi, A., Forstermann, U., Heldt, R., and Hertting, G. (1985). The formation and regional distribution of prostaglandins D2 and F2 alpha in the brain of spontaneously convulsing gerbils. *Brain Res, 337*(1), 171–174.

Shaw, K. N., Commins, S., and O'Mara, S. M. (2003). Deficits in spatial learning and synaptic plasticity induced by the rapid and competitive broad-spectrum cyclooxygenase inhibitor ibuprofen are reversed by increasing endogenous brain-derived neurotrophic factor. *Eur J Neurosci, 17*(11), 2438–2446.

Shie, F. S., Montine, K. S., Breyer, R. M., and Montine, T. J. (2005). Microglial EP2 as a new target to increase amyloid beta phagocytosis and decrease amyloid beta-induced damage to neurons. *Brain Pathol, 15*(2), 134–138.

Stewart, W. F., Kawas, C., Corrada, M., and Metter, E. J. (1997). Risk of Alzheimer's disease and duration of NSAID use. *Neurology, 48*, 626–632.

Stock, J. L., Shinjo, K., Burkhardt, J., Roach, M., Taniguchi, K., Ishikawa, T., et al. (2001). The prostaglandin E2 EP1 receptor mediates pain perception and regulates blood pressure. *J Clin Invest, 107*(3), 325–331.

Sugaya, K., Uz, T., Kumar, V., and Manev, H. (2000). New anti-inflammatory treatment strategy in Alzheimer's disease. *Jpn J Pharmacol, 82*(2), 85–94.

Sugimoto, Y., Shigemoto, R., Namba, T., Negishi, M., Mizuno, N., Narumiya, S., et al. (1994). Distribution of the messenger RNA for the prostaglandin E receptor subtype EP3 in the mouse nervous system. *Neuroscience, 62*(3), 919–928.

Sugimoto, Y., Yamasaki, A., Segi, E., Tsuboi, K., Aze, Y., Nishimura, T., et al. (1997). Failure of parturition in mice lacking the prostaglandin F receptor. *Science, 277*(5326), 681–683.

Teather, L. A., Packard, M. G., and Bazan, N. G. (2002). Post-training cyclooxygenase-2 (COX-2) inhibition impairs memory consolidation. *Learn Mem, 9*(1), 41–47.

Teismann, P., Tieu, K., Choi, DK., Wu, D. C., Naini, A., Hunot, S., Vila, M., Jackson-Lewis, V., and Przedborski, S. (2003). Cyclooxygenase-2 is instrumental in Parkinson's disease neurodegeneration. *Proc Natl Acad Sci U S A, 100*(9), 5473–5478.

Thomas, D., Mannon, R., Mannon, P., Latour, A., Oliver, J., Hoffman, M., Smithies, O., Koller, B., and Coffman, T. (1998). Coagulation defects and altered hemodynamic responses in mice lacking receptors for thromboxane A2. *J Clin Invest, 102*, 1994–2001.

Treffkorn, L., Scheibe, R., Maruyama, T., and Dieter, P. (2004). PGE2 exerts its effect on the LPS-induced release of TNF-alpha, ET-1, IL-1alpha, IL-6 and IL-10 via the EP2 and EP4 receptor in rat liver macrophages. *Prostaglandins Other Lipid Mediat, 74*(1–4), 113–123.

Tuppo, E. E., Forman, L. J., Spur, B. W., Chan-Ting, R. E., Chopra, A., and Cavalieri, T. A. (2001). Sign of lipid peroxidation as measured in the urine of patients with probable Alzheimer's disease. *Brain Res Bull, 54*(5), 565–568.

Ushikubi, F., Segi, E., Sugimoto, Y., Murata, T., Matsuoka, T., Kobayashi, T., et al. (1998). Impaired febrile response in mice lacking the prostaglandin E receptor subtype EP3. *Nature, 395*(6699), 281–284.

Vassiliou, E., Jing, H., and Ganea, D. (2003). Prostaglandin E2 inhibits TNF production in murine bone marrow-derived dendritic cells. *Cell Immunol, 223*(2), 120–132.

Watanabe, Y., Tokumoto, H., Yamashita, A., Narumiya, S., Mizuno, N., and Hayaishi, O. (1985). Specific bindings of prostaglandin D2, E2 and F2 alpha in postmortem human brain. *Brain Res, 342*(1), 110–116.

Watanabe, Y., Matsumura, K., Takechi, H., Kato, K., Morii, H., Bjorkman, M., Langstrom, B., Noyori, R., Suzuki, M., and Watanabe, Y. (1999). A novel subtype of prostacyclin receptor in the central nervous system. *J Neurochem, 72*, 2583–2592.

Weggen, S., Eriksen, J. L., Das, P., Sagi, S. A., Wang, R., Pietrzik, C. U., et al. (2001). A subset of NSAIDs lower amyloidogenic Abeta42 independently of cyclooxygenase activity. *Nature, 414*(6860), 212–216.

Weggen, S., Eriksen, J. L., Sagi, S. A., Pietrzik, C. U., Golde, T. E., and Koo, E. H. (2003a). Abeta42-lowering nonsteroidal anti-inflammatory drugs preserve intramembrane cleavage of the amyloid precursor protein (APP) and ErbB-4 receptor and signaling through the APP intracellular domain. *J Biol Chem, 278*(33), 30748–30754.

Weggen, S., Eriksen, J. L., Sagi, S. A., Pietrzik, C. U., Ozols, V., Fauq, A., et al. (2003b). Evidence that nonsteroidal anti-inflammatory drugs decrease amyloid beta 42 production by direct modulation of gamma-secretase activity. *J Biol Chem, 278*(34), 31831–31837.

Wong, P. C., Pardo, C. A., Borchelt, D. R., Lee, M. K., Copeland, N. G., Jenkins, N. A., et al. (1995). An adverse property of a familial ALS-linked SOD1 mutation causes motor neuron disease characterized by vacuolar degeneration of mitochondria. *Neuron, 14*(6), 1105–1116.

Yamagata, K., Andreasson, K. I., Kaufmann, W. E., Barnes, C. A., and Worley, P. F. (1993). Expression of a mitogen-inducible cyclooxygenase in brain neurons: regulation by synaptic activity and glucocorticoids. *Neuron, 11*(2), 371–386.

Yan, Q., Zhang, J., Liu, H., Babu-Khan, S., Vassar, R., Biere, A. L., et al. (2003). Anti-inflammatory drug therapy alters beta-amyloid processing and deposition in an animal model of Alzheimer's disease. *J Neurosci, 23*(20), 7504–7509.

Yasojima, K., Tourtellotte, W. W., McGeer, E. G., and McGeer, P. L. (2001). Marked increase in cyclooxygenase-2 in ALS spinal cord: implications for therapy. *Neurology, 57*(6), 952–956.

Yiangou, Y., Facer, P., Durrenberger, P., Chessell, I. P., Naylor, A., Bountra, C., et al. (2006). COX-2, CB2 and P2X7-immunoreactivities are increased in activated microglial cells/macrophages of multiple sclerosis and amyotrophic lateral sclerosis spinal cord. *BMC Neurol, 6*, 12.

Zhang, J. and Rivest, S. (1999). Distribution, regulation and colocalization of the genes encoding the EP2- and EP4-PGE2 receptors in the rat brain and neuronal responses to systemic inflammation. *Eur J Neurosci, 11*(8), 2651–2668.

7
Pattern Recognition Receptors in CNS Disease

Pamela A. Carpentier, D'Anne S. Duncan, and Stephen D. Miller

1 Introduction

Pattern recognition receptors (PRRs) are germline encoded receptors utilized by cells of the innate immune system for pathogen recognition. PRRs are classically activated by pathogen-associated molecular patterns (PAMPs) present in whole classes of pathogens, but not in mammalian cells, termed the "infectious nonself model" (Medzhitov and Janeway, 2002). It has also been recent appreciated that there are self-derived products released upon tissue injury or necrotic cell death that can activate PRRs (Morgan et al., 2005). PRR activation leads to opsonization, activation of complement and coagulation cascades, phagocytosis, activation of proinflammatory signaling pathways, and the induction of apoptosis (Janeway and Medzhitov, 2002; Matzinger, 2002). Additionally, activation of the innate immune system is crucial for the induction of adaptive immune responses and the eventual clearance of pathogens (Janeway and Medzhitov, 2002).

The CNS has long been considered to be an immunologically 'privileged' organ, largely due to early studies showing that rejection of allografts placed within the CNS was delayed or deficient in comparison to rejection of grafts placed at peripheral sites (Geyer et al., 1985; Head and Griffin, 1985). In recent years however, it has become clear that the CNS should be more accurately described as an immunologically 'specialized' organ. The CNS is constitutively monitored by the immune system, but at low levels compared to other sites (Hickey, 2001), making it crucial that CNS-resident cells be capable of rapidly recognizing and responding to infection. In this chapter, we discuss PRRs expressed by CNS resident cells and how they contribute to pro-inflammatory responses during CNS infection. We also discuss recent studies demonstrating that PRRs may contribute to neurodegenerative diseases when chronically activated, but their controlled activation may actually have neuroprotective functions.

T.E. Lane et al. (eds.), *Central Nervous System Diseases and Inflammation.*
© Springer 2008

2 Expression and Regulation of Pattern Recognition Receptors by CNS Resident Cells

2.1 Toll-like Receptors (TLRs)

TLRs are a large family of PRRs in the vertebrate immune system and are highly evolutionarily conserved, providing evidence of their importance in host defense (Roach et al., 2005). At least 13 TLR genes have been identified in mice, and the family is even larger in other vertebrate species, although ligands have not been identified for all of them and some may be pseudogenes in various species (Kawai and Akira, 2005; Roach et al., 2005). All known TLRs with the possible exception of TLR3 are believed to signal through the adaptor protein MyD88, which also is critical for interleukin-1 (IL-1) receptor signaling, and all lead to the activation of the transcription factor nuclear factor κ B (NFκB) (Kawai and Akira, 2005). TLR3 and TLR4 can signal through an MyD88-independent pathway dependent on the adaptor TRIF and activate both NFκB and interferon response factor 3 (IRF3) (Kawai and Akira, 2005). Many of the most well known PAMPs signal through TLRs, including lipopolysaccharide (LPS) and double stranded RNA (dsRNA) or its synthetic mimic polyinosinic-polycytidylic acid (poly I:C) signal through TLRs. The complete signaling mechanisms have been reviewed elsewhere (Roach et al., 2005; Zuany-Amorim et al., 2002) and the ligands for the known TLRs are summarized in Table 7.1.

In the resting CNS, TLRs1-9 have been detected by quantitative real time PCR, with particularly strong expression of TLR3 (Bottcher et al., 2003; McKimmie et al., 2005). In order to determine regional variation in TLR expression, in situ hybridizations have been performed for TLR2 and TLR4 and demonstrated constitutive expression primarily in circumventricular organs (CVOs) and meninges, areas with direct access to the circulation (Laflamme and Rivest, 2001; Laflamme et al., 2001). More recent studies suggest that TLR4 may also be expressed in the parenchyma throughout the brain as well, although at lower levels (Chakravarty and Herkenham, 2005). Expression of TLRs are upregulated in the CNS in response to circulating LPS, bacterial and viral infections, and other neuroinflammatory diseases, although

Table 7.1 Summary of mammalian TLRs with some known ligands [summarized from (Kawai and Akira, 2005; Zuany-Amorim et al., 2002)]

Toll-like receptor	Ligand(s)
TLR1/2, TLR2/6	Peptidoglycans, diacyl and triacyl lipopeptides, LPS, zymosan
TLR3	dsRNA, poly I:C
TLR4	LPS, respiratory syncitial virus proteins, saturated and unsaturated fatty acids, hyaluronic acid fragments
TLR5	Flagellin
TLR7, TLR8	Imidazoquinolines, ssRNA
TLR9	Unmethylated CpG DNA
TLR11	Uropathogenic bacteria

the extent to which this is due to the CNS infiltration of TLR-expressing immune cells is unknown (Bottcher et al., 2003; Bsibsi et al., 2002; Laflamme and Rivest, 2001; Laflamme et al., 2001; McKimmie et al., 2005; Zekki et al., 2002).

Studies using purified cultured CNS cells have demonstrated TLR expression on multiple cell types, most commonly microglia and astrocytes. Primary murine microglia in vitro constitutively express TLRs1-9, while human microglia express robust levels of TLRs1-8 but low TLR9 levels (Bsibsi et al., 2002; Dalpke et al., 2002; Kielian et al., 2002; Olson and Miller, 2004). In comparison, astrocytes express much lower levels of TLRs, but also express a wide variety (Bowman et al., 2003; Carpentier et al., 2005; Olson and Miller, 2004). Human astrocytes express TLRs1-5 and 9 and primary murine astrocytes reportedly express TLRs1-9, although a fair amount of controversy still exists with regards to the particular TLRs expressed by astrocytes and at what levels (Bsibsi et al., 2002; Carpentier et al., 2005; Farina et al., 2005; Jack et al., 2005; McKimmie and Fazakerley, 2005). It is apparent however that astrocytes express particularly high levels of TLR3 (Bsibsi et al., 2002; Carpentier et al., 2005; Farina et al., 2005), suggesting that astrocytes may be particularly important for anti-viral responses in the CNS. Microglia and astrocytes in vitro significantly upregulate TLRs upon treatment with cytokines, TLR agonists, and following infection with various pathogens (Bowman et al., 2003; Carpentier et al., 2005; Esen et al., 2004; Kielian et al., 2002; McKimmie and Fazakerley, 2005; Olson and Miller, 2004), providing a mechanism for amplification of inflammatory responses to pathogens infecting the CNS. This amplification of receptor expression may particularly important in astrocytes and in the CNS parenchyma, where constitutive expression of TLRs is quite low.

Although several studies describe TLR expression on astrocytes and microglia, few report TLR expression on other CNS resident cells. Human oligodendrocytes constitutively express TLR2 and 3 in vitro (Bsibsi et al., 2002). Human NT2-N cells, a neuronal cell line, express TLRs1-4 in vitro, and upregulate TLR3, interleukin-6 (IL-6), interferon-β (IFN-β), tumor necrosis factor (TNF), and CCL5 transcripts in response to viral infection or treatment with poly I:C (Prehaud et al., 2005), but this observation has yet to be confirmed in primary cells or in the intact CNS. A murine cerebral endothelial cell line (MB114EN) constitutively expresses TLRs 2, 4, and 9 (Constantin et al., 2004), and human brain microvessel endothelial cells in vitro are able to respond to LPS with the upregulation of adhesion molecules and chemokines, suggesting a key role of TLR signaling in promoting leukocyte extravasation into the CNS (Shukaliak and Dorovini-Zis, 2000; Wong and Dorovini-Zis, 1992, 1995).

2.2   Non-TLR Recognition of Virus

All cells express machinery to recognize and respond to viral infection through the activation of ubiquitously expressed viral PRRs. The serine/threonine kinase dsRNA-activated protein kinase R (PKR) is constitutively expressed at low levels in most cell types including CNS resident cells under normal physiological conditions and is activated by dsRNA, which may be present in viral genomes or produced during viral replication (Williams, 1999). Upon dsRNA binding, PKR autophosphorylates and also phosphorylates a number of substrates, the best characterized of which is the α subunit of eukaryotic initiation factor 2 (EIF-2α), leading to translation inhibition in the infected cell. Other functions of PKR include the induction type I IFNs, cytokines, and chemokines and the promotion of apoptosis (Williams, 1999). DExD/H box-containing RNA helicases retinoic acid inducible gene 1 (RIG-1) and melanoma differentiation associated gene 5 (Mda5) are also ubiquitously expressed PRRs activated by dsRNA and have reported roles in viral-induced upregulation of type I interferons (Yoneyama et al., 2004).

2.3   Non-TLR Recognition of Bacteria

There are also non-TLR PRRs which recognize various components of bacteria. CD14, a receptor which coordinates with TLR4 and TLR2 for signaling, is expressed on microglia in vitro and in vivo (Becher et al., 1996; Esen and Kielian, 2005; Laflamme and Rivest, 2001). The mannose receptor is a type I transmembrane C-type lectin that recognizes mannose-containing carbohydrate structures found on bacteria (Apostolopoulos and McKenzie, 2001). Its functions include receptor-mediated endocytosis and phagocytosis, microbicidal activity, and induction of cytokines, cell adhesion molecules, and major histocompatibility complex class II (MHC II) molecules (Apostolopoulos and McKenzie, 2001). Both primary rodent microglia and astrocytes constitutively express mannose receptors in vitro, and mannose receptors can be detected in the adult CNS in perivascular microglia/macrophages as well as a subset of neurons and astrocytes (Burudi and Regnier-Vigouroux, 2001; Burudi et al., 1999; Marzolo et al., 1999). Interestingly, mannose receptor expression on perivascular macrophages is upregulated in the CNS after injury, but its expression on microglia and astrocytes in vitro is downregulated in response to pro-inflammatory cytokines or LPS (Burudi et al., 1999; Galea et al., 2005; Marzolo et al., 1999). The functional importance of the downregulation of this receptor in the presence of inflammatory and infectious stimuli is unclear.

Originally characterized on macrophages for their ability to bind low-density lipoproteins, scavenger receptors type A and BI (SR-A and SR-BI) and CD36 also recognize a variety of PAMPs including LPS and lipoteichoic acid, as well as whole bacteria (Husemann et al., 2002). The main roles of SRs in the immune system are in mediating phagocytosis and uptake of pathogens, although their activation may also induce antigen presentation functions and reactive oxygen species (ROS) formation (Mukhopadhyay and Gordon, 2004). Murine microglia express SR-A, SR-BI, and CD36 in vitro although resting expression in vivo is low or undetectable (Husemann et al., 2002). However, these receptors can be upregulated by microglia in response to LPS or CNS injury (Bell et al., 1994; Husemann et al., 2002). Astrocytes may also express SR-BI, but lack SR-A and CD36 expression (Husemann et al., 2002; Husemann and Silverstein, 2001).

Nucleotide-binding oligomerization domain (NOD) molecules recognize bacterial peptidoglycans (PGNs) and LPS and are able to induce apoptosis and regulate inflammatory responses (Inohara and Nunez, 2003). Primary murine astrocytes constitutively express low levels of NOD1 and robust NOD2 mRNA. Following exposure to LPS, flagellin, CpG DNA, and bacterial pathogens astrocytes upregulate NOD2 in vitro (Sterka et al., 2006). Primary microglia express low levels of NOD1 mRNA, but NOD2 expression on microglia is undetermined (Sterka et al., 2006).

3 Role of CNS Pattern Recognition Receptors in Response to Infection

The limited immune surveillance of the CNS makes it crucial that resident cells be able to rapidly recognize and respond to infection. Glial cells express TLRs and other PRRs and respond to their ligation by upregulating a variety of pro-inflammatory functions. The immune functions of these cells are important for the early control of pathogen replication and direct the recruitment and activation of cells of the adaptive immune system.

3.1 PRR Stimulation Activates Innate Immune Functions of Glia Cells

It is well known that both astrocytes and microglia have the potential to contribute to innate immune functions in the CNS following exposure to pathogens or PAMPs, and many of these functions have direct anti-microbial consequences. LPS, poly I: C as well as viral infection induce type I IFN expression by microglia and astrocytes, important for control of infection through the inhibition of translation, degradation of dsRNA and upregulation of MHC class I and class II antigens required for antigen presentation in the CNS (Carpentier et al., 2005; Olson et al., 2001; Olson and Miller, 2004; Palma et al., 2003; Samuel, 2001). Engagement of TLR ligands also induce cultured glia to produce iNOS, leading to high levels of nitric oxide production which has microbicidal activity (Boje and Arora, 1992; Carpentier et al., 2005; Dalpke et al., 2002; Galea et al., 1994; Olson et al., 2001; Olson and Miller, 2004). Both microglia and astrocytes express iNOS in vivo in response to CNS infection as well (Mack et al., 2003; Oleszak et al., 1997; Sun et al., 1995). Additionally, LPS can induce human astrocytes to express β-defensin, a small anti-microbial molecule effective at killing Gram-negative bacteria, some viruses, and fungi (Hao et al., 2001).

In response to TLR stimulation, glia also express chemokines, adhesion molecules, and cytokines important for the recruitment, infiltration and activation of peripheral leukocytes. The anatomic proximity of astrocytic endfeet to CVEs producing chemokines may render astrocytic production of chemokines particularly potent. Pathogens, as well as isolated PAMPs LPS, poly I:C, PGN and CpG DNA can stimulate the production of the chemokines CCL2, CCL3, CCL4, CCL5 and CXCL10 in astrocytes and microglia (Aravalli et al., 2005; Carpentier et al., 2005; Esen et al., 2004; Hayashi et al., 1995; Hua and Lee, 2000; Kielian et al., 2002; Olson et al., 2001; Olson and Miller, 2004; Palma and Kim, 2001; Takeshita et al., 2001). Importantly, both astrocytes and microglia have been implicated as a major source of these chemokines in vivo during a variety of neuroinflammatory diseases [reviewed in (Ambrosini and Aloisi, 2004)].

Astrocytes also promote leukocyte invasion of the CNS via their expression of VCAM-1 and ICAM-1, which are known to be crucial for leukocyte infiltration (Ransohoff et al., 2003). Astrocytes in vitro express low constitutive levels of these molecules, that can be upregulated by LPS, poly I:C, or CpG DNA (Carpentier et al., 2005; Lee et al., 2004; Pang et al., 2001) Interestingly, when CVE expression of VCAM-1 is intact, but astrocytic expression is selectively lost, T cells accumulate perivascularly but do not penetrate into the CNS parenchyma (Gimenez et al., 2004) suggesting that astrocytic expression of adhesion molecules is crucial for the penetration of leukocytes into the CNS parenchyma. Microglia can also express high levels of adhesion molecules after stimulation via TLRs, although it is unknown how they contribute to leukocyte invasion (Dalpke et al., 2002; Olson and Miller, 2004).

Glial cells are also sources of cytokines that may impact developing adaptive immune responses in the CNS. Stimulation of astrocytes with flagellin, PGN, LPS, CpG, or poly I:C is effective in inducing IL-6, IL-1 and TNF production, the cytokines most frequently expressed by activated astrocytes (Bowman et al., 2003; Carpentier et al., 2005; Esen et al., 2004; Farina et al., 2005; Sharif et al., 1993; Takeshita et al., 2001). These cytokines are also produced by astrocytes in vivo during neuroinflammation (Maimone et al., 1997; Sun et al., 1995). Microglia also express TNF, IL-6, and IL-1β in response to the TLR agonists LPS, poly I: C, PGN and CpG DNA (Kielian et al., 2002; Lee et al., 1993; Olson and Miller, 2004; Takeshita et al., 2001). All of these molecules have the potential to activate a pro-inflammatory response in the both the CNS and the periphery and have also been implicated in mediating BBB damage (de Vries et al., 1996) and promoting efficient leukocyte entry into the CNS. Microglia in vitro produce IL-12, IL-18, and IL-23, potent cytokines involved in Th1 differentiation, in response to many of these stimuli (Constantinescu et al., 1996; Dalpke et al., 2002; Li et al., 2003; Olson and Miller, 2004; Stalder et al., 1997; Takeshita et al., 2001). Astrocytes have been reported to express these molecules as well (Constantinescu et al., 1996, 2005; Stalder et al., 1997), although it is generally thought that microglia are the major CNS resident source of IL-12 and IL-23 (Wheeler and Owens, 2005). The complete roles of cytokines during CNS inflammation is discussed elsewhere in this volume.

TLR stimuli also promote the phagocytic activity of microglia which become bactericidal upon stimulation with CpG DNA, *E. coli* DNA, and *S. aureus* in vitro (Dalpke et al., 2002; Kielian et al., 2002). Increased phagocytic activity promotes the processing of antigens for presentation to T cells. Upon LPS, poly I:C, or CpG DNA stimulation, murine microglia upregulate cell surface expression of both costimulatory molecules and MHC I and II, which are critical for T cell activation (Dalpke et al., 2002; Kielian et al., 2002; Olson and Miller, 2004). Microglia become capable of processing and presenting antigen to CD4+ and CD8+ T cells following TLR ligand and IFN-γ exposure, indicating microglia may be relevant for activating CD4+ and CD8+ T cells infiltrating the CNS (Dalpke et al., 2002; Dhib-Jalbut et al., 1990; Frei et al., 1987; McMahon et al., 2005; Olson and Miller, 2004).

Although PAMPs are efficient at inducing innate immune functions of astrocytes, they are not effective at inducing antigen presenting cell functions of astrocytes (Carpentier et al., 2005). LPS or poly I:C stimulated astrocytes are not able to efficiently activate CD4+ T cells due to a deficiency of MHC class II expression, indicating that astrocytes are likely not involved in activation of T cells infiltrating the CNS early after infection. In contrast, astrocytes constitutively express low levels of MHC I which is increased by poly I:C stimulation or virus infection and these cells can therefore possibly activate or be lysed by CD8+ T cells (Carpentier et al., 2007; Cornet et al., 2000; Suzumura et al., 1986).

Several investigators have confirmed the necessity of TLRs for responses of glial cells to these ligands as well as to pathogens. As expected, CD14 and TLR4 are required for glial responses to LPS (Kitamura et al., 2001; Qin et al., 2005). Similarly, TLR3 is necessary for full glial activation by poly I:C (Park et al., 2006; So et al., 2006; Town et al., 2006) and TLR2 mediates glial responses to PGN (Esen et al., 2004; Kielian et al., 2005a). Interestingly, TLR2 also mediates astrocyte responses to intact *S. aureus*, but it is dispensable for microglial activation by the same bacteria (Esen et al., 2004; Kielian et al., 2005a), indicating that microglia, but not astrocytes, have multiple mechanisms by which recognition of this bacteria occurs. We have also noted that while TLR3$^{-/-}$ astrocytes display reduced responses to poly I:C, their responses to direct infection with Theiler's virus are normal. Rather, it appears that PKR is the more important mediator of inflammation in virus-infected astrocytes (Carpentier et al., 2007). It is not surprising that responses to intact pathogens do not depend on a single TLR, but rather multiple pathways can be activated and compensate for loss or lack of another. Such redundancy is necessary for protective immunity, since so many pathogens have evolved mechanisms to block innate immune signaling pathways as a mechanism of immune evasion.

3.2 CNS Cells are Activated by Peripherally Administered PAMPs

It has long been recognized that peripheral infection has profound effects in the CNS, resulting in fever and stimulation of the hypothalamic-pituitary-adrenal axis. Peripheral infection or injection of LPS also causes a set of behavioral symptoms collectively termed "sickness behavior", including anorexia, modification of sleep patterns, decreased locomotor activity, libido, social and exploratory behavior (Roth et al., 2004). Peripheral injection of LPS results in a variety of molecular changes in the CNS, particularly the upregulation of pro-inflammatory cytokines like IL-6, IL-1β, and TNF, (Breder et al., 1994; Chakravarty and Herkenham, 2005; Vallieres and Rivest, 1997). Many of these effects can be mimicked by the intravenous injection of high levels of the recombinant cytokines, and blocking cytokine action in the periphery or CNS can partially inhibit the induction of fever and sickness behavior (Roth et al., 2004), suggesting that many of the infection-induced CNS effects result from the production of pro-inflammatory cytokines by peripheral immune cells. However, cytokine-targeting strategies to prevent LPS-induced fever are only partially effective, and the early response to LPS is often intact (Roth et al., 1998), indicating that other mechanisms must also be involved. The CNS response to pyrogens is thought to be initiated in the CVOs, which lack a blood–brain barrier and therefore are accessible to large circulating molecules. CVOs express high levels of TLR4 and CD14 in comparison to the rest of the CNS parenchyma (Chakravarty and Herkenham, 2005; Laflamme and Rivest, 2001), demonstrating the possibility that these organs can directly respond to circulating LPS. Recently, TLR4 expression on CNS resident cells was shown to be crucial for the induction and maintenance of pro-inflammatory cytokine expression in the CNS using bone marrow chimeras in which TLR4 was intact in the periphery, but absent on CNS resident cells (Chakravarty and Herkenham, 2005). Although best characterized for LPS, peripheral administration of other TLR ligands also induces sickness behavior and activation of CNS resident cells (Cremeans-Smith and Newberry, 2003; Katafuchi et al., 2003; Zhang et al., 2005).

3.3   In vivo Roles of PRRs in Control of CNS Infection

Recently, use of mice genetically deficient for various PRRs has allowed the study of their in vivo role during CNS infection. This field of study is still in its infancy, and the potential differential contributions of PRR signaling in the periphery vs. the CNS has not yet been thoroughly assessed. Even so, there appear to be some intriguing differences in how the CNS and periphery use PRRs in response to infection. MyD88$^{-/-}$ or TLR2$^{-/-}$ mice are more susceptible to fatal meningitis induced by *Streptococcus pneumoniae* or *Listeria meningitis* (Echchannaoui et al., 2002; Koedel et al., 2003, 2004). Two separate groups have reported increased clinical severity of S. pneumoniae-induced meningitis in TLR2$^{-/-}$ mice, associated with increased bacterial load in the CNS, massive although delayed leukocyte infiltration, and increased BBB permeability. In the first study by Echchannaoui et al. (2002), the authors found no differences in bacterial titer or cytokine production in the periphery of TLR2$^{-/-}$ and wild type mice, while Koedel et al. (2003) noted an increase in blood, but not splenic bacterial load in TLR2$^{-/-}$ mice. The differences between these two studies could potentially be explained by the route of infection: the former used an intracerebral route of infection with lower levels of bacteria, while the latter infected intracisternally with higher numbers of bacteria, which led to higher peripheral bacterial load and faster onset of clinical symptoms.

Studies of the role of TLR2 in *Staphylococcus aureus* infection also demonstrate differences between peripheral and CNS responses. TLR2$^{-/-}$ mice are highly susceptible to intravenous *S. aureus* infection, resulting in increased mortality and increased bacterial titers in the blood (Takeuchi et al., 2000). In contrast, there was no difference between TLR2$^{-/-}$ and wild type mice in the survival, clinical symptoms or bacterial load associated with *S. aureus*-induced brain abscess (Kielian et al., 2005b). Despite the lack of clinical effect, TLR2$^{-/-}$ mice in this model displayed decreased TNF, CXCL2 and iNOS production in the brain, with a paradoxical increase in IL-17 (Kielian et al., 2005b). In accordance with these observations, it has been reported that TLR2 is important for the induction of cytokines from macrophages, but not microglia, incubated with intact *S. aureus* (Kielian et al., 2005a; Takeuchi et al., 2000). The additional mechanisms by which microglia recognize *S. aureus* have not yet been determined.

TLR2 is also important for mediating responses to herpes simplex virus 1 (HSV-1), as TLR2$^{-/-}$ mice show decreased peripheral inflammatory responses and serum cytokine levels after HSV-1 infection (Kurt-Jones et al., 2004). Interestingly, these mice are protected from lethal encephalitis which correlates with decreased CNS inflammation. Similarly, TLR3$^{-/-}$ mice are protected from lethal West Nile virus encephalitis (Wang et al., 2004). In the periphery, these mice have increased viral loads and decreased inflammatory responses. The decreased inflammatory response in the periphery failed to damage the BBB, and the virus was unable to access the CNS and cause fatal disease. In contrast, intracerebral infection of West Nile virus had identical effects in wild type and TLR3$^{-/-}$ mice, indicating that TLR3 in the CNS may not be important for responses to this virus (Wang et al., 2004). Mice which constitutively express a dominant negative PKR also show protection from fatal poliovirus infection associated with decreased inflammation but normal viral replication (Scheuner et al., 2003). These studies demonstrated that host inflammatory responses in the CNS during infection may be more responsible for self tissue damage than direct effects of the pathogen itself. Determination of the roles of TLRs specifically in CNS-resident cells during infection will require the use of bone marrow chimeric animals in which CNS resident cells and peripheral immune cells are mismatched with respect to TLR expression, studies which are ongoing in our and other laboratories.

4 Role of CNS Pattern Recognition Receptors in Neurodegeneration

4.1 PRR Stimulation Induces Neuron and Oligodendrocyte Death

Although the stimulation of TLRs on glial cells activate functions that are important for the elimination of pathogens, these same functions can be toxic to cells of the CNS that have limited regenerative capacity. Injection of poly I:C into the CNS causes strong glial activation and profound neurodegeneration in the surrounding tissue

(Melton et al., 2003). Injection of LPS also causes profound and long lasting glial activation (Hartlage-Rubsamen et al., 1999; Herber et al., 2006) associated with oligodendrocyte death, demyelination and increased vulnerability of neurons to injury, dependent on TLR4, MyD88 and CD14 (Lehnardt et al., 2002, 2003; Milatovic et al., 2004). Neural precursor cells appear to be exquisitely sensitive to the effects of LPS, as peripheral injection of even low levels of LPS reduces neurogenesis in the hippocampus and olfactory bulb and induces apoptosis of progenitor cells in the rostral migratory stream (Monje et al., 2003; Mori et al., 2005). The effects of LPS on neurogenesis can be blocked by the anti-inflammatory drug indomethacin, indicating the disruption of neurogenesis is due to inflammation in the CNS (Monje et al., 2003).

Many of these studies have implicated a role for glial cell-derived inflammatory mediators in the CNS damage induced by TLR stimuli. Cytokines produced directly by activated glial cells or by peripheral immune cells activated in the CNS can be toxic to CNS resident cells, which is highlighted by the observation that mice with transgenic overexpression of IL-6, TNF, or IFN-γ in the CNS develop severe neurologic disease (Wang et al., 2002). IFN-γ and TNF are also directly toxic to oligodendrocytes in culture (Selmaj and Raine, 1988; Vartanian et al., 1995).

The production of iNOS by TLR-stimulated glia has potential to be toxic in the CNS. Peroxynitrite (ONOO$^-$), the product of NO reacting with superoxide anions, is a potent oxidizing and nitrating agent, thought to mediate most of the damage observed in the presence of high levels of NO (Smith et al., 1999). Reactive oxygen species (ROS) are particularly damaging to neurons and oligodendrocytes, due to their low levels of antioxidant defenses and the high lipid to protein ratio (Smith et al., 1999). As previously discussed, iNOS is induced in microglia or astrocytes treated with a variety of TLR stimuli which subsequently produce high levels of NO. TLR-stimulated glial cultures are therefore toxic to cultured neurons and oligodendrocytes, and this toxicity can be blocked by pharmacologic inhibitors of NOS (Bal-Price and Brown, 2001; Boje and Arora, 1992; Iliev et al., 2004; Merrill et al., 1993). Inhibition of iNOS by aminoguanidine results in decreased inflammation, demyelination, axonal damage and necrosis after CNS viral infection (Rose et al., 1998), but whether this is a direct effect on cell death or an indirect effect on the immune response, BBB permeability, or other factors is unclear.

Cytokine and NO production can also enhance CNS injury from glutamate excitotoxicity. Levels of extracellular glutamate are normally tightly controlled by astrocytes which express high levels of glutamate transporters (Sonnewald et al., 2002). In oligodendrocytes and neurons, excess glutamate causes calcium overload, which leads to mitochondrial dysfunction, formation of ROS, activation of toxic proteases, and ultimately apoptosis (Arundine and Tymianski, 2004; Matute et al., 2001) Cytokines produced by TLR-stimulated glia such as IL-6 and TNF can enhance glutamate excitotoxicity in cultured neurons (Chao and Hu, 1994; Qiu et al., 1998). In vitro, glutamate uptake by astrocytes is inhibited in the presence of cytokines or poly I:C, which causes the accumulation of extracellular glutamate (Fine et al., 1996; Scumpia et al., 2005; Takahashi et al., 2003). Glutamate excitotoxicity is a major mechanism of CNS damage in vivo, since glutamate receptor antagonism can reduce virus-induced neurodegeneration (Nargi-Aizenman et al., 2004).

4.2 PRRs in Multiple Sclerosis

The neurodegeneration and demyelination induced by CNS administration of PAMPs has advanced the hypothesis that the activation of PRRs and ensuing inflammation may be involved in a variety of neurodegenerative diseases. Multiple sclerosis (MS) is believed to be a CD4+ T cell-mediated autoimmune demyelinating disease of the CNS (Sospedra and Martin, 2005). Although its exact etiology is unknown, it has been hypothesized that it may have an infectious trigger and infections are known to exacerbate disease episodes (Kurtzke, 1993). The principal murine model of MS is experimental autoimmune encephalomyelitis (EAE), in which an autoimmune response is primed in the periphery by the administration of myelin peptides or proteins emulsified in complete Freund's adjuvant (CFA), which contains heat killed *Mycobacteria tuberculosis*, or by adoptive transfer of activated myelin-specific CD4+ T cells. In EAE and MS lesions, the expression of multiple TLRs are increased (Bsibsi et al., 2002; Prinz et al., 2006; Zekki et al., 2002), probably due to both increases in expression by CNS-resident cells as well by CNS-infiltrating peripheral TLR-expressing leukocytes. TLR and/or IL-1 signaling is crucial for the development of EAE because MyD88$^{-/-}$ mice do not develop EAE under the typical immunization protocol (Prinz et al., 2006). TLR9$^{-/-}$ and TLR4$^{-/-}$ mice also show decreased clinical disease after peripheral priming, although the effect is less dramatic than the loss of all MyD88-dependent pathways (Prinz et al., 2006). At least some of this effect is probably due to the inability to effectively prime Th1 responses, which is severely defective in mice with compromised TLR signaling (Schnare et al., 2001). Similar deficiency in priming likely occurs in encephalitogenic Th17 cells. Additionally, it was recently demonstrated that TLR4 is required for the action of pertussis toxin, which promotes leukocyte rolling and adhesion to CNS microvessels and CNS infiltration (Kerfoot et al., 2004). Not all roles for TLR signaling in EAE, however, are due to effects on the peripheral immune system, as MyD88$^{-/-}$ mice also developed decreased clinical disease after the transfer of activated wild-type, myelin-specific, CD4+ T cells and because chimeric mice specifically deficient in CNS expression of MyD88 also exhibit reduced clinical disease (Prinz et al., 2006). Mice deficient specifically in TLR9 in the CNS also develop delayed and reduced disease compared to wild type mice, with decreases in leukocyte infiltration as well as axonal and myelin damage (Prinz et al., 2006). These studies define important roles for both TLR9- and other MyD88-dependent pathways in CNS-resident cells during EAE-associated neurologic damage.

4.3 PRRs in Alzheimer's Disease

Alzheimer's disease (AD) is a devastating neurodegenerative disease characterized by progressive memory loss, cognitive decline and neuronal loss. Although there is no known infectious component, neuroinflammation is a major feature of disease (McGeer and McGeer, 2003). One pathologic hallmark of AD is the amyloid plaque, which consists of deposits of Aβ peptide, activated microglia and astrocytes, and dystrophic neurites (Nagele et al., 2004). A great deal of research has focused on how Aβ deposits may lead to glial activation, and a variety of PRRs have been hypothesized to play a role.

Although no TLR to date has been identified which is directly activated by Aβ, CD14, which coordinates with either TLR4 or TLR2 for signaling, can bind to fibrillar Aβ and mediate its induction of nitrates, IL-6, and TNF (Fassbender et al., 2004). Whether TLR4 also contributes to CD14 signaling in response to Aβ has not yet been determined, but it is interesting to note that a human polymorphism in TLR4, but not CD14, has been linked to AD risk (Combarros et al., 2005; Minoretti et al., 2006).

The murine formyl peptide receptor 2 (mFPR2) is classically expressed in neutrophils and other phagocytic leukocytes and is activated by N-formylated peptides produced by bacteria (Le et al., 2002). This PRR is also expressed in microglia and CNS-infiltrating leukocytes of AD patients and in vitro is important for chemotaxis and oxidative stress induced by Aβ peptides (Le et al., 2001; Tiffany et al., 2001). Interestingly, while resting microglia express little or no mFPR2, the activation of TLRs on microglia by LPS, PGN or CpG induces its upregulation and function (Chen et al., 2006; Cui et al., 2002; Iribarren et al., 2005).

Finally, scavenger receptors are also upregulated in AD brains and have been reported to be expressed by both microglia and astrocytes (Christie et al., 1996; Husemann and Silverstein, 2001). SR-A was the first SR demonstrated to promote microglial adhesion to and uptake of Aβ fibrils and mediate production of ROS (El Khoury et al., 1996; Paresce et al., 1996). However, the generation of transgenic mice which develop AD and are deficient in SR-A did not reveal any differences in plaque formation or neurodegeneration (Huang et al., 1999), leading to questions of the in vivo relevance of these observations. Shortly thereafter, it was shown that SR-BI can mediate Aβ adhesion and internalization by microglia, although its ability to do so was only revealed in the absence of SR-A (Husemann et al., 2001; Paresce et al., 1996). CD36 may also bind fibrillar Aβ and is crucial for the induction of ROS, cytokines and chemokine in microglial cell lines or primary microglia stimulated with fibrillar Aβ (Bamberger et al., 2003; Coraci et al., 2002; El Khoury et al., 2003). CD36 also mediates Aβ-induced chemotaxis of microglia in vitro and in vivo after intracerebral injection of fibrillar Aβ, the first demonstration of the in vivo role of PRRs in microglial responses to Aβ (El Khoury et al., 2003).

5   Neuroprotection Mediated by PRR activation

Although chronic or dysregulated innate immunity in the CNS is potentially damaging to this tissue, its controlled activation may also have neuroprotective effects. Astrocytes stimulated with poly I:C or LPS induce the expression of ciliary neurotrophic factor (Bsibsi et al., 2006), a potent oligodendrocyte survival factor and a major protective factor in CNS demyelinating disease (Barres et al., 1993; Linker et al., 2002). Astrocytes stimulated with poly I:C, but not LPS, express a variety of other neutrophic factors, including vascular endothelial growth factor, neurotrophin-4, brain derived neurotrophic factor, and glial growth factor 2, as well as the anti-inflammatory cytokines transforming growth factor-β and IL-10 (Bsibsi et al., 2006). Accordingly, conditioned media from poly I:C-stimulated astrocytes can promote endothelial cell growth and survival of brain slices in vitro.

Additionally, IL-6 and TNF produced by astrocytes and microglia stimulated with PAMPs, although potently pro-inflammatory in the periphery, also have been demonstrated to have neuroprotective functions in the CNS (Wang et al., 2002). Cerebral infusion of IL-6 is neuroprotective in models of ischemia and excitotoxicity, perhaps through its induction of nerve growth factor by astrocytes (Kossmann et al., 1996; Loddick et al., 1998; Toulmond et al., 1992). TNF can also promote neuronal survival in vitro, even after a variety of challenges by glutamate or kainic acid, glucose deprivation, excess iron or Aβ (Barger et al., 1995; Bruce et al., 1996; Cheng et al., 1994). Mice deficient in TNF and lymphotoxin-α show worse outcome after traumatic brain injury, kainic acid lesions, and middle cerebral artery occlusion (Bruce et al., 1996; Scherbel et al., 1999). These mice also develop enhanced EAE, although this appears to be dependent on the genetic background and/or immunizing antigen (Eugster et al., 1999; Frei et al., 1997; Liu et al., 1998). Interestingly, the neuroprotective effects of TNF appear to be mediated specifically by the p75 receptor, since p75-deficient mice, but not p55 deficient mice, show exacerbated EAE and cuprizone-induced demyelination (Arnett et al., 2001; Eugster et al., 1999; Suvannavejh et al., 2000). This could be due to a direct effect on oligodendrocytes, since p75-deficient mice exhibit decreased oligodendrocyte progenitor proliferation and decreased remyelination after cuprizone withdrawal (Arnett et al., 2001). In accordance with a regenerative function, MS patients who underwent anti-TNF therapy, an effective treatment against other autoimmune disorders, exhibited exacerbations of their clinical disease (Lenercept Multiple Sclerosis Study Group, 1999).

Another surprising role for innate immunity in myelin repair has emerged from recent studies co-infusing LPS with neurotoxins. Ethidium bromide injection into the corpus callosum causes demyelination and loss of oligodendrocytes. LPS co-injection promoted the survival and/or regeneration of oligodendrocytes as measured by an increase in the number and broader distribution of cells expressing oligodendrocyte-specific genes (Glezer et al., 2006). LPS co-injection with Tween into the brain also limited tissue damage and promoted debris clearance from the site of injury (Glezer et al., 2006). Although the exact mechanism is unknown, it is possible that LPS activates phagocytic activities of microglia, induces cytokines with pro-regenerative properties, and/or neurotrophin expression by astrocytes. Collectively, these studies highlight that a properly controlled and limited innate immune response can be beneficial to protect cells of the CNS.

6   Conclusions

CNS resident cells, particularly microglia and to a lesser extent astrocytes, express a variety of PRRs that allow them to respond to almost any pathogen invading the CNS. The ligation of PRRs activates a host of pro-inflammatory responses from these cells, including the production of type I interferons, nitric oxide, cytokines and chemokines. Additionally, microglia exposed to pathogens upregulate their ability to acquire, process and present antigen resulting in more efficient activation of activated T cells. In the CNS, PRRs are important for pro-inflammatory responses to pathogens. Interestingly, the inhibition of these pro-inflammatory responses can be beneficial or detrimental to host survival, depending on the particular pathogen and disease. Similarly, the chronic activation of PRRs may lead to neurodegeneration, but limited or controlled activation may be neuroprotective. These results suggest that the manipulation of PRR activation in the CNS may be a viable approach both in the treatment of CNS infections and neurodegenerative diseases, while dysregulation of this process may contribute to neuronal or oligodendrocyte cell death. This is a burgeoning field of research, and clearly more research is needed to determine the full range of PRR functions in the CNS.

Acknowledgments The authors wish to acknowledge the support of grants from the National Institutes of Health, the National Multiple Sclerosis Society, and the Myelin Repair Foundation.

References

Ambrosini, E. and Aloisi, F. (2004). Chemokines and glial cells: a complex network in the central nervous system. *Neurochem. Res.* 29:1017–1038.

Apostolopoulos, V. and McKenzie, I. F. (2001). Role of the mannose receptor in the immune response. *Curr. Mol. Med.* 1:469–474.

Aravalli, R. N., Hu, S., Rowen, T. N., Palmquist, J. M., and Lokensgard J. R. (2005). Cutting edge: TLR2-mediated proinflammatory cytokine and chemokine production by microglial cells in response to herpes simplex virus. *J Immunol* 175:4189–93.

Arnett, H. A., Mason, J., Marino, M., Suzuki, K., Matsushima, G. K., and Ting, J. P. (2001). TNF alpha promotes proliferation of oligodendrocyte progenitors and remyelination. *Nat. Neurosci.* 4:1116–1122.

Arundine, M. and Tymianski, M. (2004). Molecular mechanisms of glutamate-dependent neurodegeneration in ischemia and traumatic brain injury. *Cell. Mol. Life Sci.* 61:657–668.

Bal-Price, A. and Brown, G. C. (2001). Inflammatory neurodegeneration mediated by nitric oxide from activated glia-inhibiting neuronal respiration, causing glutamate release and excitotoxicity. *J. Neurosci.* 21:6480–6491.

Bamberger, M. E., Harris, M. E., McDonald, D. R., Husemann, J., and Landreth, G. E. (2003). A cell surface receptor complex for fibrillar beta-amyloid mediates microglial activation. *J. Neurosci.* 23:2665–2674.

Barger, S. W., Horster, D., Furukawa, K., Goodman, Y., Krieglstein, J., and Mattson, M. P. (1995). Tumor necrosis factors alpha and beta protect neurons against amyloid beta-peptide toxicity: evidence for involvement of a kappa B-binding factor and attenuation of peroxide and Ca^{2+} accumulation. *Proc. Natl. Acad. Sci. U.S.A.* 92:9328–9332.

Barres, B. A., Schmid, R., Sendnter, M., and Raff, M. C. (1993). Multiple extracellular signals are required for long-term oligodendrocyte survival. *Development* 118:283–295.

Becher, B., Fedorowicz, V., and Antel, J. P. (1996). Regulation of CD14 expression on human adult central nervous system-derived microglia. *J. Neurosci. Res.* 45:375–381.

Bell, M. D., Lopez-Gonzalez, R., Lawson, L., Hughes, D., Fraser, I., Gordon, S., and Perry, V. H. (1994). Upregulation of the macrophage scavenger receptor in response to different forms of injury in the CNS. *J. Neurocytol.* 23:605–613.

Boje, K. M. and Arora, P. K. (1992). Microglial-produced nitric oxide and reactive nitrogen oxides mediate neuronal cell death. *Brain Res.* 587:250–256.

Bottcher, T., von Mering, M., Ebert, S., Meyding-Lamade, U., Kuhnt, U., Gerber, J., and Nau, R. (2003). Differential regulation of Toll-like receptor mRNAs in experimental murine central nervous system infections. *Neurosci. Lett.* 344:17–20.

Bowman, C. C., Rasley, A., Tranguch, S. L., and Marriott, I. (2003). Cultured astrocytes express toll-like receptors for bacterial products. *Glia* 43:281–291.

Breder, C. D., Hazuka, C., Ghayur, T., Klug, C., Huginin, M., Yasuda, K., Teng, M., and Saper, C. B. (1994). Regional induction of tumor necrosis factor alpha expression in the mouse brain after systemic lipopolysaccharide administration. *Proc. Natl. Acad. Sci. U.S.A.* 91:11393–11397.

Bruce, A. J., Boling, W., Kindy, M. S., Peschon, J., Kraemer, P. J., Carpenter, M. K., Holtsberg, F. W., and Mattson, M. P. (1996). Altered neuronal and microglial responses to excitotoxic and ischemic brain injury in mice lacking TNF receptors. *Nat. Med.* 2:788–794.

Bsibsi, M., Ravid, R., Gveric, D., and van Noort, J. M. (2002). Broad expression of Toll-like receptors in the human central nervous system. *J. Neuropathol. Exp. Neurol.* 61:1013–1021.

Bsibsi, M., Persoon-Deen, C., Verwer, R. W., Meeuwsen, S., Ravid, R., and Van Noort, J. M. (2006). Toll-like receptor 3 on adult human astrocytes triggers production of neuroprotective mediators. *Glia* 53:688–695.

Burudi, E. M. and Regnier-Vigouroux, A. (2001). Regional and cellular expression of the mannose receptor in the post-natal developing mouse brain. *Cell Tissue Res.* 303:307–317.

Burudi, E. M., Riese, S., Stahl, P. D., and Regnier-Vigouroux, A. (1999). Identification and functional characterization of the mannose receptor in astrocytes. *Glia* 25:44–55.

Carpentier, P. A., Begolka, W. S., Olson, J. K., Elhofy, A., Karpus, W. J., and Miller, S. D. (2005). Differential activation of astrocytes by innate and adaptive immune stimuli. *Glia* 49:360–374.

Carpentier, P. A., Williams, B. R., and Miller, S. D. (2007). Distinct roles of protein kinase R and toll-like receptor 3 in the activation of astrocytes by viral stimuli. *Glia* 2007 Feb;55(3):239–52.

Chakravarty, S. and Herkenham, M. (2005). Toll-like receptor 4 on nonhematopoietic cells sustains CNS inflammation during endotoxemia, independent of systemic cytokines. *J. Neurosci.* 25:1788–1796.

Chao, C. C. and Hu, S. (1994). Tumor necrosis factor-alpha potentiates glutamate neurotoxicity in human fetal brain cell cultures. *Dev. Neurosci.* 16:172–179.

Chen, K., Iribarren, P., Hu, J., Chen, J., Gong, W., Cho, E. H., Lockett, S., Dunlop, N. M., and Wang, J. M. (2006). Activation of Toll-like receptor 2 on microglia promotes cell uptake of Alzheimer disease-associated amyloid beta peptide. *J. Biol. Chem.* 281:3651–3659.

Cheng, B., Christakos, S., and Mattson, M. P. (1994). Tumor necrosis factors protect neurons against metabolic-excitotoxic insults and promote maintenance of calcium homeostasis. *Neuron* 12:139–153.

Christie, R. H., Freeman, M., and Hyman, B. T. (1996). Expression of the macrophage scavenger receptor, a multifunctional lipoprotein receptor, in microglia associated with senile plaques in Alzheimer's disease. *Am. J. Pathol.* 148:399–403.

Combarros, O., Infante, J., Rodriguez, E., Llorca, J., Pena, N., Fernandez-Viadero, C., and Berciano, J. (2005). CD14 receptor polymorphism and Alzheimer's disease risk. *Neurosci. Lett.* 380:193–196.

Constantin, D., Cordenier, A., Robinson, K., Ala'Aldeen, D. A., and Murphy, S. (2004). Neisseria meningitidis-induced death of cerebrovascular endothelium: mechanisms triggering transcriptional activation of inducible nitric oxide synthase. *J. Neurochem.* 89:1166–1174.

Constantinescu, C. S., Frei, K., Wysocka, M., Trinchieri, G., Malipiero, U., Rostami, A., and Fontana, A. (1996). Astrocytes and microglia produce interleukin-12 p40. *Ann. N. Y. Acad. Sci.* 795:328–333.

Constantinescu, C. S., Tani, M., Ransohoff, R. M., Wysocka, M., Hilliard, B., Fujioka, T., Murphy, S., Tighe, P. J., Sarma, J. D., Trinchieri, G., and Rostami, A. (2005). Astrocytes as antigen-presenting cells: expression of IL-12/IL-23. *J. Neurochem.* 95:331–340.

Coraci, I. S., Husemann, J., Berman, J. W., Hulette, C., Dufour, J. H., Campanella, G. K., Luster, A. D., Silverstein, S. C., and El-Khoury, J. B. (2002). CD36, a class B scavenger receptor, is expressed on microglia in Alzheimer's disease brains and can mediate production of reactive oxygen species in response to beta-amyloid fibrils. *Am. J. Pathol.* 160:101–112.

Cornet, A., Bettelli, E., Oukka, M., Cambouris, C., Avellana-Adalid, V., Kosmatopoulos, K., and Liblau, R. S. (2000). Role of astrocytes in antigen presentation and naive T-cell activation. *J. Neuroimmunol.* 106:69–77.

Cremeans-Smith, J. K. and Newberry, B. H. (2003). Zymosan: induction of sickness behavior and interaction with lipopolysaccharide. *Physiol. Behav.* 80:177–184.

Cui, Y. H., Le, Y., Gong, W., Proost, P., Van Damme, J., Murphy, W. J., and Wang, J. M. (2002). Bacterial lipopolysaccharide selectively up-regulates the function of the chemotactic peptide receptor formyl peptide receptor 2 in murine microglial cells. *J. Immunol.* 168:434–442.

Dalpke, A. H., Schafer, M. K., Frey, M., Zimmermann, S., Tebbe, J., Weihe, E., and Heeg, K. (2002). Immunostimulatory CpG-DNA activates murine microglia. *J. Immunol.* 168:4854–4863.

de Vries, H. E., Blom-Roosemalen, M. C., van Oosten, M., de Boer, A. G., van Berkel, T. J., Breimer, D. D., and Kuiper, J. (1996). The influence of cytokines on the integrity of the blood–brain barrier in vitro. *J. Neuroimmunol.* 64:37–43.

Dhib-Jalbut, S., Kufta, C. V., Flerlage, M., Shimojo, N., and McFarland, H. F. (1990). Adult human glial cells can present target antigens to HLA- restricted cytotoxic T-cells. *J. Neuroimmunol.* 29:203–211.

Echchannaoui, H., Frei, K., Schnell, C., Leib, S. L., Zimmerli, W., and Landmann, R. (2002). Toll-like receptor 2-deficient mice are highly susceptible to *Streptococcus pneumoniae* meningitis because of reduced bacterial clearing and enhanced inflammation. *J. Infect. Dis.* 186:798–806.

El Khoury, J., Hickman, S. E., Thomas, C. A., Cao, L., Silverstein, S. C., and Loike, J. D. (1996). Scavenger receptor-mediated adhesion of microglia to beta-amyloid fibrils. *Nature* 382:716–719.

El Khoury, J. B., Moore, K. J., Means, T. K., Leung, J., Terada, K., Toft, M., Freeman, M. W., and Luster, A. D. (2003). CD36 mediates the innate host response to beta-amyloid. *J. Exp. Med.* 197:1657–1666.

Esen, N. and Kielian, T. (2005). Recognition of *Staphylococcus aureus*-derived peptidoglycan (PGN) but not intact bacteria is mediated by CD14 in microglia. *J. Neuroimmunol.* 170:93–104.

Esen, N., Tanga, F. Y., DeLeo, J. A., and Kielian, T. (2004). Toll-like receptor 2 (TLR2) mediates astrocyte activation in response to the Gram-positive bacterium *Staphylococcus aureus*. *J. Neurochem.* 88:746–758.

Eugster, H. P., Frei, K., Bachmann, R., Bluethmann, H., Lassmann, H., and Fontana, A. (1999). Severity of symptoms and demyelination in MOG-induced EAE depends on TNFR1. *Eur. J. Immunol.* 29:626–632.

Farina, C., Krumbholz, M., Giese, T., Hartmann, G., Aloisi, F., and Meinl, E. (2005). Preferential expression and function of Toll-like receptor 3 in human astrocytes. *J. Neuroimmunol.* 159:12–19.

Fassbender, K., Walter, S., Kuhl, S., Landmann, R., Ishii, K., Bertsch, T., Stalder, A. K., Muehlhauser, F., Liu, Y., Ulmer, A. J., et al. (2004). The LPS receptor (CD14) links innate immunity with Alzheimer's disease. *FASEB J.* 18:203–205.

Fine, S. M., Angel, R. A., Perry, S. W., Epstein, L. G., Rothstein, J. D., Dewhurst, S., and Gelbard, H. A. (1996). Tumor necrosis factor alpha inhibits glutamate uptake by primary human astrocytes. Implications for pathogenesis of HIV-1 dementia. *J. Biol. Chem.* 271: 15303–15306.

Frei, K., Siepl, C., Groscurth, P., Bodmer, S., Schwerdel, C., and Fontana, A. (1987). Antigen presentation and tumor cytotoxicity by interferon-gamma- treated microglial cells. *Eur. J. Immunol.* 17:1271–1278.

Frei, K., Eugster, H. P., Bopst, M., Constantinescu, C. S., Lavi, E., and Fontana, A. (1997). Tumor necrosis factor alpha and lymphotoxin alpha are not required for induction of acute experimental autoimmune encephalomyelitis. *J. Exp. Med.* 185:2177–2182.

Galea, E., D. Reis, J., and Feinstein, D. L. (1994). Cloning and expression of inducible nitric oxide synthase from rat astrocytes. *J. Neurosci. Res.* 37:406–414.

Galea, I., Palin, K., Newman, T. A., Van Rooijen, N., Perry, V. H., and Boche, D. (2005). Mannose receptor expression specifically reveals perivascular macrophages in normal, injured, and diseased mouse brain. *Glia* 49:375–384.

Geyer, S. J., Gill, T. J., III, Kunz, H. W., and Moody, E. (1985). Immunogenetic aspects of transplantation in the rat brain. *Transplant.* 39:244–247.

Gimenez, M. A., Sim, J. E., and Russell, J. H. (2004). TNFR1-dependent VCAM-1 expression by astrocytes exposes the CNS to destructive inflammation. *J. Neuroimmunol.* 151:116–125.

Glezer, I., Lapointe, A., and Rivest, S. (2006). Innate immunity triggers oligodendrocyte progenitor reactivity and confines damages to brain injuries. *FASEB J.*

Hao, H. N., Zhao, J., Lotoczky, G., Grever, W. E., and Lyman, W. D. (2001). Induction of human beta-defensin-2 expression in human astrocytes by lipopolysaccharide and cytokines. *J. Neurochem.* 77:1027–1035.

Hartlage-Rubsamen, M., Lemke, R., and Schliebs, R. (1999). Interleukin-1beta, inducible nitric oxide synthase, and nuclear factor-kappaB are induced in morphologically distinct microglia after rat hippocampal lipopolysaccharide/interferon-gamma injection. *J Neurosci Res* 57:388–398.

Hayashi, M., Luo, Y., Laning, J., Strieter, R. M., and Dorf, M. E. (1995). Production and function of monocyte chemoattractant protein-1 and other beta-chemokines in murine glial cells. *J. Neuroimmunol.* 60:143–150.

Head, J. R. and Griffin, W. S. (1985). Functional capacity of solid tissue transplants in the brain: evidence for immunological privilege. *Proc. Roy. Soc. Lond. Ser. B: Biol. Sci.* 224:375–387.

Herber, D. L., Maloney, J. L., Roth, L. M., Freeman, M. J., Morgan, D., and Gordon, M. N. (2006). Diverse microglial responses after intrahippocampal administration of lipopolysaccharide. *Glia* 53:382–391.

Hickey, W. F. (2001). Basic principles of immunological surveillance of the normal central nervous system. *Glia* 36:118–124.

Hua, L. L., and Lee, S. C. (2000). Distinct patterns of stimulus-inducible chemokine mRNA accumulation in human fetal astrocytes and microglia. *Glia* 30:74–81.

Huang, F., Buttini, M., Wyss-Coray, T., McConlogue, L., Kodama, T., Pitas, R. E., and Mucke, L. (1999). Elimination of the class A scavenger receptor does not affect amyloid plaque formation or neurodegeneration in transgenic mice expressing human amyloid protein precursors. *Am. J. Pathol.* 155:1741–1747.

Husemann, J. and Silverstein, S. C. (2001). Expression of scavenger receptor class B, type I, by astrocytes and vascular smooth muscle cells in normal adult mouse and human brain and in Alzheimer's disease brain. *Am. J. Pathol.* 158:825–832.

Husemann, J., Loike, J. D., Kodama, T., and Silverstein, S. C. (2001). Scavenger receptor class B type I (SR-BI) mediates adhesion of neonatal murine microglia to fibrillar beta-amyloid. *J. Neuroimmunol.* 114:142–150.

Husemann, J., Loike, J. D., Anankov, R., Febbraio, M., and Silverstein, S. C. (2002). Scavenger receptors in neurobiology and neuropathology: their role on microglia and other cells of the nervous system. *Glia* 40:195–205.

Iliev, A. I., Stringaris, A. K., Nau, R., and Neumann, H. (2004). Neuronal injury mediated via stimulation of microglial toll-like receptor-9 (TLR9). *FASEB J.* 18:412–414.

Inohara, N. and Nunez, G. (2003). NODs: intracellular proteins involved in inflammation and apoptosis. *Nat. Rev. Immunol.* 3:371–382.

Iribarren, P., Chen, K., Hu, J., Gong, W., Cho, E. H., Lockett, S., Uranchimeg, B., and Wang, J. M. (2005). CpG-containing oligodeoxynucleotide promotes microglial cell uptake of amyloid beta 1-42 peptide by up-regulating the expression of the G-protein- coupled receptor mFPR2. *FASEB J.* 19:2032–2034.

Jack, C. S., Arbour, N., Manusow, J., Montgrain, V., Blain, M., McCrea, E., Shapiro, A., and Antel, J. P. (2005). TLR signaling tailors innate immune responses in human microglia and astrocytes. *J. Immunol.* 175:4320–4330.

Janeway, C. A., Jr. and Medzhitov, R. (2002). Innate immune recognition. *Annu. Rev. Immunol.* 20:197–216.

Katafuchi, T., Kondo, T., Yasaka, T., Kubo, K., Take, S., and Yoshimura, M. (2003). Prolonged effects of polyriboinosinic:polyribocytidylic acid on spontaneous running wheel activity and brain interferon-alpha mRNA in rats: a model for immunologically induced fatigue. *Neuroscience* 120:837–845.

Kawai, T. and Akira, S. (2005). Pathogen recognition with Toll-like receptors. *Curr. Opin. Immunol.* 17:338–344.

Kerfoot, S. M., Long, E. M., Hickey, M. J., Andonegui, G., Lapointe, B. M., Zanardo, R. C., Bonder, C., James, W. G., Robbins, S. M., and Kubes, P. (2004). TLR4 contributes to disease-inducing mechanisms resulting in central nervous system autoimmune disease. *J. Immunol.* 173:7070–7077.

Kielian, T., Mayes, P., and Kielian, M. (2002). Characterization of microglial responses to *Staphylococcus aureus*: effects on cytokine, costimulatory molecule, and Toll-like receptor expression. *J. Neuroimmunol.* 130:86–99.

Kielian, T., Esen, N., and Bearden, E. D. (2005a). Toll-like receptor 2 (TLR2) is pivotal for recognition of S. aureus peptidoglycan but not intact bacteria by microglia. *Glia* 49:567–576.

Kielian, T., Haney, A., Mayes, P. M., Garg, S., and Esen, N. (2005b). Toll-like receptor 2 modulates the proinflammatory milieu in *Staphylococcus aureus*-induced brain abscess. *Infect. Immun.* 73:7428–7435.

Kitamura, Y., Kakimura, J., Koike, H., Umeki, M., Gebicke-Haerter, P. J., Nomura, Y., and Taniguchi, T. (2001). Effects of 15-deoxy-delta(12,14) prostaglandin J(2) and interleukin-4 in Toll-like receptor-4-mutant glial cells. *Eur. J. Pharmacol.* 411:223–230.

Koedel, U., Angele, B., Rupprecht, T., Wagner, H., Roggenkamp, A., Pfister, H. W., and Kirschning, C. J. (2003). Toll-like receptor 2 participates in mediation of immune response in experimental pneumococcal meningitis. *J. Immunol.* 170:438–444.

Koedel, U., Rupprecht, T., Angele, B., Heesemann, J., Wagner, H., Pfister, H. W., and Kirschning, C. J. (2004). MyD88 is required for mounting a robust host immune response to *Streptococcus pneumoniae* in the CNS. *Brain* 127:1437–1445.

Kossmann, T., Hans, V., Imhof, H. G., Trentz, O., and Morganti-Kossmann, M. C. (1996). Interleukin-6 released in human cerebrospinal fluid following traumatic brain injury may trigger nerve growth factor production in astrocytes. *Brain Res.* 713:143–152.

Kurt-Jones, E. A., Chan, M., Zhou, S., Wang, J., Reed, G., Bronson, R., Arnold, M. M., Knipe, D. M., and Finberg, R. W. (2004). Herpes simplex virus 1 interaction with Toll-like receptor 2 contributes to lethal encephalitis. *Proc. Natl. Acad. Sci. U.S.A.* 101:1315–1320.

Kurtzke, J. F. (1993). Epidemiologic evidence for multiple sclerosis as an infection. *Clin. Microbiol. Rev.* 6:382–427.

Laflamme, N. and Rivest, S. (2001). Toll-like receptor 4: the missing link of the cerebral innate immune response triggered by circulating gram-negative bacterial cell wall components. *FASEB J.* 15:155–163.

Laflamme, N., Soucy, G., and Rivest, S. (2001). Circulating cell wall components derived from gram-negative, not gram-positive, bacteria cause a profound induction of the gene-encoding Toll-like receptor 2 in the CNS. *J. Neurochem.* 79:648–657.

Le, Y., Gong, W., Tiffany, H. L., Tumanov, A., Nedospasov, S., Shen, W., Dunlop, N. M., Gao, J. L., Murphy, P. M., Oppenheim, J. J., and Wang, J. M. (2001). Amyloid (beta)42 activates a G-protein-coupled chemoattractant receptor, FPR-like-1. *J. Neurosci.* 21:RC123.

Le, Y., Murphy, P. M., and Wang, J. M. (2002). Formyl-peptide receptors revisited. *Trends Immunol.* 23:541–548.

Lee, S. C., Liu, W., Dickson, D. W., Brosnan, C. F., and Berman, J. W. (1993). Cytokine production by human fetal microglia and astrocytes. Differential induction by lipopolysaccharide and IL-1 beta. *J. Immunol.* 150:2659–2667.

Lee, S., Hong, J., Choi, S. Y., Oh, S. B., Park, K., Kim, J. S., Karin, M., and Lee, S. J. (2004). CpG oligodeoxynucleotides induce expression of proinflammatory cytokines and chemokines in astrocytes: the role of c-Jun N-terminal kinase in CpG ODN-mediated NF-kappaB activation. *J. Neuroimmunol.* 153:50–63.

Lehnardt, S., Lachance, C., Patrizi, S., Lefebvre, S., Follett, P. L., Jensen, F. E., Rosenberg, P. A., Volpe, J. J., and Vartanian, T. (2002). The toll-like receptor TLR4 is necessary for lipopolysaccharide-induced oligodendrocyte injury in the CNS. *J. Neurosci.* 22:2478–2486.

Lehnardt, S., Massillon, L., Follett, P., Jensen, F. E., Ratan, R., Rosenberg, P. A., Volpe, J. J., and Vartanian, T. (2003). Activation of innate immunity in the CNS triggers neurodegeneration through a Toll-like receptor 4-dependent pathway. *Proc. Natl. Acad. Sci. U.S.A.* 100: 8514–8519.

Lenercept Multiple Sclerosis Study Group and The University of British Columbia MS/MRI Analysis Group. (1999). TNF neutralization in MS: results of a randomized, placebo-controlled multicenter study. *Neurology* 53:457–465.

Li, J., Gran, B., Zhang, G. X., Ventura, E. S., Siglienti, I., Rostami, A., and Kamoun, M. (2003). Differential expression and regulation of IL-23 and IL-12 subunits and receptors in adult mouse microglia. *J. Neurol. Sci.* 215:95–103.

Linker, R. A., Maurer, M., Gaupp, S., Martini, R., Holtmann, B., Giess, R., Rieckmann, P., Lassmann, H., Toyka, K. V., Sendtner, M., and Gold, R. (2002). CNTF is a major protective factor in demyelinating CNS disease: a neurotrophic cytokine as modulator in neuroinflammation. *Nat. Med.* 8:620–624.

Liu, J., Marino, M. W., Wong, G., Grail, D., Dunn, A., Bettadapurs, J., Slavin, A. J., Old, L., and Bernard, C. C. A. (1998). TNF is a potent anti-inflammatory cytokine in autoimmune-mediated demyelination. *Nat. Med.* 4:78–83.

Loddick, S. A., Turnbull, A. V., and Rothwell, N. J. (1998). Cerebral interleukin-6 is neuroprotective during permanent focal cerebral ischemia in the rat. *J. Cereb. Blood Flow Metab.* 18:176–179.

Mack, C. L., Neville, K. L., and Miller, S. D. (2003). Microglia are activated to become competent antigen presenting and effector cells in the inflammatory environment of the Theiler's virus model of multiple sclerosis. *J. Neuroimmunol.* 144:68–79.

Maimone, D., Guazzi, G. C., and Annunziata, P. (1997). IL-6 detection in multiple sclerosis brain. *J. Neurol. Sci.* 146:59–65.

Marzolo, M. P., von Bernhardi, R., and Inestrosa, N. C. (1999). Mannose receptor is present in a functional state in rat microglial cells. *J. Neurosci. Res.* 58:387–395.

Matute, C., Alberdi, E., Domercq, M., Perez-Cerda, F., Perez-Samartin, A., and Sanchez-Gomez, M. V. (2001). The link between excitotoxic oligodendroglial death and demyelinating diseases. *Trends Neurosci.* 24:224–230.

Matzinger, P. (2002). The danger model: a renewed sense of self. *Science* 296:301–305.

McGeer, E. G. and McGeer, P. L. (2003). Inflammatory processes in Alzheimer's disease. *Prog. Neuropsychopharmacol. Biol. Psychiatry* 27:741–749.

McKimmie, C. S. and Fazakerley, J. K. (2005). In response to pathogens, glial cells dynamically and differentially regulate Toll-like receptor gene expression. *J. Neuroimmunol.* 169:116–125.

McKimmie, C. S., Johnson, N., Fooks, A. R., and Fazakerley, J. K. (2005). Viruses selectively upregulate Toll-like receptors in the central nervous system. *Biochem. Biophys. Res. Commun.* 336:925–933.

McMahon, E. J., Bailey, S. L., Castenada, C. V., Waldner, H., and Miller, S. D. (2005). Epitope spreading initiates in the CNS in two mouse models of multiple sclerosis. *Nat. Med.* 11:335–339.

Medzhitov, R. and Janeway, C. A., Jr. (2002). Decoding the patterns of self and nonself by the innate immune system. *Science* 296:298–300.

Melton, L. M., Keith, A. B., Davis, S., Oakley, A. E., Edwardson, J. A., and Morris, C. M. (2003). Chronic glial activation, neurodegeneration, and APP immunoreactive deposits following acute administration of double-stranded RNA. *Glia* 44:1–12.

Merrill, J. E., Ignarro, L. J., Sherman, M. P., Melinek, J., and Lane, T. E. (1993). Microglial cell cytotoxicity of oligodendrocytes is mediated through nitric oxide. *J. Immunol.* 151:2132–2141.

Milatovic, D., Zaja-Milatovic, S., Montine, K. S., Shie, F. S., and Montine, T. J. (2004). Neuronal oxidative damage and dendritic degeneration following activation of CD14-dependent innate immune response in vivo. *J. Neuroinflammation* 1:20.

Minoretti, P., Gazzaruso, C., Vito, C. D., Emanuele, E., Bianchi, M., Coen, E., Reino, M., and Geroldi, D. (2006). Effect of the functional toll-like receptor 4 Asp299Gly polymorphism on susceptibility to late-onset Alzheimer's disease. *Neurosci. Lett.* 391:147–149.

Monje, M. L., Toda, H., and Palmer, T. D. (2003). Inflammatory blockade restores adult hippocampal neurogenesis. *Science* 302:1760–1765.

Morgan, M. E., van Bilsen, J. H., Bakker, A. M., Heemskerk, B., Schilham, M. W., Hartgers, F. C., Elferink, B. G., van der Zanden, L., de Vries, R. R., Huizinga, T. W., et al. (2005). Expression of FOXP3 mRNA is not confined to CD4+ CD25+ T regulatory cells in humans. *Hum. Immunol.* 66:13–20.

Mori, K., Kaneko, Y. S., Nakashima, A., Nagatsu, I., Takahashi, H., and Ota, A. (2005). Peripheral lipopolysaccharide induces apoptosis in the murine olfactory bulb. *Brain Res.* 1039:116–129.

Mukhopadhyay, S. and Gordon, S. (2004). The role of scavenger receptors in pathogen recognition and innate immunity. *Immunobiology* 209:39–49.

Nagele, R. G., Wegiel, J., Venkataraman, V., Imaki, H., Wang, K. C., and Wegiel, J. (2004). Contribution of glial cells to the development of amyloid plaques in Alzheimer's disease. *Neurobiol. Aging* 25:663–674.

Nargi-Aizenman, J. L., Havert, M. B., Zhang, M., Irani, D. N., Rothstein, J. D., and Griffin, D. E. (2004). Glutamate receptor antagonists protect from virus-induced neural degeneration. *Ann. Neurol.* 55:541–549.

Oleszak, E. L., Katsetos, C. D., Kuzmak, J., and Varadhachary, A. (1997). Inducible nitric oxide synthase in Theiler's murine encephalomyelitis virus infection. *J. Virol.* 71:3228–3235.

Olson, J. K. and Miller, S. D. (2004). Microglia initiate central nervous system innate and adaptive immune responses through multiple TLRs. *J. Immunol.* 173:3916–3924.

Olson, J. K., Girvin, A. M., and Miller, S. D. (2001). Direct activation of innate and antigen presenting functions of microglia following infection with Theiler's virus *J. Virol.* 75:9780–9789.

Palma, J. P., Kwon, D., Clipstone, N. A., and Kim, B. S. (2003). Infection with Theiler's murine encephalomyelitis virus directly induces proinflammatory cytokines in primary astrocytes via NF-kappaB activation: potential role for the initiation of demyelinating disease. *J. Virol.* 77:6322–6331.

Pang, Y., Cai, Z., and Rhodes, P. G. (2001). Analysis of genes differentially expressed in astrocytes stimulated with lipopolysaccharide using cDNA arrays. *Brain Res.* 914:15–22.

Paresce, D. M., Ghosh, R. N., and Maxfield, F. R. (1996). Microglial cells internalize aggregates of the Alzheimer's disease amyloid beta-protein via a scavenger receptor. *Neuron* 17:553–565.

Park, C., Lee, S., Cho, I. H., Lee, H. K., Kim, D., Choi, S. Y., Oh, S. B., Park, K., Kim, J. S., and Lee, S. J. (2006). TLR3-mediated signal induces proinflammatory cytokine and chemokine gene expression in astrocytes: differential signaling mechanisms of TLR3-induced IP-10 and IL-8 gene expression. *Glia* 53:248–256.

Prehaud, C., Megret, F., Lafage, M., and Lafon, M. (2005). Virus infection switches TLR-3-positive human neurons to become strong producers of beta interferon. *J. Virol.* 79:12893–12904.

Prinz, M., Garbe, F., Schmidt, H., Mildner, A., Gutcher, I., Wolter, K., Piesche, M., Schroers, R., Weiss, E., Kirschning, C. J., et al. (2006). Innate immunity mediated by TLR9 modulates pathogenicity in an animal model of multiple sclerosis. *J. Clin. Invest.* 116:456–464.

Qin, L., Li, G., Qian, X., Liu, Y., Wu, X., Liu, B., Hong, J. S., and Block, M. L. (2005). Interactive role of the toll-like receptor 4 and reactive oxygen species in LPS-induced microglia activation. *Glia* 52:78–84.

Qiu, Z., Sweeney, D. D., Netzeband, J. G., and Gruol, D. L. (1998). Chronic interleukin-6 alters NMDA receptor-mediated membrane responses and enhances neurotoxicity in developing CNS neurons. *J. Neurosci.* 18:10445–10456.

Ransohoff, R. M., Kivisakk, P., and Kidd, G. (2003). Three or more routes for leukocyte migration into the central nervous system. *Nat. Rev. Immunol.* 3:569–581.

Roach, J. C., Glusman, G., Rowen, L., Kaur, A., Purcell, M. K., Smith, K. D., Hood, L. E., and Aderem, A. (2005). The evolution of vertebrate Toll-like receptors. *Proc. Natl. Acad. Sci. U.S.A.* 102:9577–9582.

Rose, J. W., Hill, K. E., Wada, Y., Kurtz, C. B., Tsunoda, I., Fujinami, R. S., and Cross, A. H. (1998). Nitric oxide synthase inhibitor, aminoguanidine, reduces inflammation and demyelination produced by Theiler's virus infection. *J. Neuroimmunol.*:82–89.

Roth, J., Martin, D., Storr, B., and Zeisberger, E. (1998). Neutralization of pyrogen-induced tumour necrosis factor by its type 1 soluble receptor in guinea-pigs: effects on fever and interleukin-6 release. *J. Physiol.* 509:267–275.

Roth, J., Harre, E. M., Rummel, C., Gerstberger, R., and Hubschle, T. (2004). Signaling the brain in systemic inflammation: role of sensory circumventricular organs. *Front. Biosci.* 9:290–300.

Samuel, C. E. (2001). Antiviral actions of interferons. *Clin. Microbiol. Rev.* 14:778–809.

Scherbel, U., Raghupathi, R., Nakamura, M., Saatman, K. E., Trojanowski, J. Q., Neugebauer, E., Marino, M. W., and McIntosh, T. K. (1999). Differential acute and chronic responses of tumor necrosis factor-deficient mice to experimental brain injury. *Proc. Natl. Acad. Sci. U.S.A.* 96:8721–8726.

Scheuner, D., Gromeier, M., Davies, M. V., Dorner, A. J., Song, B., Patel, R. V., Wimmer, E. J., McLendon, R. E., and Kaufman, R. J. (2003). The double-stranded RNA-activated protein kinase mediates viral-induced encephalitis. *Virology* 317:263–274.

Schnare, M., Barton, G. M., Holt, A. C., Takeda, K., Akira, S., and Medzhitov, R. (2001). Toll-like receptors control activation of adaptive immune responses. *Nat. Immunol.* 2:947–950.

Scumpia, P. O., Kelly, K. M., Reeves, W. H., and Stevens, B. R. (2005). Double-stranded RNA signals antiviral and inflammatory programs and dysfunctional glutamate transport in TLR3-expressing astrocytes. *Glia* 52:153–162.

Selmaj, K. and Raine, C. S. (1988). Tumor necrosis factor mediates myelin and oligodendrocyte damage in vitro. *Ann. Neurol.* 23:339–346.

Sharif, S. F., Hariri, R. J., Chang, V. A., Barie, P. S., Wang, R. S., and Ghajar, J. B. (1993). Human astrocyte production of tumour necrosis factor-alpha, interleukin-1 beta, and interleukin-6 following exposure to lipopolysaccharide endotoxin. *Neurol. Res.* 15:109–112.

Shukaliak, J. A. and Dorovini-Zis, K. (2000). Expression of the beta-chemokines RANTES and MIP-1 beta by human brain microvessel endothelial cells in primary culture. *J. Neuropathol. Exp. Neurol.* 59:339–352.

Smith, K. J., Kapoor, R., and Felts, P. A. (1999). Demyelination: the role of reactive oxygen and nitrogen species. *Brain Pathol.* 9:69–92.

So, E. Y., Kang, M. H., and Kim, B. S. (2006). Induction of chemokine and cytokine genes in astrocytes following infection with Theiler's murine encephalomyelitis virus is mediated by the Toll-like receptor 3. *Glia* 53:858–867.

Sonnewald, U., Qu, H., and Aschner, M. (2002). Pharmacology and toxicology of astrocyte-neuron glutamate transport and cycling. *J. Pharmacol. Exp. Ther.* 301:1–6.

Sospedra, M. and Martin, R. (2005). Immunology of multiple sclerosis. *Annu. Rev. Immunol.* 23:683–747.

Stalder, A. K., Pagenstecher, A., Yu, N. C., Kincaid, C., Chiang, C. S., Hobbs, M. V., Bloom, F. E., and Campbell, I. L. (1997). Lipopolysaccharide-induced IL-12 expression in the central nervous system and cultured astrocytes and microglia. *J. Immunol.* 159:1344–1351.

Sterka, D., Jr., Rati, D. M., and Marriott, I. (2006). Functional expression of NOD2, a novel pattern recognition receptor for bacterial motifs, in primary murine astrocytes. *Glia* 53:322–330.

Sun, N., Grzybicki, D., Castro, R. F., Murphy, S., and Perlman, S. (1995). Activation of astrocytes in the spinal cord of mice chronically infected with a neurotropic coronavirus. *Virology* 213:482–493.

Suvannavejh, G. C., Lee, H. O., Padilla, J., Dal Canto, M. C., Barrett, T. A., and Miller, S. D. (2000). Divergent roles for p55 and p75 tumor necrosis factor receptors in the pathogenesis of MOG(35-55)-induced experimental autoimmune encephalomyelitis. *Cell. Immunol.* 205:24–33.

Suzumura, A., Lavi, E., Weiss, S. R., and Silberberg, D. H. (1986). Coronavirus infection induces H-2 antigen expression on oligodendrocytes and astrocytes. *Science* 232:991–993.

Takahashi, J. L., Giuliani, F., Power, C., Imai, Y., and Yong, V. W. (2003). Interleukin-1beta promotes oligodendrocyte death through glutamate excitotoxicity. *Ann. Neurol.* 53:588–595.

Takeshita, S., Takeshita, F., Haddad, D. E., Janabi, N., and Klinman, D. M. (2001). Activation of microglia and astrocytes by CpG oligodeoxynucleotides. *Neuroreport* 12:3029–3032.

Takeuchi, O., Hoshino, K., and Akira, S. (2000). Cutting edge: TLR2-deficient and MyD88-deficient mice are highly susceptible to *Staphylococcus aureus* infection. *J. Immunol.* 165:5392–5396.

Tiffany, H. L., Lavigne, M. C., Cui, Y. H., Wang, J. M., Leto, T. L., Gao, J. L., and Murphy, P. M. (2001). Amyloid-beta induces chemotaxis and oxidant stress by acting at formylpeptide receptor 2, a G protein-coupled receptor expressed in phagocytes and brain. *J. Biol. Chem.* 276:23645–23652.

Toulmond, S., Vige, X., Fage, D., and Benavides, J. (1992). Local infusion of interleukin-6 attenuates the neurotoxic effects of NMDA on rat striatal cholinergic neurons. *Neurosci. Lett.* 144:49–52.

Town, T., Jeng, D., Alexopoulou, L., Tan, J., and Flavell, R. A. (2006). Microglia recognize double-stranded RNA via TLR3. *J. Immunol.* 176:3804–3812.

Vallieres, L. and Rivest, S. (1997). Regulation of the genes encoding interleukin-6, its receptor, and gp130 in the rat brain in response to the immune activator lipopolysaccharide and the proinflammatory cytokine interleukin-1beta. *J. Neurochem.* 69:1668–1683.

Vartanian, T., Li, Y., Zhao, M., and Stefansson, K. (1995). Interferon-gamma-induced oligodendrocyte cell death: implications for the pathogenesis of multiple sclerosis. *Mol. Med.* 1:732–743.

Wang, J., Asensio, V. C., and Campbell, I. L. (2002). Cytokines and chemokines as mediators of protection and injury in the central nervous system assessed in transgenic mice. *Curr. Top. Microbiol. Immunol.* 265:23–48.

Wang, T., Town, T., Alexopoulou, L., Anderson, J. F., Fikrig, E., and Flavell, R. A. (2004). Toll-like receptor 3 mediates West Nile virus entry into the brain causing lethal encephalitis. *Nat. Med.* 10:1366–1373.

Wheeler, R. D. and Owens, T. (2005). The changing face of cytokines in the brain: perspectives from EAE. *Curr. Pharm. Des.* 11:1031–1037.

Williams, B. R. (1999). PKR; a sentinel kinase for cellular stress. *Oncogene* 18:6112–6120.

Wong, D. and Dorovini-Zis, K. (1992). Upregulation of intercellular adhesion molecule-1 (ICAM-1) expression in primary cultures of human brain microvessel endothelial cells by cytokines and lipopolysaccharide. *J. Neuroimmunol.* 39:11–21.

Wong, D. and Dorovini-Zis, K. (1995). Expression of vascular cell adhesion molecule-1 (VCAM-1) by human brain microvessel endothelial cells in primary culture. *Microvasc. Res.* 49:325–339.

Yoneyama, M., Kikuchi, M., Natsukawa, T., Shinobu, N., Imaizumi, T., Miyagishi, M., Taira, K., Akira, S., and Fujita, T. (2004). The RNA helicase RIG-I has an essential function in double-stranded RNA-induced innate antiviral responses. *Nat Immunol* 5:730–737.

Zekki, H., Feinstein, D. L., and Rivest, S. (2002). The clinical course of experimental autoimmune encephalomyelitis is associated with a profound and sustained transcriptional activation of the genes encoding toll-like receptor 2 and CD14 in the mouse CNS. *Brain Pathol.* 12:308–319.

Zhang, Z., Trautmann, K., and Schluesener, H. J. (2005). Microglia activation in rat spinal cord by systemic injection of TLR3 and TLR7/8 agonists. *J. Neuroimmunol.* 164:154–160.

Zuany-Amorim, C., Hastewell, J., and Walker, C. (2002). Toll-like receptors as potential therapeutic targets for multiple diseases. *Nat. Rev. Drug Discov.* 1:797–807.

8
Central Nervous System Diseases and Inflammation

The Complement System in the CNS: Thinking again

Andrea J. Tenner and Karntipa Pisalyaput

The immune system has evolved numerous mechanisms to protect the host against perceived danger, as well as regulatory mechanisms for the resolution of inflammation and/or repair of the host. While the adaptive immune responses that include specific antibody and T cells are critical in eliminating many pathogenic organisms, the first immune responses are provided by the innate immune system. This immediate response is critical in assessing the level of "danger" or injury and consequently directing the subsequent recruitment of other immune system components. These responses are rapid, and include both cellular elements (phagocytic cells, natural killer cells) and protein elements (the complement system, defensins). Some components of the host response to danger if insufficiently regulated can result excessive inflammation and tissue damage including neurotoxicity in the central nervous system (CNS). Thus, a balance between generating a toxic environment for pathogens while providing reparative functions to the tissues must be maintained, and thus requires monitoring systems and appropriate modulation of immune response mediators. This is certainly the case for the generally proinflammatory complement cascade, a powerful effective mechanism of the immune system.

The complement system has been implicated in several neurological diseases (reviewed in (Gasque et al., 2000)). In this chapter we review the basis for the damaging role of complement in diseases of the CNS, but also focus on the more recent evidence for neuroprotective responses triggered by complement proteins and activation fragments (Wyss-Coray and Mucke, 2002; Rus et al., 2006; Dhib-Jalbut et al., 2006; Tenner and Fonseca, 2006). Such mechanisms appear to be induced to offset the effects of neuronal injury and facilitate repair. Since these neuroprotective mechanisms are induced rapidly in response to neurotoxic events, it is sometimes difficult to separate the components of the detrimental events from those signaling pathways inducing neuroprotection. Thus, precise knowledge of the signaling molecules, cell surface receptors and biochemical pathways involved are necessary to permit design of therapeutic interventions to inhibit the detrimental responses while promoting beneficial or reparative functions.

T.E. Lane et al. (eds.), *Central Nervous System Diseases and Inflammation.*
© Springer 2008

1   The Complement System

The complement (C') system is a powerful effector mechanism of the immune system that protects the host by generating activation fragments and molecular assemblies that result in elimination of pathogens, immune complexes or other activating species (Gasque, 2004; Rus et al., 2005b). The complement system can be activated via three distinct recognition pathways: the classical pathway via C1q, the lectin pathway via mannose binding lectin (MBL), or the alternative pathway via C3 (two of which are depicted in Fig. 8.1). All three pathways converge at C3. In addition to their functions in the C' activation cascade, C3b, C4b and C1q can opsonize (target to professional phagocytes for ingestion) pathogens or cellular debris, facilitating their rapid clearance (Bobak et al., 1987; Taylor et al., 2000; Sahu and Lambris, 2001). C' activation products of enzymatic cleavage, C3a and C5a, initiate the local inflammatory response through both increasing the permeability of the local vasculature and recruitment of leukocytes to the site of infection or injury. These recruited cells are often activated to produce potent inflammatory mediators and toxic free radicals or macromolecules, ideally targeted to the pathogen. The terminal stage of complement activation is the assembly of the membranolytic membrane attack complex (MAC or C5b-9), a membrane pore that results in the demise of pathogens via membrane lysis. While host cell damage from the complement system is inhibited by cell surface complement regulatory molecules (CD59, CD46, CD55, CR1) (Song, 2004) and extracellular inhibitors and inactivators (Factor H, Factor I, C4bBP) (Pangburn, 2000; Blom et al., 2001), tissue damage can result from chronic activation of the complement system by antibody-antigen complexes, specific pathogens, cellular debris or misfolded proteins which may persist due to overproduction or reduced clearance (Walport, 2001; Van Beek et al., 2003). In the brain, activation of complement can result in recruitment to the site of damage and stimulation of microglia and astrocytes (Yao et al., 1990). These activated glial cells can be phagocytic but can also secrete proinflammatory cytokines as well as reactive oxygen species and nitric oxide, essentially creating a local inflammatory reaction that can ultimately accelerate the pathology and neuronal dysfunction (Streit, 2004) seen in human disease (Akiyama et al., 2000; Kulkarni et al., 2004).

There has been a surge in research into the role of complement in age-related degenerative diseases since the 2005 description of the association of complement Factor H (and subsequently Factor B as well) polymorphisms with 50–70% of the risk for age related macular degeneration (Maller et al., 2006; Haines et al., 2005; Hageman et al., 2005; Klein et al., 2005). It had been reported earlier that complement protein (C1q and C3 cleavage products) deposition was associated with the drusen deposits in the eye of patients with age-related macular degeneration (Baudouin et al., 1992; Johnson et al., 2002). While effects of complement activation are likely secondary to the initiating events of this disorder (Hageman et al., 2001), these data suggest that the magnitude of local inflammation caused by inefficiently controlled complement activation (via Factor H polymorphisms) or more efficient activation (via Factor B polymorphisms) can have serious consequences in chronic degenerative disease.

Color Plates

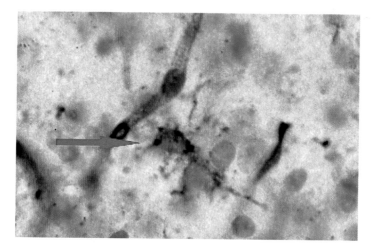

Fig. 1.1 A typical parenchymal microglia extending processes to all elements of its environment. Microglia and blood vessels are visualized in brown using tomato lectin. Nuclei are visualized in blue using hematoxylin.

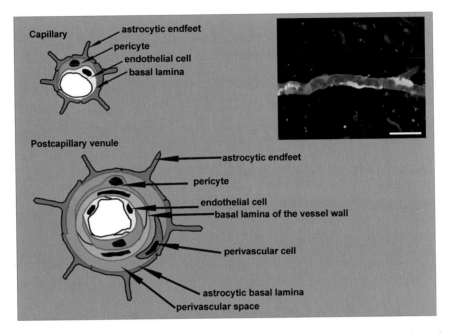

Fig. 3.1 *Left side:* Topography of the blood-brain interface: Difference between capillaries and venule. *Upper panel:* By definition, capillaries lack a "media" of smooth muscle cells. The pericyte (green) is thus separated from the endothelial cells (beige) by a basement membrane (red) only. The outer vascular basement membrane and the one on top of the glia limitans (blue) are fused to a common "gliovascular" membrane. Note the intimate contact of the astrocytic endfeet and the pericytes and endothelial cells which may underlie certain aspects of blood–brain barrier differentiation. *Lower panel:* Less pronounced barrier function of the endothelium may be explained by the following differences between venules and capillaries: Astrocytic endfeet cannot contact the pericytes and the endothelium due to the perivascular space and the smooth muscle cells of the "media" (orange); the latter also separates the pericytes from the endothelium. Pericytes are often mixed up with perivascular cells (pink), a heterogeneous population of leptomeningeal cells, macrophages, and other leukocytes. The brain parenchyma proper (neuropil) is delineated by the astrocytic endfeet forming the glia limitans (blue). It is not clear if all pericytes are located at the outermost portion of the vascular wall (upper green cell). Some may engulf the endothelium or be part of the media (lower green cell). *Right side:* Pericytes (green) oriented along the longitudinal axis of a capillary. The fused gliovascular membrane is labeled with an antibody to laminin (red). Scale bar: 20 μm.

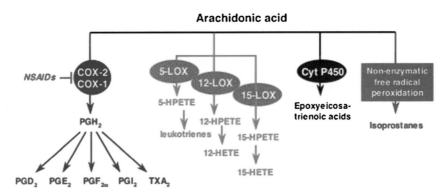

Fig. 6.1 Arachidonic acid is metabolized by three enzyme systems: (1) the cyclooxygenases COX-1 and COX-2 to yield PGH$_2$, which is converted by specific synthases to the five prostanoids, (2) the lipoxygenases 5-LOX, 12-LOX, and 15-LOX to yield leukotrienes and HETEs, and (3) the cytochrome P450 epoxygenases to yield epoxyeicosanotrienoic acids. AA is also non-enzymatically oxidized by free radicals to form a large family of isoprostanes.

Complement Cascades

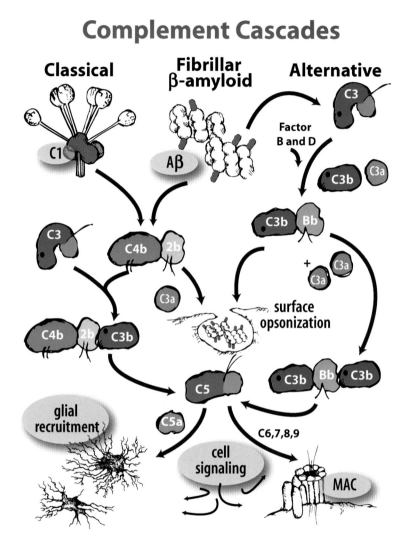

Fig. 8.1 Schematic diagram of complement activation by fibrillar β-amyloid. Fibrillar Aβ can activate the classical complement pathway in the absence of antibody, or the alternative pathway. These two pathways converge at C3 activation with end results including opsonization (C3b deposits on fibrillar Aβ) and engulfment (of Aβ), leukocyte recruitment (by anaphylatoxins C3a and C5a) and potential bystander lysis by the membrane attack complex (MAC).

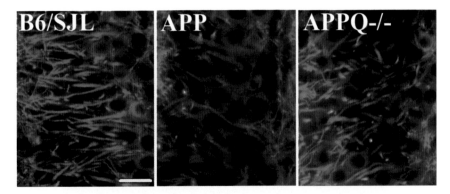

Fig. 8.2 APPQ–/– mice show less neuronal degeneration. The hippocampus from 16-month-old wild type mice (B6/SJL), mice transgenic for the human mutant APP protein which confers amyloid plaque accumulation and neuropathology (APP), and APP with a genetically ablated C1q gene (APPQ–/–) were immunostained with antibody to MAP2, a neuronal microtubule associated protein indicative of neuronal integrity. APPQ–/– mice show higher MAP2 immunostaining in the pyramidal neurons of the CA3c area of hippocampus than the complement sufficient APP transgenic mouse. Additional photomicrographs and image analysis of results from multiple animals and ages have been published (Fonseca et al., 2004). Scale bar: 20 μm.

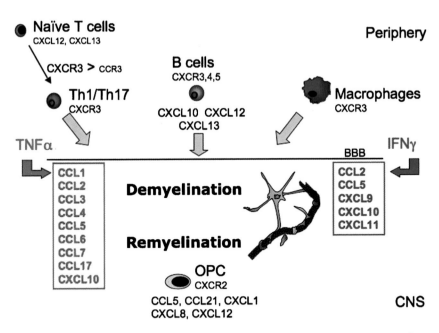

Fig. 9.1 Schematic to summarize the major points covered in this chapter. Demyelinating disease results primarily from entry of leukocytes to the CNS, which is directed by chemokines acting on receptors specifically expressed by leukocyte subsets such as T cells, B cells and macrophages. Neutrophils are not shown in the figure, for greater simplicity. The cytokines TNFα and IFNγ induce selected chemokines, these are indicated in boxes for each cytokine. Remyelination is portrayed as an intra-CNS event, the role of chemokines being to direct migration of oligodendrocyte precursor cells (OPC).

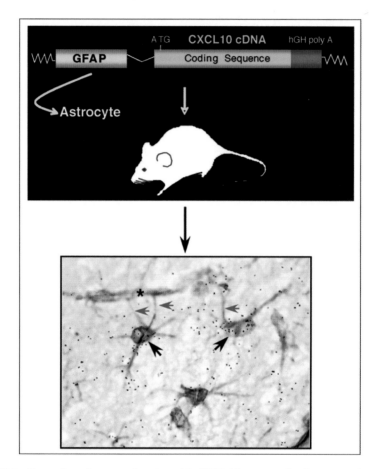

Fig. 10.1 Generation of transgenic mice with CXCL10 gene production targeted to astrocytes. A cDNA encoding murine CXCL10 was inserted downstream of the murine GFAP promoter and upstream of a human growth hormone (hGH) polyadenylation signal sequence. This transgene construct was microinjected into the germ line of mice to generate a stable transgenic line. The presence of CXCL10 gene expression in astrocytes (lower panel) was confirmed by in situ hybridization for CXCL10 RNA combined with immunohistochemistry for GFAP protein. Note that some CXCL10 RNA positive astrocytes (black arrows) extend processes (blue arrows) to a blood vessel (asterisk) that is close by.

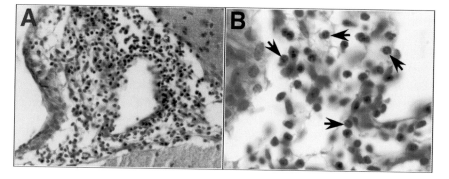

Fig. 10.2 Histological evaluation of the brain from GFAP-CXCL10 transgenic mice. A. appearance of a perivascular leukocyte infiltrate in the meninges (haematoxylin and eosin stained section; original magnification ×100). B. numerous neutrophils were observed (arrows) in the leukocyte infiltrates, (haematoxylin & eosin stained section; original magnification ×1,000)

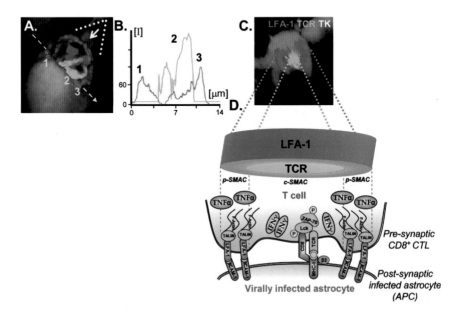

Fig. 14.1 Immunological synapses in the brain in vivo during an antiviral immune response. This figure shows an immunological synapse formed between CD8 T cells and an adenovirally infected astrocyte. (**a**) shows the synapse as seen under the confocal microscope; the infected cell is detected through its expression of a marker gene expressed from the viral genome (i.e. HSV1-TK; white), and the CD8 + T cell is detected through its expression of LFA-1 (red) and TCR (green). (**b**) shows the relative quantification of fluorescence across the yellow arrow in (**a**); 1,2, 3 are the areas indicated in the fluorescence intensity graph in (**b**). Notice the typical distribution of increased intensity of LFA-1 at the p-SMAC (in red), and the peak of TCR intensity at the c-SMAC. (**c**) shows the view from the 3-D reconstruction of the image stack illustrated in (**a**). The view shown in (**c**) is viewed from the white triangle and white arrow in (**a**). (**d**) illustrates schematically the distribution of molecules at the p-SMAC and c-SMAC, and also illustrates the intracellular re-distribution and potential secretion of T cell effector molecules either towards the immunological synapse (i.e. IFN-γ), or outside the immunological synapse (i.e. TNF-α). (**d**) is based on results from Huse et al. (2006), and Lowenstein et al. (2007).

Complement Cascades

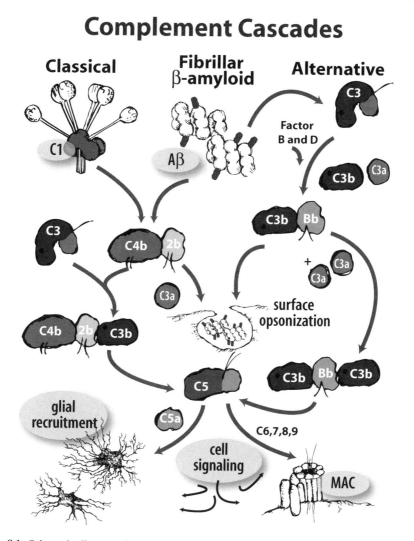

Fig. 8.1 Schematic diagram of complement activation by fibrillar β-amyloid. Fibrillar Aβ can activate the classical complement pathway in the absence of antibody or the alternative pathway. These two pathways converge at C3 activation with end results including opsonization (C3b deposits on fibrillar Aβ) and engulfment (of Aβ), leukocyte recruitment (by anaphylatoxins C3a and C5a) and potential bystander lysis by the membrane attack complex (MAC) (*See also color plates*).

2 Synthesis of Complement Proteins

While the liver was originally considered the primary site of synthesis for the complement proteins, it is now apparent that these proteins are synthesized by multiple cell types, particularly of myeloid origin, including induced synthesis of C' proteins in the CNS in certain disorders (Walker and McGeer, 1992; Barnum, 1995; Morgan and Gasque, 1997; Strohmeyer et al., 2000). For example, synthesis of most complement factors has been shown to occur within the AD brain (Johnson et al., 1992; Shen et al., 1997), and receptors for complement activation products are found on neurons as well as on microglia and astrocytes (reviewed in (Nataf et al., 1999)). Thus, there is no requirement for a compromised blood–brain barrier to initiate complement activity in the brain.

Data from several studies have indicated that C1q can be synthesized in the absence of subcomponents C1r and C1s (components that are required for classical complement activation) (Bensa et al., 1983; Reboul et al., 1985). In an in vivo approach employing microarray analysis of mRNA induction in a murine model of oxidative stress during aging (Lee et al., 2000), C1q (and C4) was among the most highly induced genes detected in both neocortex and cerebellum as a consequence of aging. While C1q was upregulated, there was no evidence for induction of other C1 subcomponents, C1r and C1s, or C3. Interestingly, this induction of C1q was not seen in aged mice that were maintained on caloric restriction diet (and thus presumably under less oxidative stress). In other injury models, CNS synthesis of C1q has been observed following viral infection, kainic acid treatment, cuprizone model of demyelination and remyelination and ischemia/reperfusion (Dandoy-Dron et al., 1998; Arnett et al., 2003; Goldsmith et al., 1997; Lampert-Etchells et al., 1993; Dietzschold et al., 1995; Huang et al., 1999). Synthesis of C1q has also been detected in Alzheimer's disease brain (Afagh et al., 1996; Shen et al., 1997; Johnson et al., 1992) and other human neurodegenerative diseases, such as Huntington's disease (Singhrao et al., 1999), or mouse models of diseases such as Sanfilippo syndrome (Ohmi et al., 2003). These observations suggest that C1q may be upregulated as a response to injury and may possibly serve additional functions in addition to complement activation and the induction of proinflammatory events that often follow.

3   Complement Proteins Induce and Regulate Inflammatory Events

As indicated above, C1q is the recognition component of the classical complement cascade, which can lead to a proinflammatory environment upon activation. However, C1q has also been shown to enhance phagocytic activity of monocytes, macrophages and microglia both independently (as reviewed in (Bohlson et al., 2007) and in conjunction with deposited C3 activation products, C3b and iC3b. In addition, C1q plays a role in the clearance of apoptotic cells both in vivo and in vitro (Botto et al., 1998; Bohlson et al., 2007; Ogden et al., 2001). C1q binds to apoptotic cells and cellular debris through its globular heads (Korb and Ahearn, 1997; Navratil et al., 2001), directly facilitating the rapid removal of damaged cells and thus limiting the release of potentially damaging intracellular components. While there are several apoptotic cell markers (such as phosphatidylserine) and proteins that bind to apoptotic cells and signal through receptors on phagocytic cells to induce rapid clearance [as described in reviews (Savill and Fadok, 2000; Grimsley and Ravichandran, 2003)], in mammalian systems only a few (for example, C1q, SP-D, CD93, MFG-E8, MBL) have been confirmed thus far to have a nonredundant role in this process in vivo as demonstrated by gene knock out experiments (Stuart et al., 2005; Hanayama et al., 2004; Scott et al., 2001; Botto et al., 1998; Norsworthy et al., 2004). Since C1q enhances clearance of apoptotic cells, one possible protective function for C1q induction following neuronal injury as described above, may be to flag damaged neurons and/or neuronal blebs for rapid clearance and thus prevent release of neurotoxic levels of intracellular components (such as glutamate) in initial stages of CNS injury.

More recently, C1q has been shown to modulate proinflammatory cytokine expression in peripheral monocytes (Fraser et al., 2006), down regulating the level of mRNA and protein of inflammatory cytokines including IL-1α and IL-1β, while increasing mRNA and protein secretion of anti-inflammatory IL-10. While signaling pathways induced in monocytes by C1q involve the generation of the "decoy" p50p50 NFkB dimer and CREB phosphorylation (Fraser et al., 2007), which are known to be involved in the regulation of cytokine gene expression, whether these factors lead to the C1q-induced regulation of cytokines and to what extent these responses are similar in microglia or brain macrophages remain to be formally demonstrated. A regulatory pathway that results in suppression of proinflammatory cytokines would be beneficial in resolving inflammation and/or avoiding autoimmunity to cell constituents in the CNS. Thus, a hypothesis under current investigation is that induction of C1q synthesis may be a response to injury that promotes rapid clearance of apoptotic cells and/or concomitant suppression of inflammation by microglia.

The small peptides C3a and C5a (78 and 74 amino acids, respectively) that are produced by activation of all complement pathways (Fig. 8.1) diffuse from the site of activation and bind to G protein-coupled receptors C3aR and C5aR (CD88), respectively, on neutrophils, endothelial cells, glial cells, and neurons (O'Barr et al., 2001), inducing directed chemotaxis (Yao et al., 1990), Ca++signaling (Möller et al., 1997), and other cell type-specific responses (reviewed in (Ember et al., 1998)). In vivo, C5a promotes the margination, extravasation, migration, and activation of leukocytes. It also can induce increases in vascular permeability, and ultimately vascular hemorrhage. Many of the detrimental inflammatory effects of complement can result from the influx and activation of inflammatory myeloid-derived cells (neutrophils, macrophages, microglia).

Seemingly paradoxically, C5-deficient mice showed enhanced neurodegeneration in response to some excitotoxic stimuli (Pasinetti et al., 1996; Tocco et al., 1997), and the presence of C5 limited detrimental responses to neurodegenerative stimuli in other injury models (Rus et al., 2005a). In addition, Kohl and colleagues describe the C5a downregulation of LPS-stimulated IL-12p70 in murine macrophages, which correlates with priming (or directing) for a limited Th1 and enhanced Th2 response to TLR-4 mediated cell triggering. This response was dependent on C5aR/CD88 (i.e., C5a had no effect on this signaling in CD88−/− mice) (Hawlisch et al., 2005). Furthermore, C5a, as well as C3a, has been reported to provide direct neuroprotection (Van Beek et al., 2001; O'Barr et al., 2001; Benard et al., 2004; Osaka et al., 1999; Mukherjee and Pasinetti, 2001) suggesting that this peptide may have neuroprotective roles as well as proinflammatory activity.

An early report that C3a induces NGF expression (Heese et al., 1998) in human microglial cell cultures suggested that this activation fragment may facilitate repair as well as contribute to damage in the brain (see below). Fontaine and colleagues reported that both C3a and C5a triggered increases in the nerve growth factor (NGF) mRNA levels in astrocytes that resulted in increased NGF secretion upon costimulation with IL-1β (Jauneau et al., 2006). In a separate report, C3a and C5a primed cells for responsiveness to reparative/regenerative growth factors (Markiewski et al., 2004), further supporting the possibility that complement activation fragments influence neurotrophin function, and thus survival of neurons.

Three specific examples of central nervous system disorders in which complement appears to influence disease progression will be discussed here to demonstrate the need for clear understanding of both the detrimental and beneficial effects of complement system so that therapies may be optimized to inhibit the detrimental events and enhance the protective or reparative activities of these proteins. However, there are other CNS disease scenarios which have been reviewed recently (Mocco et al., 2006) that also support the notion that complement activation in the CNS can be advantageous as well as detrimental.

4   Complement and Inflammation in Alzheimer Disease

Several hypotheses for the underlying causes of the cognitive loss that is identified as the major characteristic of Alzheimer Disease have been proposed and are consistent with observations described over the past two decades for both human AD and transgenic mouse models of AD. One hypothesis states that Aβ, which is a 39–42 amino acid peptide derived from the cleavage of amyloid precursor protein (APP), plays a major role in the progressive pathology and cognitive dysfunction seen in AD. Genetic evidence for this "amyloid hypothesis" includes the presence of mutations in various genes that result in the accumulation of amyloid or APP gene duplication in cases of familial AD (Golde, 2005; Tanzi and Bertram, 2005). However, there are distinct pools of amyloid peptide, including intraneuronal accumulations, soluble oligomeric structures, as well as prominent fibrillar extracellular plaques, and the precise relationships between different pools of amyloid and the major learning and memory deficits are still being investigated. Various amyloid peptide preparations are correlated with cognitive deficiencies in mice and have been shown to induce neuronal degeneration in primary neuron cultures and LTP dysfunction upon exogenous injection into rodent brain (Walsh et al., 2002, 2005). Several transgenic mouse models have shown synaptic and cognitive deficits prior to plaque detection suggesting that soluble multimeric Aβ assembly states and/or intraneuronal Aβ are responsible for early synaptic loss (Walsh et al., 2002; Takahashi et al., 2002; Koistinaho et al., 2001; Oddo et al., 2003).

Extensive in vitro studies by this laboratory and others have shown that C1q, both alone and in the classical complement initiation complex C1 (Fig. 8.1), binds to fibrillar Aβ (beta sheet conformation) and in C1 activates the classical complement pathway (Rogers et al., 1992; Jiang et al., 1994; Chen et al., 1996). Activation of the alternative complement pathway by interaction of Aβ with C3 has also been documented in vitro (Bradt et al., 1998; Watson et al., 1997). Importantly, colocalization of C1q and C3 with neuritic plaques was the first suggestion that complement may contribute to the progression of Alzheimer Disease (Eikelenboom et al., 1989) particularly at late stages of the disease when the amyloid plaques contain the fibrillar (complement-activating) form of the peptide (Afagh et al., 1996). Detection of the complement membranolytic complex, C5b-9, in areas containing the fibrillar plaques in AD (Webster et al., 1997) indicate that the complete cascade is activated in the AD brain. Reactive microglia and astrocytes also surround these fibrillar amyloid containing plaques, as would be predicted as a result of recruitment by complement C5a. Elevated pro-inflammatory cytokines including IL-1β, TNF-α, IL-6, and IL-8 have been found in the AD brain, with IL-1β perhaps being the earliest to be induced (Akiyama et al., 2000; Griffin and Mrak, 2002). The proinflammatory complement activation products recruiting in glia to the plaque could initiate a secondary wave of inflammatory cytokines as well as toxic free radicals that accelerate local neuronal damage, loss, and decline of cognitive function in AD. Interestingly, in transgenic mouse models of AD, C1q is also associated with fibrillar amyloid plaques (Fonseca et al., 2004; Matsuoka et al., 2001). The contribution of complement to amyloid-induced inflammation and pathology in animal models has been convincingly demonstrated in transgenic mice expressing mutant APP (Tg2576) in the presence and absence of classical complement pathway activation. Tg2576 mice deficient in C1q, demonstrated a 50% reduction in microgliosis and astrocytic activation and significantly less decrease (60–65%) in neuronal integrity (Fig. 8.2) relative to the complement sufficient transgenic APP mice (Fonseca et al., 2004). Further supporting the hypothesis of complement-mediated neuropathology, Bergamaschini, et al. tested the effect of Enoxaparin, a low molecular weight heparin that inhibits Aβ-induced complement activation and activation of the contact system, in the APP23 transgenic mouse, another model of amyloid toxicity/AD. Treatment with this compound 3 times a week for 6 months, resulted in a greater than 2-fold decrease in amyloid deposits, total amyloid load and decreased astrogliosis at the end of treatment (Bergamaschini et al., 2004). As in human AD, it is evident that an inflammatory environment exists in these animal models and that combinations of cytokines, costimulatory ligands, and innate signaling molecules synergize in this disease to produce a toxic environment which likely contributes to neuronal functional/synaptic deficits (Akiyama et al., 2000; Todd et al., 2004; Tan et al., 1999). Fig. 8.3 illustrates a model in which complement mediated enhancement of both glial inflammation and direct neurotoxicity may be contributing to progression of Alzheimer Disease at a stage in which fibrillar amyloid plaques accumulate.

In contrast to the detrimental effects of complement in this mouse model of AD, work in a different murine model of amyloid induced neuropathology suggests that

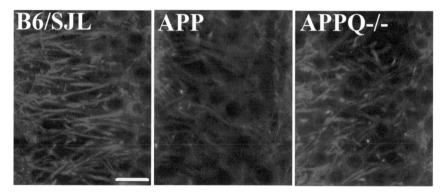

Fig. 8.2 APPQ–/– mice show less neuronal degeneration. The hippocampus from 16 month old wild type mice (B6/SJL), mice transgenic for the human mutant APP protein which confers amyloid plaque accumulation and neuropathology (APP), and APP with a genetically ablated C1q gene (APPQ–/–) were immunostained with antibody to MAP2, a neuronal microtubule associated protein indicative of neuronal integrity. APPQ–/– mice show higher MAP2 immunostaining in the pyramidal neurons of the CA3c area of hippocampus than the complement sufficient APP transgenic mouse. Additional photomicrographs and image analysis of results from multiple animals and ages have been published (Fonseca et al., 2004). Scale bar: 20 μm (*See also color plates*).

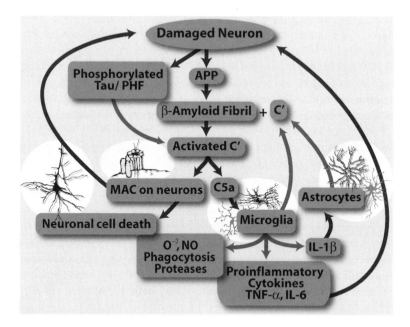

Fig. 8.3 Model of complement activation and inflammation induced damage in Alzheimer Disease. Complement activation in AD has a variety of consequences and activators. C' activation results in activation of glia, which can be induced to produce proinflammatory cytokines and neurotoxic substances, and formation of the lytic MAC complex, which may contribute to neuron damage and death.

at least some of the complement factors can decrease neuropathology (Wyss-Coray et al., 2002). That is, when the cleavage of complement component C3 was inhibited in a transgenic APP model by over expressing the soluble Crry protein (which inhibits formation of the enzyme that amplifies the cleavage of C3 to C3b), Aβ deposition was higher than in comparable mice that did not over-express Crry and was accompanied by less microgliosis but elevated loss of NeuN relative to the APP wild type (Wyss-Coray et al., 2002). One explanation provided by the investigators was that the lack of microglia (and/or lack of C3b as an opsonin to enhance phagocytosis) prevented clearance of Aβ and thus ultimately led to an increase in the number of degenerating neurons. The caveats of this model are the systemic nature of the complement inhibition and the lifetime inhibition of the transgenic over expression of Crry (Wyss-Coray et al., 2002). Nevertheless, at this point the mechanisms resulting in the difference in outcome between the APP C1q–/– mouse described by us and that of the Crry transgene are unknown. However, the data do suggest that a balance of complex interactions is involved.

Interestingly, in at least two other dementias, familial British dementia (FBD) and familial Danish dementia (FDD) (the Chromosome 13 dementias), deposits of fibrillar conformations of abnormal proteins are also associated with complement components C1q, C4d and C5b-9, glial recruitment and inflammation (Rostagno et al., 2002, 2005). In vitro studies demonstrated that both classical and alternative complement pathways are activated by the abnormal proteins (which are distinct from the beta amyloid in AD) (Rostagno et al., 2002). Similar to the Alzheimer Disease pathologic peptides, the oligomeric form of these peptides induce neuronal death directly in in vitro models, but the contribution of this toxicity to the dementia seen in patients has yet to be determined (Gibson et al., 2004; El-Agnaf et al., 2001). As with AD, there may be multiple pathways leading to cognitive dysfunction, and thus therapeutic intervention including regulation of complement-induced inflammation may improve patient outcome.

5   Complement in Multiple Sclerosis

Multiple sclerosis (MS) is a demyelinating CNS disorder which is considered an autoimmune disease with myelin and oligodendrocytes as the target. Many studies into the mechanisms responsible for MS pathology have relied on use of the experimental autoimmune encephalitis (EAE) mouse model in which immunization with myelin oligodendrocyte glycoprotein (MOG) is used to induce demyelination. Early studies using the EAE mouse model implicated a role for complement in pathology, since animals treated with Cobra Venom Factor (CVF), which depletes serum complement, demonstrated an attenuated form of the disease when compared to controls (reviewed in (Barnum and Szalai, 2006)). Pioneering studies by Moon Shin demonstrated in vitro that myelin basic protein activated the classical complement pathway in the absence of antibody, suggesting that complement may amplify tissue injury, as well as participate in clearance of damaged cells (Vanguri et al., 1982). Use of sCR1 to inhibit complement activation by inhibiting C3 and C5 convertases resulted in attenuated disease in EAE mouse (reviewed in (Barnum and Szalai, 2006)) demonstrating that in this model the detrimental effects of complement outweighed any protective effect. However these early studies, which involved inhibiting the all downstream components of complement, were broad and therefore may be difficult to apply as a clinical treatment.

More recently several researchers have utilized a targeted approach to studying the role of complement in the development of EAE through the use of mice genetically altered for complement protein expression or activity. The contribution of C3a, a cleavage product of C3 generated upon complement activation, was examined using C3a receptor knock-out mouse (C3aR−/−). EAE was induced in these animals by injection of MOG, and while onset of disease was not affected by deletion of C3aR, disease severity in terms of clinical score and macrophage and T cell infiltration, was reduced in C3aR−/− when compared to that of control mice (Boos et al., 2004). In the same study, transgenic mice expressing C3a in the CNS (under the control of the GFAP promoter) were generated and EAE induced. Again, onset of disease was similar in the C3a expressing transgenic as in the control mice, but here disease severity increased significantly over controls with a larger mortality rate in the C3a expressing transgenic animals. Given that neither C3aR deletion nor C3a expression affected disease onset and differences were only observed around 3 weeks post-immunization, it has been suggested that the contribution of C3aR/C3a is during the chronic phase of EAE. Similar attenuation of disease was previously observed in C3−/− mice (Nataf et al., 2000). Possible functions for C3a/C3aR include modulation of cytokine profile and or/chemoattraction of encephalitogenic cells into the CNS, two factors which are capable of influencing the extent of disease. In contrast to the protective effects observed in C3−/−, no effect on EAE onset or progression has been observed in C4−/− mice (Boos et al., 2005), suggesting that the alternative pathway of complement can be activated in this model.

C5a is another cleavage product resulting from complement activation which has been examined for a possible role in EAE. This fragment of C5 is extensively studied in CNS disorders due to its chemoattractant and inflammatory properties. The classical C5a receptor (C5aR/CD88) is up-regulated in EAE on neurons, glial cells and infiltrating T-cells, however inhibition of C5aR with a small molecule C5a receptor antagonist (Morgan et al., 2004) or use of a C5aR −/− (Reiman et al., 2002) did not affect the development of EAE. A transgenic mouse that produces C5a exclusively in the brain was also studied, but expression of C5a was found to have no effect on the progression of EAE (Reiman et al., 2005). Together, these data suggest that neither C5a nor C5aR are essential for EAE development and that neither contribute significantly to disease severity. In separate studies, the role of C5 was examined using a C5 knockout and no difference in EAE development of clinical signs was observed in these mice when compared to controls (Barnum and Szalai, 2006).

While these studies are consistent with those involving C5aR−/− and C5a expressing transgenic mice, other studies focused on the role of the C5b-9 suggest that the other cleaved fragment of C5, C5b, does have a role in EAE. The C5b-9 complex is the end result of complement activation and its formation on a cell membrane can result in lysis of the target cell. Colocalization of C5b-9 with oligodendrocytes in MS lesions has been observed and it has been suggested that the lytic complex contributes to demyelination in vivo. However, in vitro studies have shown that sublytic amounts of C5b-9 can induce cell cycle activation in oligodendrocytes and prevent them from dying by apoptosis (Rus et al., 2005a). In vivo, greater remyelination were observed in C5 sufficient mice, compared to C5 deficient mice (Niculescu et al., 2004). Also, fewer TUNEL + oligodendrocytes were observed in C5 sufficient mice than in C5 deficient mice, findings which are consistent with the in vitro studies demonstrating inhibition of oligodendrocyte apoptosis. This inhibition of apoptosis by sublytic complement was shown to involve activation of signaling pathways that phosphorylate BAD and reduce its association with Bcl-X$_L$ (reviewed in (Rus et al., 2005a)). These data suggest that C5b-9 could play a role in neuroprotection by preventing oligodendrocyte cell death (via the mitochondrial pathway of apoptosis), thereby promoting remyelination and allowing damaged cells to recover.

6 Complement in Spinal Cord Injury/Demyelination

Previous studies have demonstrated that inflammation plays a central role in contusion-induced spinal cord injury (Carlson et al., 1998; Streit et al., 1998; Popovich et al., 1997) and that inhibition of inflammation with methylprednisolone after SCI results in an improved function outcome for patients (Bracken, 2000). As part of the inflammatory process, complement activation has been implicated in spinal cord demyelination. This hypothesis was supported by studies which demonstrate that serum complement is required to induce demyelination in an immunological intraspinal infusion protocol involving infusion of myelin-specific IgG with serum complement (Dyer et al., 1998, 2005). Use of complement deficient serum in this model demonstrated that while activation of the classical complement cascade is sufficient for demyelination, activation of the alternative pathway is not necessary. Additionally, demyelination was unaffected when formation of MAC was eliminated by using C5 or C6 deficient serum. Since the terminal component of complement activation is not required for injury, it can be suggested that a large contribution to injury resulting from complement activation lies in induction of the inflammatory response. That is, anaphylatoxins (C3a, atleast) produced following complement activation may recruit and activate microglia, which can then secrete cytotoxic substances themselves and/or phagocytose opsonized myelin.

In an experimental rat model of SCI, complement activation was observed 1 day following SCI and then again 6 weeks after injury (Anderson et al., 2004). Immunohistochemistry showed reactivity for C1q, C4, Factor B, and C5b-9 in the gray and white matter of injured spinal cords at 1–42 days following injury. In a follow up study using the same experimental paradigm, increased staining for factor H and clusterin were observed at 1–42 days following injury (Anderson et al.,

2005). While the biological significance of the presence of these proteins has not been determined, factor H and clusterin are regulators of complement activity and therefore may be important targets for therapies aimed at regulation of complement activation following injury. The results from these immunohistochemical studies are supported by data from a gene expression profiling of contused rat spinal cords (Aimone et al., 2004). Microarray analysis of tissue samples from the injury epicenter and surrounding tissue show that C1q mRNA was induced by 24 h post-SCI and persisted through 35 days post-SCI. C3 mRNA was also increased at 7 days post-SCI. This observed up-regulation of complement RNA and evidence of complement activation following SCI strongly implicate a role for complement in spinal cord injury. However, it is not clear what the functional consequences of this complement up-regulation are. While complement activation can lead to damaging results, such as formation of the lytic MAC complex and induction of inflammatory phagocytes, there exists an unexamined possibility that induced synthesis of complement proteins in these SCI models may be occurring as a protective response to injury. As mentioned above, C1q has been shown to enhance uptake of apoptotic cells (Vandivier et al., 2002; Botto et al., 1998) and its upregulation may be important to help clear cells damaged by injury and prevent further inflammation from occurring. Additionally, recent reports have implicated that complement activation products, such as C3a, may be important in stimulating neuroregeneration (Rahpeymai et al., 2006). Thus, the increase in C3 cleavage via complement activation following SCI may aid in repair of damaged tissue by providing a source for C3a. In conclusion, while it is clear that complement activation can exacerbate spinal cord injury and play a role in demyelination, it still remains to be determined what other consequences, possibly beneficial, follow complement induction. Delineating the precise mechanisms involved in SCI induced pathology will help direct strategies for therapeutic intervention.

7   Therapeutic Strategies

In each case described here and reviewed by others (Mocco et al., 2006; Guo and Ward, 2005), there is evidence that complement activation has both proinflammatory and neuroprotective roles. In view of evidence indicating varying outcomes, complete and long term inhibition of complement may not be optimal. Drugs that target specific effector functions and/or that modulate complement activation may be more advantageous, with the goal of preserving and/or enhancing the potential protective/reparative effects of specific complement components.

Since different complement activation fragments may induce detrimental effects in different diseases, more precise analysis of the role of isolated components is required. Thus far, tissue-specific responses and the differentiation state of the cell have been used to explain the proinflammatory vs. anti-inflammatory effects of complement, but the molecular pathways involved (critical for identifying therapeutic targets) are only beginning to be defined. One approach to discerning these disease-specific components is to use the combination of ligand and receptor knockout mice (tissue specific ablation being optimal) and antagonists or inhibitors of the different complement activation fragments or enzyme complexes respectively, similar to that described above in animal models of EAE.

Developing inhibitors of C5a activation of myeloid cells and/or receptor antagonists of C5a has been targeted as a major mechanism for inhibiting acute C5a-induced inflammatory disorders as well as chronic peripheral disorders (systemic lupus erythematosus [SLE], rheumatoid arthritis, and adult respiratory disease syndrome) (Konteatis et al., 1994). If early complement activation products (C3b, C3a) have beneficial effects (such as enhanced clearance and/or priming for neurotrophin activity), C5a receptor antagonists that provide the inhibition of the complement system downstream of these events would be beneficial. Such inhibitors are beginning to be explored in CNS disorders. For example, a C5aR antagonist (AcF-[OPdChaWR] (AcPhe [Orn-Pro-D-cyclohexylalanine-Trp-Arg])), has been shown to reduce disease activity in several rat models [for example, (Proctor et al., 2004; Woodruff et al., 2003, 2004)], including a brain trauma model (Sewell et al., 2004), and most recently an in vivo model of Huntington's Disease (Woodruff et al., 2006). This C5a receptor antagonist has additional favorable characteristics in that this compound was effective in a model of inflammatory bowel disease when given orally to rats (Woodruff et al., 2004), and most recently its transdermal administration inhibited C5a-mediated responses in a rat model (Proctor et al., 2006), suggesting the possibility of an oral route or a "C5aR antagonist patch" for drug delivery if proven efficacious. Alternatively, a C5a mutant peptide constructed by Kohl and colleagues has also been shown to be effective in an animal model of ischemia/reperfusion and immune complex disease (Heller et al., 1999). While the blood–brain barrier may hinder access of this peptide to the brain, this peptide can be valuable in defining pathogenic mechanisms.

Recently, a novel transmembrane C5a binding protein (C5L2) was identified and originally proposed to be a "decoy" receptor that binds or "scavenges" C5a, but does not signal cell activation (Ohno et al., 2000; Okinaga et al., 2003) (Gao et al., 2005). Consistent with this discovery, a C5L2 knockout mouse has been generated and shown to have greatly increased inflammatory responses to immune complex-induced injury (Gerard et al., 2005). Inhibition of C5L2 expression has recently been shown to lead to iNOS upregulation (Gavrilyuk et al., 2005), presumably due to the increased availability of C5a to its signaling receptor, CD88. Thus, some of the effects of C5a may be modified by modulated C5L2 expression, as well as possible modulation of C5aR (CD88) expression, providing additional opportunities for fine-tuning of therapeutic targeting (Ohno et al., 2000; Gerard et al., 2005; Gavrilyuk et al., 2005). Thus, it will be essential to have knowledge of the expression of these receptors in healthy and diseased brain and of the factors that modulate their expression.

Two additional strategies involve targeting complement inhibitors specifically to areas of injury. One example of this approach is CR2-Crry, a chimeric molecule composed of the mouse C3 convertase inhibitor, Crry, fused to an extracellular portion of complement receptor 2, CR2. CR2 is normally a membrane receptor that binds the cleavage fragments of C3 that become deposited at sites of complement activation. In this soluble fusion protein, the CR2 localizes the inhibitor to the activated C3 complex that is to be inhibited, and the Crry portion of the molecule prevents amplification of the complement cascade at the site. This inhibitor has proven effective in reducing neutrophil infiltration and tissue pathology and improving locomotion in a mouse model of spinal cord injury (Qiao et al., 2006). Since Crry is the functional homolog of the human CR1, these data suggest comparable results may be feasible in humans. A second approach (also being tested in mice) involves using a retroviral vector to express in vivo a fusion protein consisting of Crry and a single chain monoclonal antibody to a tissue specific antigen (which can also be placed under the control of a tissue specific promoter) (Spitzer et al., 2004). In a proof of principal in vivo experiment, Crry fused to a single chain anti mouse glycophorin A (a red cell marker) antibody was shown to improve survival of IgM coated red blood cells in a Crry knock out mouse, thereby demonstrating that this fusion protein can be synthesized and localized to the target cell in levels that functionally prevent complement-mediated cell lysis (Spitzer et al., 2006).

Finally, recent reports have implicated C3a and possibly C5a in stem cell regeneration and directed migration of stem cells (Reca et al., 2003; Mastellos et al., 2001; Markiewski et al., 2004). A more recent report provides evidence for positive effects of C3a on neurogenesis in adult and injured brain (Rahpeymai et al., 2006). While further investigations are needed, the possibility of using complement to influence regeneration in the injured brain is intriguing.

8 Summary

Complement activation, and the subsequent induction of inflammation, does occur in the CNS. In addition however, as summarized here, recent data support protective functions for complement which are distinct from the detrimental consequences of dysregulated complement activation. What remains to be determined is the relative contribution of complement to inflammation and subsequent loss of neuronal function, and additionally the specific biochemical pathways through which the different complement activation products play counterbalancing roles in injury and repair in each disease. Therapeutic interventions in murine models of Alzheimer's Disease targeting inflammation and oxidative stress have proven successful in reducing amyloid plaque burden/pathology as well as the proinflammatory cytokine IL-1β and other indicators of oxidative stress (Yao et al., 2004). If complement does contribute to the exacerbation of inflammation and neuronal loss as suggested in murine models of the disease, specific complement inhibitors that would block pathogenic activation but not the systemic protective functions of the complement system could be valuable therapeutics. While there are multiple potential targets for therapeutic intervention in neurodegenerative studies, different approaches will likely be beneficial at different stages of each disease, and thus a cocktail of therapeutic reagents, including targeted complement inhibitors rather than a single drug, may be more successful in preventing or slowing CNS degeneration.

Acknowledgements Authors thank Cheryl Cotman for illustrations in Figs. 8.1 and 8.3 and Dr. Maria Fonseca for Fig. 8.2. The work in the authors' lab was supported by NIH NS 35144 and AG 00538 and NIH training grant AG00096-21.

References

Afagh, A., Cummings, B.J., Cribbs, D.H., Cotman, C.W., and Tenner, A.J. (1996). Localization and cell association of C1q in Alzheimer's disease brain. *Exp. Neurol. 138*, 22–32.

Aimone, J.B., Leasure, J.L., Perreau, V.M., and Thallmair, M. (2004). Spatial and temporal gene expression profiling of the contused rat spinal cord. *Exp. Neurol. 189*, 204–221.

Akiyama, H., Barger, S., Barnum, S., Bradt, B., Bauer, J., Cole, G.M., Cooper, N.R., Eikelenboom, P., Emmerling, M., Fiebich, B.L., Finch, C.E., Frautschy, S., Griffin, W.S., Hampel, H., Hull, M., Landreth, G., Lue, L., Mrak, R., Mackenzie, I.R., McGeer, P.L., O'Banion, M.K., Pachter, J., Pasinetti, G., Plata-Salaman, C., Rogers, J., Rydel, R., Shen, Y., Streit, W., Strohmeyer, R., Tooyoma, I., Van Muiswinkel, F.L., Veerhuis, R., Walker, D., Webster, S., Wegrzyniak, B., Wenk, G., and Wyss-Coray, T. (2000). Inflammation and Alzheimer's disease. *Neurobiol. Aging 21*, 383–421.

Anderson, A.J., Robert, S., Huang, W., Young, W., and Cotman, C.W. (2004). Activation of complement pathways after contusion-induced spinal cord injury. *J. Neurotrauma 21*, 1831–1846.

Anderson, A.J., Najbauer, J., Huang, W., Young, W., and Robert, S. (2005). Upregulation of complement inhibitors in association with vulnerable cells following contusion-induced spinal cord injury. *J. Neurotrauma 22*, 382–397.

Arnett, H.A., Wang, Y., Matsushima, G.K., Suzuki, K., and Ting, J.P. (2003). Functional genomic analysis of remyelination reveals importance of inflammation in oligodendrocyte regeneration. *J. Neurosci. 23*, 9824–9832.

Barnum, S.R. (1995). Complement biosynthesis in the central nervous system. *Crit. Rev. Oral Biol. Med. 6*, 132–146.

Barnum, S.R. and Szalai, A.J. (2006). Complement and demyelinating disease: no MAC needed? *Brain Res. Brain Res. Rev. 52*, 58–68.

Baudouin, C., Peyman, G.A., Fredj-Reygrobellet, D., Gordon, W.C., Lapalus, P., Gastaud, P., and Bazan, N.G. (1992). Immunohistological study of subretinal membranes in age-related macular degeneration. *Jpn. J. Ophthalmol. 36*, 443–451.

Benard, M., Gonzalez, B.J., Schouft, M.T., Falluel-Morel, A., Vaudry, D., Chan, P., Vaudry, H., and Fontaine, M. (2004). Characterization of C3a and C5a receptors in rat cerebellar granule neurons during maturation. Neuroprotective effect of C5a against apoptotic cell death. *J. Biol. Chem. 279*, 43487–43496.

Bensa, J.C., Reboul, A., and Colomb, M.G. (1983). Biosynthesis in vitro of complement subcomponents C1q, C1s and C1 inhibitor by resting and stimulated human monocytes. *Biochem. J. 216*, 385–392.

Bergamaschini, L., Rossi, E., Storini, C., Pizzimenti, S., Distaso, M., Perego, C., De Luigi, A., Vergani, C., and Grazia, D.S. (2004). Peripheral treatment with enoxaparin, a low molecular weight heparin, reduces plaques and beta-amyloid accumulation in a mouse model of Alzheimer's disease. *J. Neurosci. 24*, 4181–4186.

Blom, A.M., Kask, L., and Dahlback, B. (2001). Structural requirements for the complement regulatory activities of C4BP. *J. Biol. Chem. 276*, 27136–27144.

Bobak, D.A., Gaither, T.G., Frank, M.M., and Tenner, A.J. (1987). Modulation of FcR function by complement: subcomponent C1q enhances the phagocytosis of IgG-opsonized targets by human monocytes and culture-derived macrophages. *J. Immunol. 138*, 1150–1156.

Bohlson, S.S., Fraser, D.A., and Tenner, A.J. (2007). Complement proteins C1q and MBL are pattern recognition molecules that signal immediate and long-term protective immune functions. *Mol. Immunol. 44*, 33–43.

Boos, L., Campbell, I.L., Ames, R., Wetsel, R.A., and Barnum, S.R. (2004). Deletion of the complement anaphylatoxin c3a receptor attenuates, whereas ectopic expression of c3a in the brain exacerbates, experimental autoimmune encephalomyelitis. *J. Immunol. 173*, 4708–4714.

Boos, L.A., Szalai, A.J., and Barnum, S.R. (2005). Murine complement C4 is not required for experimental autoimmune encephalomyelitis. *Glia 49*, 158–160.

Botto, M., Dell'agnola, C., Bygrave, A.E., Thompson, E.M., Cook, H.T., Petry, F., Loos, M., Pandolfi, P.P., and Walport, M.J. (1998). Homozygous C1q deficiency causes glomerulonephritis associated with multiple apoptotic bodies. *Nat. Genet. 19*, 56–59.

Bracken, M.B. (2000). The use of methylprednisolone. *J. Neurosurg. 93*, 340–341.

Bradt, B.M., Kolb, W.P., and Cooper, N.R. (1998). Complement-dependent proinflammatory properties of the Alzheimer's disease β-peptide. *J. Exp. Med. 188*, 431–438.

Carlson, S.L., Parrish, M.E., Springer, J.E., Doty, K., and Dossett, L. (1998). Acute inflammatory response in spinal cord following impact injury. *Exp. Neurol. 151*, 77–88.

Chen, S., Frederickson, R.C., and Brunden, K.R. (1996). Neuroglial-mediated immunoinflammatory responses in Alzheimer's disease: complement activation and therapeutic approaches. *Neurobiol. Aging 17*, 781–787.

Dandoy-Dron, F., Guillo, F., Benboudjema, L., Deslys, J.P., Lasmezas, C., Dormont, D., Tovey, M.G., and Dron, M. (1998). Gene expression in scrapie. Cloning of a new scrapie-responsive gene and the identification of increased levels of seven other mRNA transcripts. *J. Biol. Chem. 273*, 7691–7697.

Dhib-Jalbut, S., Arnold, D.L., Cleveland, D.W., Fisher, M., Friedlander, R.M., Mouradian, M.M., Przedborski, S., Trapp, B.D., Wyss-Coray, T., and Yong, V.W. (2006). Neurodegeneration and neuroprotection in multiple sclerosis and other neurodegenerative diseases. *J. Neuroimmunol. 176*, 198–215.

Dietzschold, B., Schwaeble, W., Schäfer, M.K.H., Hooper, D.C., Zehng, Y.M., Petry, F., Sheng, H., Fink, T., Loos, M., Koprowski, H., and Weihe, E. (1995). Expression of C1q, a subcomponent of the rat complement system, is dramatically enhanced in brains of rats with either Borna disease or experimental allergic encephalomyelitis. *J. Neurol. Sci. 130*, 11–16.

Dyer, J.K., Bourque, J.A., and Steeves, J.D. (1998). Regeneration of brainstem-spinal axons after lesion and immunological disruption of myelin in adult rat. *Exp. Neurol. 154*, 12–22.

Dyer, J.K., Bourque, J.A., and Steeves, J.D. (2005). The role of complement in immunological demyelination of the mammalian spinal cord. *Spinal Cord 43*, 417–425.

Eikelenboom, P., Hack, C.E., Rozemuller, J.M., and Stam, F.C. (1989). Complement activation in amyloid plaques in Alzheimer's dementia. *Virchows Arch. B Cell Pathol. 56*, 259–262.

El-Agnaf, O.M., Nagala, S., Patel, B.P., and Austen, B.M. (2001). Non-fibrillar oligomeric species of the amyloid ABri peptide, implicated in familial British dementia, are more potent at inducing apoptotic cell death than protofibrils or mature fibrils. *J. Mol. Biol. 310*, 157–168.

Ember, J.A., Jagels, M.A., and Hugli, T. (1998). Characterization of complement anaphylatoxins and biological responses. In *The Human Complement System in Health and Disease*, J.E. Volanakis and M.M. Frank, eds. (New York, NY: Marcel Dekker), pp. 241–284.

Fonseca, M.I., Zhou, J., Botto, M., and Tenner, A.J. (2004). Absence of C1q leads to less neuropathology in transgenic mouse models of Alzheimer's disease. *J. Neurosci. 24*, 6457–6465.

Fraser, D.A., Bohlson, S.S., Jasinskiene, N., Rawal, N., Palmarini, G., Ruiz, S., Rochford, R., and Tenner, A.J. (2006). C1q and MBL, components of the innate immune system, influence monocyte cytokine expression. *J. Leukoc. Biol. 80*, 107–116.

Fraser, D.A., Arora, M., Bohlson, S.S., Lozano, E., and Tenner, A.J. (2007). Generation of inhibitory NFkappa B complexes and pCREB correlates with the anti-inflammatory activity of complement protein C1q in human monocytes. *J. Biol. Chem. 282*, 7360–7367.

Gao, H., Neff, T.A., Guo, R.F., Speyer, C.L., Sarma, J.V., Tomlins, S., Man, Y., Riedemann, N.C., Hoesel, L.M., Younkin, E., Zetoune, F.S., and Ward, P.A. (2005). Evidence for a functional role of the second C5a receptor C5L2. *FASEB J. 19*, 1003–1005.

Gasque, P. (2004). Complement: a unique innate immune sensor for danger signals. *Mol. Immunol.* *41*, 1089–1098.

Gasque, P., Dean, Y.D., McGreal, E.P., VanBeek, J., and Morgan, B.P. (2000). Complement components of the innate immune system in health and disease in the CNS. *Immunopharmacology* *49*, 171–186.

Gavrilyuk, V., Kalinin, S., Hilbush, B.S., Middlecamp, A., McGuire, S., Pelligrino, D., Weinberg, G., and Feinstein, D.L. (2005). Identification of complement 5a-like receptor (C5L2) from astrocytes: characterization of anti-inflammatory properties. *J. Neurochem. 92*, 1140–1149.

Gerard, N.P., Lu, B., Liu, P., Craig, S., Fujiwara, Y., Okinaga, S., and Gerard, C. (2005). An anti-inflammatory function for the complement anaphylatoxin C5a-binding protein, C5L2. *J. Biol. Chem. 280*, 39677–39680.

Gibson, G., Gunasekera, N., Lee, M., Lelyveld, V., El-Agnaf, O.M., Wright, A., and Austen, B. (2004). Oligomerization and neurotoxicity of the amyloid ADan peptide implicated in familial Danish dementia. *J. Neurochem. 88*, 281–290.

Golde, T.E. (2005). The Abeta hypothesis: leading us to rationally-designed therapeutic strategies for the treatment or prevention of Alzheimer disease. *Brain Pathol. 15*, 84–87.

Goldsmith, S.K., Wals, P., Rozovsky, I., Morgan, T.E., and Finch, C.E. (1997). Kainic acid and decorticating lesions stimulate the synthesis of C1q protein in adult rat brain. *J. Neurochem. 68*, 2046–2052.

Griffin, W.S. and Mrak, R.E. (2002). Interleukin-1 in the genesis and progression of and risk for development of neuronal degeneration in Alzheimer's disease. *J. Leukoc. Biol. 72*, 233–238.

Grimsley, C. and Ravichandran, K.S. (2003). Cues for apoptotic cell engulfment: eat-me, don't eat-me and come-get-me signals. *Trends Cell Biol. 13*, 648–656.

Guo, R.F. and Ward, P.A. (2005). Role of C5a in inflammatory responses. *Annu. Rev. Immunol. 23*, 821–852.

Hageman, G.S., Luthert, P.J., Victor Chong, N.H., Johnson, L.V., Anderson, D.H., and Mullins, R.F. (2001). An integrated hypothesis that considers drusen as biomarkers of immune-mediated processes at the RPE-Bruch's membrane interface in aging and age-related macular degeneration. *Prog. Retin. Eye Res. 20*, 705–732.

Hageman, G.S., Anderson, D.H., Johnson, L.V., Hancox, L.S., Taiber, A.J., Hardisty, L.I., Hageman, J.L., Stockman, H.A., Borchardt, J.D., Gehrs, K.M., Smith, R.J., Silvestri, G., Russell, S.R., Klaver, C.C., Barbazetto, I., Chang, S., Yannuzzi, L.A., Barile, G.R., Merriam, J.C., Smith, R.T., Olsh, A.K., Bergeron, J., Zernant, J., Merriam, J.E., Gold, B., Dean, M., and Allikmets, R. (2005). A common haplotype in the complement regulatory gene factor H (HF1/CFH) predisposes individuals to age-related macular degeneration. *Proc. Natl. Acad. Sci. U.S.A 102*, 7227–7232.

Haines, J.L., Hauser, M.A., Schmidt, S., Scott, W.K., Olson, L.M., Gallins, P., Spencer, K.L., Kwan, S.Y., Noureddine, M., Gilbert, J.R., Schnetz-Boutaud, N., Agarwal, A., Postel, E.A., and Pericak-Vance, M.A. (2005). Complement factor H variant increases the risk of age-related macular degeneration. *Science 308*, 419–420.

Hanayama, R., Tanaka, M., Miyasaka, K., Aozasa, K., Koike, M., Uchiyama, Y., and Nagata, S. (2004). Autoimmune disease and impaired uptake of apoptotic cells in MFG-E8-deficient mice. *Science 304*, 1147–1150.

Hawlisch, H., Belkaid, Y., Baelder, R., Hildeman, D., Gerard, C., and Kohl, J. (2005). C5a negatively regulates toll-like receptor 4-induced immune responses. *Immunity 22*, 415–426.

Heese, K., Hock, C., and Otten, U. (1998). Inflammatory signals induce neurotrophin expression in human microglial cells. *J. Neurochem. 70*, 699–707.

Heller, T., Hennecke, M., Baumann, U., Gessner, J.E., zu Vilsendorf, A.M., Baensch, M., Boulay, F., Kola, A., Klos, A., Bautsch, W., and Kohl, J. (1999). Selection of a C5a receptor antagonist from phage libraries attenuating the inflammatory response in immune complex disease and ischemia/reperfusion injury. *J. Immunol. 163*, 985–994.

Huang, J., Kim, L.J., Mealey, R., Marsh, Jr.H.C., Zhang, Y., Tenner, A.J., Connolly, E.S., Jr., and Pinsky, D.J. (1999). Neuronal protection in stroke by an sLe^x-glycosylated complement inhibitory protein. *Science 285*, 595–599.

Jauneau, A.C., Ischenko, A., Chatagner, A., Benard, M., Chan, P., Schouft, M.T., Patte, C., Vaudry, H., and Fontaine, M. (2006). Interleukin 1b and anaphylatoxins exert a synergistic effect on NGF expression by astrocytes. *J. Neuroinflammation. 3*, 8.

Jiang, H., Burdick, D., Glabe, C.G., Cotman, C.W., and Tenner, A.J. (1994). β-amyloid activates complement by binding to a specific region of the collagen-like domain of the C1q A chain. *J. Immunol. 152*, 5050–5059.

Johnson, S.A., Lampert-Etchells, M., Pasinetti, G.M., Rozovsky, I., and Finch, C. (1992). Complement mRNA in the mammalian brain: Responses to Alzheimer's disease and experimental brain lesioning. *Neurobiol. Aging 13*, 641–648.

Johnson, L.V., Leitner, W.P., Rivest, A.J., Staples, M.K., Radeke, M.J., and Anderson, D.H. (2002). The Alzheimer's A{beta}-peptide is deposited at sites of complement activation in pathologic deposits associated with aging and age-related macular degeneration. *Proc. Natl. Acad. Sci. U.S.A. 99*, 14682–14687.

Klein, R.J., Zeiss, C., Chew, E.Y., Tsai, J.Y., Sackler, R.S., Haynes, C., Henning, A.K., Sangiovanni, J.P., Mane, S.M., Mayne, S.T., Bracken, M.B., Ferris, F.L., Ott, J., Barnstable, C., and Hoh, J. (2005). Complement factor H polymorphism in age-related macular degeneration. *Science 308*, 385–389.

Koistinaho, M., Ort, M., Cimadevilla, J.M., Vondrous, R., Cordell, B., Koistinaho, J., Bures, J., and Higgins, L.S. (2001). Specific spatial learning deficits become severe with age in beta – amyloid precursor protein transgenic mice that harbor diffuse beta – amyloid deposits but do not form plaques. *Proc. Natl. Acad. Sci. U.S.A. 98*, 14675–14680.

Konteatis, Z.D., Siciliano, S.J., Van, R.G., Molineaux, C.J., Pandya, S., Fischer, P., Rosen, H., Mumford, R.A., and Springer, M.S. (1994). Development of C5a receptor antagonists. Differential loss of functional responses. *J. Immunol. 153*, 4200–4205.

Korb, L.C. and Ahearn, J.M. (1997). C1q binding directly and specifically to surface blebs of apoptotic human keratinocytes. *J. Immunol. 158*, 4525–4528.

Kulkarni, A.P., Kellaway, L.A., Lahiri, D.K., and Kotwal, G.J. (2004). Neuroprotection from complement-mediated inflammatory damage. *Ann. N.Y. Acad. Sci. 1035*, 147–164.

Lampert-Etchells, M., Pasinetti, G.M., Finch, C.E., and Johnson, S.A. (1993). Regional localization of cells containing complement C1q and C4 mRNAs in the frontal cortex during Alzheimer's disease. *Neurodegeneration 2*, 111–121.

Lee, C.K., Weindruch, R., and Prolla, T.A. (2000). Gene-expression profile of the ageing brain in mice. *Nat. Genet. 25*, 294–297.

Maller, J., George, S., Purcell, S., Fagerness, J., Altshuler, D., Daly, M.J., and Seddon, J.M. (2006). Common variation in three genes, including a noncoding variant in CFH, strongly influences risk of age-related macular degeneration. *Nat. Genet. 38*, 1055–1059.

Markiewski, M.M., Mastellos, D., Tudoran, R., DeAngelis, R.A., Strey, C.W., Franchini, S., Wetsel, R.A., Erdei, A., and Lambris, J.D. (2004). C3a and C3b activation products of the third component of complement (C3) are critical for normal liver recovery after toxic injury. *J. Immunol. 173*, 747–754.

Mastellos, D., Papadimitriou, J.C., Franchini, S., Tsonis, P.A., and Lambris, J.D. (2001). A novel role of complement: mice deficient in the fifth component of complement (C5) exhibit impaired liver regeneration. *J. Immunol. 166*, 2479–2486.

Matsuoka, Y., Picciano, M., Malester, B., LaFrancois, J., Zehr, C., Daeschner, J.M., Olschowka, J.A., Fonseca, M.I., O'Banion, M.K., Tenner, A.J., Lemere, C.A., and Duff, K. (2001). Inflammatory responses to amyloidosis in a transgenic mouse model of Alzheimer's disease. *Am. J. Pathol. 158*, 1345–1354.

Mocco, J., Sughrue, M.E., Ducruet, A.F., Komotar, R.J., Sosunov, S.A., and Connolly, E.S., Jr. (2006). The complement system: a potential target for stroke therapy. *Adv. Exp. Med. Biol. 586*, 189–201.

Möller, T., Nolte, C., Burger, R., Verkhratsky, A., and Kettenmann, H. (1997). Mechanisms of C5a and C3a complement fragment-induced $[Ca^{2+}]_i$ signaling in mouse microglia. *J. Neurosci. 17*, 615–624.

Morgan, B.P. and Gasque, P. (1997). Extrahepatic complement biosynthesis: where, when and why? *Clin. Exp. Immunol. 107*, 1–7.

Morgan, B.P., Griffiths, M., Khanom, H., Taylor, S.M., and Neal, J.W. (2004). Blockade of the C5a receptor fails to protect against experimental autoimmune encephalomyelitis in rats. *Clin. Exp. Immunol. 138*, 430–438.

Mukherjee, P. and Pasinetti, G.M. (2001). Complement anaphylatoxin C5a neuroprotects through mitogen-activated protein kinase-dependent inhibition of caspase 3. *J. Neurochem. 77*, 43–49.

Nataf, S., Stahel, P.F., Davoust, N., and Barnum, S.R. (1999). Complement anaphylatoxin receptors on neurons: new tricks for old receptors? *Trends Neurosci 22*, 397–402.

Nataf, S., Carroll, S.L., Wetsel, R.A., Szalai, A.J., and Barnum, S.R. (2000). Attenuation of experimental autoimmune demyelination in complement- deficient mice. *J. Immunol. 165*, 5867–5873.

Navratil, J.S., Watkins, S.C., Wisnieski, J.J., and Ahearn, J.M. (2001). The globular heads of C1q specifically recognize surface blebs of apoptotic vascular endothelial cells. *J. Immunol. 166*, 3231–3239.

Niculescu, T., Weerth, S., Niculescu, F., Cudrici, C., Rus, V., Raine, C.S., Shin, M.L., and Rus, H. (2004). Effects of complement C5 on apoptosis in experimental autoimmune encephalomyelitis. *J. Immunol. 172*, 5702–5706.

Norsworthy, P.J., Fossati-Jimack, L., Cortes-Hernandez, J., Taylor, P.R., Bygrave, A.E., Thompson, R.D., Nourshargh, S., Walport, M.J., and Botto, M. (2004). Murine CD93 (C1qRp) contributes to the removal of apoptotic cells in vivo but is not required for C1q-mediated enhancement of phagocytosis. *J. Immunol. 172*, 3406–3414.

O'Barr, S.A., Caguioa, J., Gruol, D., Perkins, G., Ember, J.A., Hugli, T., and Cooper, N.R. (2001). Neuronal expression of a functional receptor for the C5a complement activation fragment. *J. Immunol. 166*, 4154–4162.

Oddo, S., Caccamo, A., Shepherd, J.D., Murphy, M.P., Golde, T.E., Kayed, R., Metherate, R., Mattson, M.P., Akbari, Y., and LaFerla, F.M. (2003). Triple-transgenic model of Alzheimer's disease with plaques and tangles: intracellular Abeta and synaptic dysfunction. *Neuron 39*, 409–421.

Ogden, C.A., deCathelineau, A., Hoffmann, P.R., Bratton, D., Ghebrehiwet, B., Fadok, V.A., and Henson, P.M. (2001). C1q and mannose binding lectin engagement of cell surface calreticulin and CD91 initiates macropinocytosis and uptake of apoptotic cells. *J. Exp. Med. 194*, 781–796.

Ohmi, K., Greenberg, D.S., Rajavel, K.S., Ryazantsev, S., Li, H.H., and Neufeld, E.F. (2003). Activated microglia in cortex of mouse models of mucopolysaccharidoses I and IIIB. *Proc. Natl. Acad. Sci. U.S.A. 100*, 1902–1907.

Ohno, M., Hirata, T., Enomoto, M., Araki, T., Ishimaru, H., and Takahashi, T.A. (2000). A putative chemoattractant receptor, C5L2, is expressed in granulocyte and immature dendritic cells, but not in mature dendritic cells. *Mol. Immunol. 37*, 407–412.

Okinaga, S., Slattery, D., Humbles, A., Zsengeller, Z., Morteau, O., Kinrade, M.B., Brodbeck, R.M., Krause, J.E., Choe, H.R., Gerard, N.P., and Gerard, C. (2003). C5L2, a nonsignaling C5A binding protein. *Biochemistry 42*, 9406–9415.

Osaka, H., Mukherjee, P., Aisen, P.S., and Pasinetti, G.M. (1999). Complement-derived anaphylatoxin C5a protects against glutamate- mediated neurotoxicity. *J. Cell. Biochem. 73*, 303–311.

Pangburn, M.K. (2000). Host recognition and target differentiation by factor H, a regulator of the alternative pathway of complement. *Immunopharmacology 49*, 149–157.

Pasinetti, G.M., Tocco, G., Sakhi, S., Musleh, W.D., DeSimoni, M.G., Mascarucci, P., Schreiber, S., Baudry, M., and Finch, C.E. (1996). Hereditary deficiencies in complement C5 are associated with intensified neurodegenerative responses that implicate new roles for the C-system in neuronal and astrocytic functions. *Neurobiol. Dis. 3*, 197–204.

Popovich, P.G., Wei, P., and Stokes, B.T. (1997). Cellular inflammatory response after spinal cord injury in Sprague–Dawley and Lewis rats. *J. Comp Neurol. 377*, 443–464.

Proctor, L.M., Arumugam, T.V., Shiels, I., Reid, R.C., Fairlie, D.P., and Taylor, S.M. (2004). Comparative anti-inflammatory activities of antagonists to C3a and C5a receptors in a rat model of intestinal ischaemia/reperfusion injury. *Br. J. Pharmacol. 142*, 756–764.

Proctor, L.M., Woodruff, T.M., Sharma, P., Shiels, I.A., and Taylor, S.M. (2006). Transdermal pharmacology of small molecule cyclic C5a antagonists. *Adv. Exp. Med. Biol. 586*, 329–345.

Qiao, F., Atkinson, C., Song, H., Pannu, R., Singh, I., and Tomlinson, S. (2006). Complement plays an important role in spinal cord injury and represents a therapeutic target for improving recovery following trauma. *Am. J. Pathol. 169*, 1039–1047.

Rahpeymai, Y., Hietala, M.A., Wilhelmsson, U., Fotheringham, A., Davies, I., Nilsson, A.K., Zwirner, J., Wetsel, R.A., Gerard, C., Pekny, M., and Pekna, M. (2006). Complement: a novel factor in basal and ischemia-induced neurogenesis. *EMBO J. 25*, 1364–1374.

Reboul, A., Prandini, M.H., Bensa, J.C., and Colomb, M.G. (1985). Characterization of C1q, C1s and C1 Inh synthesized by stimulated human monocytes in vitro. *FEBS Lett. 190*, 65–68.

Reca, R., Mastellos, D., Majka, M., Marquez, L., Ratajczak, J., Franchini, S., Glodek, A., Honczarenko, M., Spruce, L.A., Janowska-Wieczorek, A., Lambris, J.D., and Ratajczak, M.Z. (2003). Functional receptor for C3a anaphylatoxin is expressed by normal hematopoietic stem/progenitor cells, and C3a enhances their homing-related responses to SDF-1. *Blood 101*, 3784–3793.

Reiman, R., Gerard, C., Campbell, I.L., and Barnum, S.R. (2002). Disruption of the C5a receptor gene fails to protect against experimental allergic encephalomyelitis. *Eur. J. Immunol. 32*, 1157–1163.

Reiman, R., Torres, A.C., Martin, B.K., Ting, J.P., Campbell, I.L., and Barnum, S.R. (2005). Expression of C5a in the brain does not exacerbate experimental autoimmune encephalomyelitis. *Neurosci. Lett. 390*, 134–138.

Rogers, J., Cooper, N.R., Webster, S., Schultz, J., McGeer, P.L., Styren, S.D., Civin, W.H., Brachova, L., Bradt, B., Ward, P., and Lieberburg, I. (1992). Complement activation by beta-amyloid in Alzheimer disease. *Proc. Natl. Acad. Sci. U.S.A. 89*, 10016–10020.

Rostagno, A., Revesz, T., Lashley, T., Tomidokoro, Y., Magnotti, L., Braendgaard, H., Plant, G., Bojsen-Moller, M., Holton, J., Frangione, B., and Ghiso, J. (2002). Complement activation in chromosome 13 dementias. Similarities with Alzheimer's disease. *J. Biol. Chem. 277*, 49782–49790.

Rostagno, A., Tomidokoro, Y., Lashley, T., Ng, D., Plant, G., Holton, J., Frangione, B., Revesz, T., and Ghiso, J. (2005). Chromosome 13 dementias. *Cell Mol. Life Sci. 62*, 1814–1825.

Rus, H., Cudrici, C., and Niculescu, F. (2005a). C5b-9 complement complex in autoimmune demyelination and multiple sclerosis: dual role in neuroinflammation and neuroprotection. *Ann. Med. 37*, 97–104.

Rus, H., Cudrici, C., and Niculescu, F. (2005b). The role of the complement system in innate immunity. *Immunol. Res. 33*, 103–112.

Rus, H., Cudrici, C., David, S., and Niculescu, F. (2006). The complement system in central nervous system diseases. *Autoimmunity 39*, 395–402.

Sahu, A. and Lambris, J.D. (2001). Structure and biology of complement protein C3, a connecting link between innate and acquired immunity. *Immunol. Rev. 180*, 35–48.

Savill, J. and Fadok, V. (2000). Corpse clearance defines the meaning of cell death. *Nature 407*, 784–788.

Scott, R.S., McMahon, E.J., Pops, S.M., Reap, E.A., Caricchio, R., Cohen, P.L., Earp, H.S., and Matsushima, G.K. (2001). Phagocytosis and clearance of apoptotic cells is mediated by mer. *Nature 411*, 207–211.

Sewell, D.L., Nacewicz, B., Liu, F., Macvilay, S., Erdei, A., Lambris, J.D., Sandor, M., and Fabry, Z. (2004). Complement C3 and C5 play critical roles in traumatic brain cryoinjury: blocking effects on neutrophil extravasation by C5a receptor antagonist. *J. Neuroimmunol. 155*, 55–63.

Shen, Y., Li, R., McGeer, E.G., and McGeer, P.L. (1997). Neuronal expression of mRNAs for complement proteins of the classical pathway in Alzheimer brain. *Brain Res. 769*, 391–395.

Singhrao, S.K., Neal, J.W., Morgan, B.P., and Gasque, P. (1999). Increased complement biosynthesis by microglia and complement activation on neurons in Huntington's disease. *Exp. Neurol. 159*, 362–376.

Song, W.C. (2004). Membrane complement regulatory proteins in autoimmune and inflammatory tissue injury. *Curr. Dir. Autoimmun. 7*, 181–199.

Spitzer, D., Unsinger, J., Bessler, M., and Atkinson, J.P. (2004). ScFv-mediated in vivo targeting of DAF to erythrocytes inhibits lysis by complement. *Mol. Immunol. 40*, 911–919.

Spitzer, D., Wu, X., Ma, X., Xu, L., Ponder, K.P., and Atkinson, J.P. (2006). Cutting edge: treatment of complement regulatory protein deficiency by retroviral in vivo gene therapy. *J. Immunol. 177*, 4953–4956.

Streit, W.J. (2004). Microglia and Alzheimer's disease pathogenesis. *J. Neurosci. Res. 77*, 1–8.

Streit, W.J., Semple-Rowland, S.L., Hurley, S.D., Miller, R.C., Popovich, P.G., and Stokes, B.T. (1998). Cytokine mRNA profiles in contused spinal cord and axotomized facial nucleus suggest a beneficial role for inflammation and gliosis. *Exp. Neurol. 152*, 74–87.

Strohmeyer, R., Shen, Y., and Rogers, J. (2000). Detection of complement alternative pathway mRNA and proteins in Alzheimer's disease brain. *Mol. Brain Res. 81*, 7–18.

Stuart, L.M., Takahashi, K., Shi, L., Savill, J., and Ezekowitz, R.A. (2005). Mannose-binding lectin-deficient mice display defective apoptotic cell clearance but no autoimmune phenotype. *J. Immunol. 174*, 3220–3226.

Takahashi, R.H., Milner, T.A., Li, F., Nam, E.E., Edgar, M.A., Yamaguchi, H., Beal, M.F., Xu, H., Greengard, P., and Gouras, G.K. (2002). Intraneuronal Alzheimer abeta42 accumulates in multivesicular bodies and is associated with synaptic pathology. *Am. J. Pathol. 161*, 1869–1879.

Tan, J., Town, T., Paris, D., Mori, T., Suo, Z., Crawford, F., Mattson, M.P., Flavell, R.A., and Mullan, M. (1999). Microglial activation resulting from CD40-CD40L interaction after β-amyloid stimulation. *Science 286*, 2352–2355.

Tanzi, R.E. and Bertram, L. (2005). Twenty years of the Alzheimer's disease amyloid hypothesis: a genetic perspective. *Cell 120*, 545–555.

Taylor, P.R., Carugati, A., Fadok, V.A., Cook, H.T., Andrews, M., Carroll, M.C., Savill, J.S., Henson, P.M., Botto, M., and Walport, M.J. (2000). A hierarchical role for classical pathway complement proteins in the clearance of apoptotic cells in vivo. *J. Exp. Med. 192*, 359–366.

Tenner, A.J. and Fonseca, M.I. (2006). The double-edged flower: roles of complement protein C1q in neurodegenerative diseases. *Adv. Exp. Med. Biol. 586*, 153–176.

Tocco, G., Musleh, W., Sakhi, S., Schreiber, S.S., Baudry, M., and Pasinetti, G.M. (1997). Complement and glutamate neurotoxicity. Genotypic influences of C5 in a mouse model of hippocampal neurodegeneration. *Mol. Chem. Neuropathol. 31*, 289–300.

Todd, R.J., Volmar, C.H., Dwivedi, S., Town, T., Crescentini, R., Crawford, F., Tan, J., and Mullan, M. (2004). Behavioral effects of CD40-CD40L pathway disruption in aged PSAPP mice. *Brain Res. 1015*, 161–168.

Van Beek, J., Nicole, O., Ali, C., Ischenko, A., MacKenzie, E.T., Buisson, A., and Fontaine, M. (2001). Complement anaphylatoxin C3a is selectively protective against NMDA-induced neuronal cell death. *Neuroreport 12*, 289–293.

Van Beek, J., Elward, K., and Gasque, P. (2003). Activation of complement in the central nervous system: roles in neurodegeneration and neuroprotection. *Ann. N.Y. Acad. Sci. 992*, 56–71.

Vandivier, R.W., Ogden, C.A., Fadok, V.A., Hoffmann, P.R., Brown, K.K., Botto, M., Walport, M.J., Fisher, J.H., Henson, P.M., and Greene, K.E. (2002). Role of surfactant proteins A, D, and C1q in the clearance of apoptotic cells in vivo and in vitro: calreticulin and CD91 as a common collectin receptor complex. *J. Immunol. 169*, 3978–3986.

Vanguri, P., Koski, C.L., Silverman, B., and Shin, M.L. (1982). Complement activation by isolated myelin: activation of the classical pathway in the absence of myelin-specific antibodies. *Proc. Natl. Acad. Sci. U.S.A. 79*, 3290–3294.

Walker, D.G. and McGeer, P.L. (1992). Complement gene expression in human brain: comparison between normal and Alzheimer disease cases. Brain Res. *Mol. Brain Res. 14*, 109–116.

Walport, M.J. (2001). Complement. Second of two parts. *N. Engl. J. Med. 344*, 1140–1144.

Walsh, D.M., Klyubin, I., Fadeeva, J.V., Cullen, W.K., Anwyl, R., Wolfe, M.S., Rowan, M.J., and Selkoe, D.J. (2002). Naturally secreted oligomers of amyloid beta protein potently inhibit hippocampal long-term potentiation in vivo. *Nature 416*, 535–539.

Walsh, D.M., Townsend, M., Podlisny, M.B., Shankar, G.M., Fadeeva, J.V., Agnaf, O.E., Hartley, D.M., and Selkoe, D.J. (2005). Certain inhibitors of synthetic amyloid beta-peptide (Abeta) fibrillogenesis block oligomerization of natural Abeta and thereby rescue long-term potentiation. *J. Neurosci. 25*, 2455–2462.

Watson, M.D., Roher, A.E., Kim, K.S., Spiegel, K., and Emmerling, M.R. (1997). Complement interactions with amyloid-β1–42: a nidus for inflammation in AD brains. Amyloid: *Int. J. Exp. Clin. Invest. 4*, 147–156.

Webster, S., Lue, L.F., Brachova, L., Tenner, A.J., McGeer, P.L., Terai, K., Walker, D.G., Bradt, B., Cooper, N.R., and Rogers, J. (1997). Molecular and cellular characterization of the membrane attack complex, C5b-9, in Alzheimer's disease. *Neurobiol. Aging 18*, 415–421.

Woodruff, T.M., Arumugam, T.V., Shiels, I.A., Reid, R.C., Fairlie, D.P., and Taylor, S.M. (2003). A potent human C5a receptor antagonist protects against disease pathology in a rat model of inflammatory bowel disease. *J. Immunol. 171*, 5514–5520.

Woodruff, T.M., Arumugam, T.V., Shiels, I.A., Reid, R.C., Fairlie, D.P., and Taylor, S.M. (2004). Protective effects of a potent C5a receptor antagonist on experimental acute limb ischemia-reperfusion in rats. *J. Surg. Res. 116*, 81–90.

Woodruff, T.M., Crane, J.W., Proctor, L.M., Buller, K.M., Shek, A.B., de, V.K., Pollitt, S., Williams, H.M., Shiels, I.A., Monk, P.N., and Taylor, S.M. (2006). Therapeutic activity of C5a receptor antagonists in a rat model of neurodegeneration. *FASEB J. 20*, 1407–1417.

Wyss-Coray, T. and Mucke, L. (2002). Inflammation in neurodegenerative disease–a double-edged sword. *Neuron 35*, 419–432.

Wyss-Coray, T., Yan, F., Lin, A.H., Lambris, J.D., Alexander, J.J., Quigg, R.J., and Masliah, E. (2002). Prominent neurodegeneration and increased plaque formation in complement-inhibited Alzheimer's mice. *Proc. Natl. Acad. Sci. U.S.A. 99*, 10837–10842.

Yao, J., Harvath, L., Gilbert, D.L., and Colton, C.A. (1990). Chemotaxis by a CNS macrophage, the microglia. *J. Neurosci. Res. 27*, 36–42.

Yao, Y., Chinnici, C., Tang, H., Trojanowski, J.Q., Lee, V.M., and Pratico, D. (2004). Brain inflammation and oxidative stress in a transgenic mouse model of Alzheimer-like brain amyloidosis. *J. Neuroinflammation 1*, 21.

9
Chemokines and Autoimmune Demyelination

Michaela Fux, Jason Millward, and Trevor Owens

1   Introduction

Autoimmune attack on the nervous system is considered the basis for multiple sclerosis (MS) (Compston and Coles, 2002), and is also implicated in peripheral neuropathies such as Guillain–Barré Syndrome (GBS) (Kiefer et al., 2002). Myelin is probably the major target of autoimmune attack in both diseases, although non-myelin antigens are also recognized by infiltrating T cells and antibodies. Destruction of myelin (demyelination) is a central feature of MS and GBS, with accompanying inflammation viz. infiltrates of T cells and macrophages, neutrophils (depending on the subtype of disease), and B cells, again depending on disease subpathology. A working definition of inflammation is the presence of leukocytes where they don't belong, and it is instructive to consider how leukocytes in demyelinating diseases get to be 'where they don't belong'. This chapter will attempt to review the role of chemokines in this inflammatory process and how they contribute to the autoimmune pathology in demyelinating diseases. We will focus primarily on central nervous system (CNS) demyelinating disease.

2   Demyelinating Disease

2.1   Multiple Sclerosis

The most common demyelinating disease, which is also the most common neurological disease of young adults, is MS. MS presents as a heterogeneous spectrum of symptoms and progressions (Compston and Coles, 2002), though the majority of cases in Europe and North America fall into the relapsing-remitting (RR-MS) or secondary-progressive (SP-MS) categories. Primary progressive MS shares many pathological features with both, while differing in progression and not showing the same gender bias. The etiology of MS is unknown although many aspects of the disease suggest a link to infection - this remains unproven. There are also genetic associations, multiple genes conferring risk for disease or severity of progression.

T.E. Lane et al. (eds.), *Central Nervous System Diseases and Inflammation.*
© Springer 2008

One of the stronger associations is with MHC (Noseworthy et al., 2000; Compston and Coles, 2002; Keegan and Noseworthy, 2002; Dyment et al., 2004). Generalizations are helpful to understanding the role of chemokines in MS, and they include that immune infiltration of the CNS accompanies demyelinating pathology and axonal damage. However, heterogeneous patterns of neuropathology have been described in MS, based on analysis of autopsy and biopsy material. The two most commonly-represented patterns show immune pathology, with infiltrates of T cells and macrophages/microglia, differing principally from each other in the degree of B cell and antibody involvement (Lucchinetti et al., 2000). Less common pathologic patterns show hypoxia-like injury or oligodendrocyte injury, and may reflect endogenous rather than immune-mediated disease processes. The relative role of immune infiltration versus glial migration and activation is of obvious relevance to discussion of the action of specific chemokines in MS.

Remyelination is evidenced in MS by so-called 'shadow plaques' containing thinly remyelinated axons (Lassmann et al., 1997; Noseworthy et al., 2000; Compston and Coles, 2002). Remyelination in adult CNS has been studied in animal models of demyelination induced by toxins such as lysolecithin and cuprizone (Ludwin, 1981; Blakemore et al., 2000). The process is dependent on oligodendrocyte precursor cells (OPC) which can be identified partly by expression of growth factor receptors (PDGF-alpha receptor) or other markers (eg. the NG2 proteoglycan) (Zhao et al., 2005). Such precursor cells, of rodent and human origin, have been shown capable of remyelinating experimental lesions in mice and rats (Keirstead et al., 2005; Zhao et al., 2005). Transplants of such human adult neural-derived precursors were effective in remyelination of mouse brain (Windrem et al., 2004). Whether OPC's originate entirely from intra-CNS sources is debated. Despite occasional reports that OPC's may derive from bone marrow (Bonilla et al., 2002), and the observation that neurosphere-derived neuronal precursors can enter the CNS from blood (Pluchino et al., 2003), the consensus view holds that OPC's originate from the subventricular zone and the hippocampal dentate gyrus, like neuronal precursors (Gage et al., 1998; Marshall et al., 2003). Importantly, both origins would require directed migration and cellular 'stay-or-go' decision-making. Populations of OPC have been shown in MS brain, proximal to demyelinated plaques, and it has been suggested that the persistence of plaques at least partly reflects failure of OPC's to remyelinate demyelinated axons (Chang et al., 2002; Dubois-Dalcq et al., 2005). Whether and how this failure of OPC's to remyelinate can be ascribed to negative effects (or lack of positive effects) of inflammatory mediators such as cytokines and chemokines will be discussed in this chapter.

2.2 Animal Models for Demyelinating Disease

The two most commonly-used animal models for MS are Experimental Autoimmune Encephalomyelitis (EAE) and Theiler's Murine Encephalomyelitis Virus (TMEV).

Both target T cell responses to the CNS, either by deliberate immunization (EAE) or by viral infection (TMEV), and in both there is immune infiltration, demyelination and axonal damage. The immune response against the picornavirus in TMEV infection spreads to recognition of myelin antigens (Olson et al., 2004). The degree to which either demyelination or axonal damage occur varies depending on species and strain of animal, and specifics of immunization. EAE in mice is intrinsically less demyelinating than in other models, and Lewis rats do not show demyelination without additional intervention (*vide infra*) (Gold et al., 2006). But, as for MS, the generalization that both EAE and TMEV are immune-mediated myelin-targeted diseases is useful. The relative merits of both as models for MS have been discussed elsewhere (Oleszak et al., 2004; Friese et al., 2006; Gold et al., 2006), and will not be discussed further in this chapter.

There are a number of spontaneous models for demyelinating disease. These include T-cell receptor transgenics in which the majority of T cells are specific for myelin epitopes (Goverman et al., 1993; Owens et al., 2001; Bettelli et al., 2003). In some cases additional stimuli have been transgenically engineered into T-cell receptor transgenic models, such as autoantibodies (Litzenburger et al., 2000), or xeno-MHC and CD4 for presentation of antigen to a human TCR (Madsen et al., 1999). Transgenic mice in which the costimulator ligand CD86 has been overexpressed on microglia show a CD8-dominated demyelinating disease (Zehntner et al., 2003). Transgenic mice in which the myelin proteolipid protein (PLP) DM20 gene has been over-expressed in oligodendrocytes develop adult onset demyelination (Mastronardi et al., 1993). Because of its non-inflammatory nature this may represent a model for pattern IV MS pathology (Lucchinetti et al., 2000). There are also a number of transgenic and knockout models in which chemokines have been targeted. In large part, these represent experimental tests of postulates deriving from the findings that will be discussed herein. In some cases, they have yielded unexpected insights, as will be discussed.

2.3   Immune Cell Infiltration to the CNS

The immune cells that infiltrate the CNS in demyelinating disease predominantly include T lymphocytes and macrophages.

2.3.1   T Cells

Both CD4 + (MHC II-restricted) and CD8 + (MHC I-restricted) T cells are found associated to MS lesions (Traugott et al., 1983). There is evidence that CD8 + cells may be more actively involved in demyelinating pathology (Babbe et al., 2000), but that is not to say that CD4 + T cells have no function. The association of MS susceptibility to HLA-DR/DQ points to a CD4 + T cell involvement, as well as a multitude of findings showing myelin specificity of CD4 + T cells in MS (Compston and Coles, 2002). It is probably most realistic to allow a role for both types of T cells in MS. In EAE, by contrast, the weight of evidence has been towards CD4 + T cells as critical for disease. However, this may reflect the use of adjuvants in immunization, and there are some reports that CD8 + T cells can induce and mediate EAE (Friese and Fugger, 2005), as well as dominance of CD8 + T cells in CD86-induced spontaneous disease in transgenic mice (Brisebois et al., 2006). Target cells of potential immune attack in the CNS include oligodendrocytes and neurons, both of which more easily upregulate or induce MHC I than MHC II. Nevertheless, it is the case that the vast majority of analyses of EAE reflect operation of CD4 + T cells, and specifically directed experiments are needed for use of EAE in analysis of the role of CD8 + T cells. The CD8 + T cell is functionally associated with antiviral cytotoxicity so it is not unexpected that there are more reports of CD8 + T cell involvement in TM EV-induced demyelination than in EAE (Oleszak et al., 2004). This is somewhat dependent on strain of mouse and virus, and mice deficient in CD8 + T cells remain susceptible to TMEV demyelination (Begolka et al., 2001; Ransohoff et al., 2002).

2.3.2 Macrophages

Co-infiltration of macrophages with T cells not only occurs in EAE but is required for T cell entry and subsequent disease (Brosnan et al., 1981; Tran et al., 1998). Macrophages deriving from blood have been shown to play a critical role in directing CNS infiltration, via their depletion using toxin-loaded liposomes (Tran et al., 1998). Essentially similar observations have been reported for TMEV demyelination, where macrophages are required for persistent infection with virus (Rossi et al., 1997). Macrophages deriving from blood monocytes are identifiable in MS lesions. Transmigration of monocytes across in vitro models of blood–brain barrier endothelia has been extensively studied (Becher et al., 2000). The cytolytic and phagocytic capability of macrophages makes them candidate mediators of actual demyelination. Targeting by myelin-specific antibodies has been suggested to play a role in directing such cytotoxicity. Microglial cells, normally resident in CNS, share bone marrow origin and myeloid lineage markers with macrophages, and can exert many of the same functions. Microglial migration within the CNS has been described, towards lesions and sites of local injury response (Raivich and Banati, 2004).

2.3.3 Neutrophils

Infiltration of neutrophils to the CNS occurs in EAE (Zehntner et al., 2005), but has not been described in RR- or SP-MS. In EAE, neutrophils co-infiltrate with other leukocytes, with a perivascular distribution in white matter, under normal circumstances, whereas in the absence of interferon-gamma (IFNγ) they disseminate through grey and white matter more indiscriminately (Tran et al., 2000b; Zehntner et al., 2005). It has been suggested that the adjuvants used in EAE promote neutrophilia (Matthys et al., 2001), although neutrophils are also found in adoptively transferred EAE without adjuvants. Neutrophils are detected in cerebrospinal fluid (CSF) in severe subtypes of MS, notably in Devic's Disease (O'Riordan et al., 1996). There are no reports of neutrophils in TMEV, although astrocyte production of the neutrophil-attracting chemokine CXCL2 correlated to genetic susceptibility (Rubio et al., 2006).

2.3.4 B Cells

B cells have been described in MS lesions (Traugott et al., 1983), as has entry of B cells to the CNS (Knopf et al., 1998). Elevated CSF immunoglobulin titer is a diagnostic for inflammatory brain disease. In the case of MS characteristic oligoclonal banding has been used in confirmation of diagnosis. The specificity of one such band, from CSF of a patient with neuromyelitis optica, was recently identified as anti-aquaporin, pointing to a pathological role for antibody against a blood–brain barrier antigen (Lennon et al., 2005). The oligoclonality of these bands demonstrates that the elevated immunoglobulin titers in CSF are not a consequence of leakage from serum eg. through a disrupted blood–brain barrier, but represent compartment-specific synthesis. This calls attention to processes guiding the entry of B cells to the CNS, as will be discussed.

Anti-myelin antibodies can enhance demyelination in certain models of EAE, notably in Lewis rat (Linington et al., 1992). Nonetheless, the pathology associated with EAE can be induced in mice which lack B cells (Oliver et al., 2003), suggesting that whatever role antibody may play in EAE, it is not required. Entry of B cells to CNS in EAE has not been frequently described. Antibody responses to TMEV do not play a major role in disease. Anti-myelin antibodies in TMEV disease are more commonly associated to therapeutic remyelinating responses than to disease progression (Warrington et al., 2000).

The above brief description of essential features of major demyelinating diseases in the CNS identifies as critical steps:

1. Transmigration of macrophages/monocytes, neutrophils, B cells, CD4 + and CD8 + T lymphocytes from blood to CNS parenchyma
2. Control of intra-CNS distribution of infiltrated leukocytes (this also includes trans-glia limitans migration)
3. Migration of glial cells within the CNS (includes neuronal glial interactions)

3 Role of Chemokines

The role of chemokines in these stages of evolution of brain disease must now be considered.

Chemokine biology has been described elsewhere and the reader is referred to excellent reviews on the topic (Cyster, 1999; Rossi and Zlotnik, 2000). The CC and CXC families contain by far the most of the known chemokines and so predominate in discussion of chemokines in inflammatory and demyelinating disease. The CC and CXC chemokines act through G-protein coupled 7-transmembrane CCR and CXCR receptors, respectively, and chemokine involvement in disease and developmental processes is divined both through analysis of receptor expression and of chemokines themselves.

The by-now classic concept underlying the role of chemokines in cellular entry to tissues includes that CC or CXC chemokines are presented on endothelium to attract specific cell types to targeted locations. These chemokines act on cells in blood via specific receptors and this induces upregulation and activation-dependent conformational change of adhesion ligands, leading to arrest and migration at the site of chemokine expression and adhesion receptor upregulation. Response to chemokines is concentration-dependent and a gradient effect 'guides' cells towards the source of their production (reviewed in (von Andrian and Mackay, 2000)).

To understand how this operates in the context of entry of immune cells to the CNS, we must first define those chemokines that are candidate players in these immune processes. Then their induction and presentation to leukocytes at the blood–brain barrier will be discussed, particularly in experimental models for this, including transgenic and knockout animals.

3.1 Chemokines that Drive T Cell Entry

The migration of T cells in inflammatory responses is predominantly guided by receptors for CC chemokines. Mature activated T cells express CCR1, CCR2 (primarily a receptor for MCP-1 or CCL2), CCR3, CCR4 and CCR5 (reviewed in (Moser et al., 2004)). It is noteworthy that the chemokine receptors expressed by T cells that participate in tissue responses tend to be those that are 'promiscuous'in their ligand binding, allowing such T cells to respond to a broad range of potential chemoattractants. Many of the CC receptors we have listed bind CCL5/RANTES, though CCR3 and CCR5 are the receptors most commonly associated to CCL5 response. CCR4 binds CCL3/MIP-1α and CCL2/MCP-1, as well as CCL5/ RANTES, and is more associated to the Th2 subset (see below).

The precise pattern of receptor expression can be used to sub-classify T cells and has functional correlates. Th1 and Th2 CD4 + T cells can be distinguished by their reciprocal expression of CXCR3 (Th1) and CCR3 (Th2) (reviewed in (Baggiolini, 1998; Moser et al., 2004)). Th1 T cells, most strongly associated with inflammation, express CXCR3, which binds the ELR-CXC chemokines CXCL10/IP-10, CXCL9/I-TAC and CXCL11/MIG. Th1 cells also express CCR1 and CCR2. The chemokines CCL2, CCL5 and CXCL10 are of particular interest because of their regulation (eg. IP-10 = interferon-regulated protein-10), and because they are upregulated in the inflamed CNS (Engelhardt and Ransohoff, 2005). The Th2-associated CCR3 binds CCL11/eotaxin, CCL8/MCP-2, CCL7/MCP-3 and CCL13/MCP-4.

The chemokines listed so far fall into a general category of inflammatory chemokines. The alternate category is homeostatic, whose members act on resting leukocytes or at initiation of responses (Moser et al., 2004). These have a potentially important role to play in autoimmunity. Two homeostatic chemokines deserve consideration in demyelinating disease. Naive T cells in circulation respond to CXCL12, also known as SDF (stromal cell-derived factor), via CXCR4. The functionally-related chemokine CXCL13/BCA-1 or BLC is best-known for effects in the lymph node where it directs CXCR5 + B and T cell migration to enter follicles. However, CXCL12 and CXCL13 have also been detected in MS CSF and may act on T cells in CNS (Krumbholz et al., 2006).

3.2   Chemokines in B Cell Entry to CNS

There are essentially no B cells in the normal CNS and relatively few even in MS. The B cells that are found in the CNS in MS include plasmablasts (CD19 + CD138+) and plasma cells (CD19-, CD138+), as well as memory phenotype B cells (eg. CD27+) and cells with a centroblast-like phenotype (CD19 + , CD38high, CD77 +, Ki67+) (Meinl et al., 2006). The latter are associated with germinal center-like follicular aggregates (Serafini et al., 2004), that are especially prominent in progressive disease. Nevertheless, because oligoclonal Ig bands are a diagnostic for MS, it is obvious that B cell entry and activation occur early. The mechanism underlying B cell migration to the CNS has not been as thoroughly examined as that for T cells, but processes analogous to those guiding T cell entry operate. TNFα and TNF-family members (BAFF, APRIL) are identified as B cell factors which promote B cell survival in the CNS. It remains to be established whether they play an analogous role in induction of B cell-tropic chemokines, as TNFα does for T cells. The chemokines CXCL12 and CXCL13 are detected at elevated levels in EAE and MS CSF (Columba-Cabezas et al., 2003; Magliozzi et al., 2004; Krumbholz et al., 2005), and their receptors CXCR5 and CXCR4 respectively are expressed on plasmablasts and on B cells. These chemokine are best-known for directing B cell migration within lymph nodes and entry to follicles, and their expression in MS CNS is consistent with such activity. The inflammatory chemokine CXCL10 or IP-10, whose expression is strongly elevated in MS and EAE, also acts on B cells, via the CXCR3 receptor, and is implicated in entry of plasmablasts to the CNS (Meinl et al., 2006).

3.3   Chemokines that Drive Macrophage and Neutrophil Entry

Monocytes and macrophages express a wide range of chemokine receptors. The very naming of macrophage chemoattractant protein's (MCP's) and macrophage inflammatory proteins (MIP's) identifies a role for CCL2/MCP-1, CCL3/MIP-1α, CCL4/MIP-1β, CCL9/MIP-1γ, to list a few of these CC chemokines, in regulating macrophage migration. Macrophages also express receptors for CCL5/RANTES, and for the CXC chemokine CXCL10. Other CXC chemokines are associated with neutrophil trafficking. Predominant among these are CXCL1, CXCL2 (or KC) and CXCL8/IL-8, all acting through the CXCR2 receptor (reviewed in Rossi and Zlotnik, 2000).

3.4   Chemokines in MS

Studies on chemokine expression in MS patients have revealed that the composition of chemokines expressed depends on the subtypes of MS. By measuring chemokine expression in PBMCs via real-time PCR and applying a multivariate statistical analysis it was possible to distinguish healthy individuals versus MS patients as well as primary progressive versus RR-MS, respectively (Furlan et al., 2005). Interestingly chemokines (CCL2/MCP-1,CCL3/MIP-1α, CCL4/MIP-1β, CCL8/ MCP-2 and CCL7/MCP-3) and their receptors (CCR2, CCR3, and CCR5) driving T cell and macrophage infiltration, were shown to be upregulated in CSF and lesions of RR-MS patients during relapse (McManus et al., 1998a; Simpson et al., 1998, 2000a; Bartosik-Psujek and Stelmasiak, 2005).

At anatomical sites like vascular endothelium, perivascular space and parenchyma, chemokines are differently expressed, (reviewed in (Muller et al., 2004)) indicating their putative functions to guide leukocytes through different possible migration routes to CNS as e.g. (1) from blood to CSF across the choroid plexus, (2) from blood to subarachnoid space and (3) from blood to parenchymal perivascular space (Ransohoff et al., 2003). The expression of CCL21/SLC on choroid plexus epithelium and the detection of CCR7 positive T cells in CSF of MS patients speaks for a direct entry of T cells to CFS from systemic circulation (Kivisakk et al., 2004), a process which is dependent on P-selectin expression (Kivisakk et al., 2003). Furthermore Simpson et al. (1998) reported selective expression of CCL5/ RANTES in blood vessel endothelium, perivascular cells and glial limitants indicating that CCL5/RANTES is functional during transmigration from blood to parenchyma via the perivascular space. In vitro studies analyzing transmigration of ex vivo leukocytes of MS patients across human brain-derived endothelial cells (HBECs) revealed that endothelial cells produce CCL2/MCP-1 which drives leukocyte movement (Prat et al., 2002; Seguin et al., 2003). CCL2/MCP-1 was shown to be present in CSF of MS patients, however at reduced levels during relapse of the disease (Mahad et al., 2002; Bartosik-Psujek and Stelmasiak, 2005). This apparent contradiction between HBECs in vitro and chemokine expression studies in MS respectively was solved by the study of Mahad et al. showing that leukocytes consume CCL2/MCP-1 during migration across the blood-brain barrier (BBB) (Mahad et al., 2006). CCL2/MCP-1 is produced by parenchymal glial cells in active lesions of MS patients as well (Simpson et al., 1998). While expression of CCL2/MCP-1 by endothelial cells may be responsible for leukocyte transmigration across BBB, its expression within the parenchyma may result in augmenting local inflammation. CCL2/MCP-1 represents therefore an example that the same chemokine fulfils different functions depending at which anatomical site it is expressed.

It is known that chemokines not only contribute in triggering leukocyte movements. This counts among others for the homeostatic chemokines CXCL12/SDF-1 and CXCL13/BCA-1. Functions of CXCL12/SDF-1 have been studied using New Zealand Black/New Zealand White mice, expressing a self-reactive repertoire resulting in lupus associated nephritis. Administration of anti-CXCL12/SDF-1 antibodies repressed lymphocyte activation and autoantibody production in these mice (Balabanian et al., 2003). Furthermore CXCL12/SDF-1 was shown to partially inhibit spontaneous apoptosis of adenotonsillar memory T cells (Pajusto et al., 2004). Considering the homeostatic functions and the presence of CXCL12/SDF-1 in active MS lesions (Calderon et al., 2006; Krumbholz et al., 2006) it may be assumed that this chemokine is not only responsible to attract leukocytes to CNS but more importantly regulates the subsequent autoimmune reaction. Coricone et al. detected CD19$^+$ CD38^{high}CD77$^+$ Bcl2$^-$ B cells in CSF of MS patients together with CXCL12/SDF-1 and CXCL13/BCA-1 expression on the outer layer of capillaries. As these B cells are normally exclusively present in secondary lymph nodes and CXCL12/SDF-1 together with CXCL13/BCA-1 is known to be key mediator of lymphoneogenesis the authors concluded that B cell differentiation occurs within the CNS (Corcione et al., 2004). This is supported by the detection of lymphoid follicle-like structures containing B cells and CXCL13/BCA positive dendritic cells within cerebral meninges of secondary progressive MS patients (Serafini et al., 2004) and in mice with EAE (Magliozzi et al., 2004). Furthermore lymphoid chemokines like CCL19/MIP-3α and CCL21/SLC have been described in EAE (Alt et al., 2002; Columba-Cabezas et al., 2003) and in MS (Pashenkov et al., 2003). Coeval expression of their receptor CCR7 was demonstrated on T cells infiltrating the parenchyma during EAE (Alt et al., 2002). However CCR7 expression on T cells within the parenchyma could not be confirmed in MS patients (Kivisakk et al., 2004). Therefore, the mechanisms of the local induction and maintenance of autoimmunity through the formation of lymphoid follicle-like structures within the CNS needs to be further analyzed.

4 Regulation of Chemokines by Inflammatory Cytokines in CNS

Given the central roles described for IFNγ and TNFα in MS and other inflammatory diseases, it is not surprising that they should emerge as key controllers of chemokine production and response. Both have been shown to be critical for chemokine-directed immune responses in the CNS. Below we will review the regulation of chemokine responses in autoimmune demyelinating disease by IFNγ and TNFα.

4.1 TNFα-Regulated Chemokines in CNS

TNFα is produced within active demyelination lesions of MS patient by T cells, microglia and subtypes of peripheral macrophages (Bitsch et al., 1998; Bitsch et al., 2000). In addition a correlation between severity and progression of disease and elevated levels of TNFα in CSF of MS patients have been found (Sharief and Hentges, 1991; Drulovic et al., 1997). Mice overexpressing TNFα in astrocytes develop spontaneous inflammatory demyelinating disease underscoring the pathological role of TNFα (Probert et al., 1995; Akassoglou et al., 1997). Furthermore induction of EAE in mice transgenic for TNFα under the control of MBP promoter resulted in a prolonged demyelinating disease (Taupin et al., 1997). TNFα production in TMEV-induced demyelination has been reported as well (Inoue et al., 1996). TNFα mRNA expression during the peak of disease is less evaluated during TMEV infection than during EAE (Sato et al., 1997). Interestingly, by *in vivo* and *in vitro* studies it has been shown that susceptible (DBA/2J, SJL/J) as well as resistant (BALB/cByJ, C57BL/6J) strains produce TNFα upon TMEV infection (Sierra and Rubio, 1993; Sato et al., 1997; Chang et al., 2000). However, while Sato et al. reported no statistical significant difference of TNFα production between susceptible and resistance strains during early disease, Chang et al. observed such a difference (Sato et al., 1997; Chang et al., 2000). Nevertheless, during late disease continuous expression versus abrogation of TNFα production in susceptible and resistant strains respectively, has been observed in both studies (Sato et al., 1997; Chang et al., 2000).

Up to now it is well accepted that TNFα is involved in demyelination (reviewed in (Kollias et al., 1999)). It has been shown that TNFα bears immunomodulatory functions as well (reviewed in (Kollias et al., 1999; Owens et al., 2001)), e.g. while TNFα deficient B6 mice immunized with MOG showed prolonged demyelination 5 weeks after immunization, wild type mice were in remission by that time point (Kassiotis and Kollias, 2001). This may be a reason why anti-TNFα treatment in MS patients not only did not have significant beneficial effect but this therapy, while generally safe, and effective against other inflammatory autoimmune diseases, was associated with incidence of encephalitis (van Oosten et al., 1996). The multifunctional effects of TNFα have been attributed to the different biological activities of its receptors p55 and p75, respectively (reviewed in (MacEwan, 2002)). Furthermore the soluble and transmembrane forms of TNFα are proposed to contribute in different biological events (Grell et al., 1995). While the transmembrane form is thought to have beneficial effects, soluble TNFα is supposed to have pro-inflammatory activities in EAE (Ruuls et al., 2001).

Induction of EAE by immunization with MOG and MBP in mice deficient for TNFα-resulted rather in a delay than an abrogation of EAE (Korner et al., 1997a; Riminton et al., 1998; Kassiotis et al., 1999) indicating that although TNFα is not required for the development of EAE, it accelerates the onset of disease. This is supported by studies using a p55-TNFα-IgG fusion protein to block the effect of TNFα. They demonstrated inhibition or reduced severity of EAE in rats only if the fusion protein was administrated prior to onset of disease (Korner et al., 1995,

1997b). In addition a failure of leukocyte movement into CNS parenchyma was observed in these TNFα-deficient mice (Korner et al., 1997a; Riminton et al., 1998; Kassiotis et al., 1999). CCL1/TCA-3, CCL2/MCP-1, CCL3/MIP-1α, CCL4/MIP-1β, CCL5/RANTES and CXCL2-3/MIP-2 are produced within the spinal cord upon TNFα, TNFβ, LTβ, INFγ and TGFβ1 expression (Glabinski et al., 2003a) during EAE. One may therefore speculate that TNFα together with other pro-inflammatory cytokines accelerates the onset of EAE via the induction of chemokines within CNS parenchyma. While intracerebral injection of TNFα induced CCL5/RANTES, CCL3/MIP1α, CCL4/MIP-1β and IL-8/MIP-2 production within CNS, leukocyte infiltration into parenchyma could not be observed (Glabinski et al., 2003b). In addition in vitro studies showed CXCL10/IP-10, CCL2/MCP-1 and CXCL8/IL8 production by HBECs upon stimulation with supernatants derived from MBP specific TH1 cells. This effect was not observed upon stimulation with TNFα or INF γ alone (Biernacki et al., 2004). These studies demonstrated that TNFα alone is not responsible for the induction of the chemokine cocktail needed for cell entry to parenchyma.

The work of Matejuk et al. provided some further insights in the dependency of chemokine production on TNFα presence. They demonstrated missing and reduced expression of CCL3/MIP-1α CXCL2-3/MIP-2 and CCL5/RANTES, CXCL10/IP-10, respectively within spinal cord of TNFα deficient EAE mice (Matejuk et al., 2002). In addition Glabinski et al. (2004) published decreased expression of CCL3/MIP-1α and CCL4/MIP-1β in murine CNS after blockade of TNFα with soluble recombinant fusion construct of TNF receptor:Fc. These studies were confirmed by in vitro studies showing CCL2/MCP-1, CCL3/MIP-1α, CCL4/MIP-1β and CCL5/RANTES production by rat astrocytes and human microglia, respectively, upon TNFα stimulation (Guo et al., 1998; McManus et al., 1998b). Furthermore, abolition of CCL2/MCP-1 expression in spinal cord of mice deficient for the TNF receptor p55 has been observed (Gimenez et al, 2006). Taken together these studies point out the interaction between TNFα and the chemokines CCL2/MCP-1, CCL3/MIP-1α, CCL4/MIP-1β, CCL5/RANTES. Interestingly, infiltrating TH1 cells and macrophages express the receptor for CCL3/MIP-1α and CCL5/RANTES, CCR1, during EAE (Glabinski et al., 2002; Sunnemark et al., 2003) and CCR1 deficient mice are resistant to the onset of EAE (Rottman et al., 2000).

By analyzing chemokine expression in TNFα deficient mice upon EAE induction at different time points Murphy et al. delivered some more comprehension of the relationship between TNFα and chemokine production. They published that 2 days before onset of disease the production of CCL1/TCA-3, CCL2/MCP-1, CCL6/C10, CCL7/MCP-3, CCL17/TARC and CXCL10/IP-10 was dependent on TNFα, while the production of other chemokines was sustained or completely independent of TNFα (CCL5/RANTES, CXCL9/Mig and CCL3/MIP-1α, CCL4/MIP-1β, ˜CCL19/MIP-3β, CCL20/MIP-3α, CCL22/MDC, XCL1/Lymphotactin, respectively). Those chemokines which were dependent on TNFα reached almost normal levels at the day of onset in TNFα deficient mice compared to wild type

(Murphy et al., 2002), indicating that the absence of TNFα is compensated during development of EAE.

Nevertheless, the ability of TNFα to induce chemokine production within CNS as well in the periphery (Goebeler et al., 1997; Ngo et al., 1999; Janatpour et al., 2001) is well accepted. However TNFα and its receptors are expressed by infiltrating leukocytes as well as by CNS resident cells (Renno et al., 1995). Analysis of the source of TNFα during the development of EAE over time has shown that early in disease development, TNFα was mainly produced by microglia with coeval T cell infiltration and CCL5/RANTES, CCL2/MCP-1 and CXCL10/IP-10 production. With ongoing development of disease a switch occurred where macrophages were the main source of TNFα. The authors concluded that TNFα produced by microglia is the critical stimulus to induce glial chemokine production and T cell infiltration (Juedes et al., 2000). However by using radiation bone marrow chimeras Murphy et al. demonstrated that TNFα produced by infiltrating peripheral leukocytes is the essential source for rapid induction of chemokine production and EAE (Murphy et al., 2002). Furthermore the publication of Gimenez et al. revealed that irradiated TNF receptor p55 deficient mice did not develop EAE after passive transfer of wild type bone marrow whereas wild type recipient of TNF receptor p55 deficient bone marrow developed EAE comparable to non- irradiated mice. The authors concluded that the TNFα receptor essential to induce EAE is expressed on glial cells (Gimenez et al., 2006). One may therefore speculate that blood-derived TNFα stimulates TNF receptor positive glial cells to produce chemokines, which results in attracting peripheral immune cells to sites of brain inflammation. However, TNFα may also synergize with other cytokines to induce the chemokines essential to trigger rapid development of EAE. One such candidate is IFNγ, whose expression in CNS is itself regulated via TNFRI signaling (Wheeler et al., 2006).

4.2   Interferon-γ Regulated Induction of Chemokines in the CNS

Interferon-γ is produced by T cells and natural killer cells, and is generally regarded as being absent from the CNS except during inflammation, when immune cells enter the CNS in appreciable numbers. In this context, IFNγ can promote inflammatory events by inducing expression of a wide variety of genes, including other cytokines, adhesion molecules and co-stimulatory molecules, as well as chemokines (Boehm et al., 1997). Binding of IFNγ to its receptor induces oligomerization and phosphorylation of JAK1 and JAK2, which then phosphorylate STAT-1α. Phosphorylated STAT-1α homodimers bind to DNA elements, the IFNγ-activated sites (GAS), upstream of target genes (Ransohoff and Tani, 1998; Huang et al., 2000). The genetic response to IFNγ signalling is rendered more complex by contributions of additional upstream regulators (interferon-stimulated response elements, ISRE) and intermediary factors (ie. interferon regulatory factor-1). These elements are directly implicated in the transcription of multiple chemokines, including CXCL10 (interferon-inducible protein of 10 kDa) (Ohmori and Hamilton, 1993), CXCL9 (monokine induced by IFNγ) (Ohmori et al., 1997) and CXCL11 (interferon-inducible T cell α chemokine) (Cole et al., 1998). There is evidence that IFNγ can also enhance expression of certain chemokine receptors, for example CXCR4 and CCR5 (Croitoru-Lamoury et al., 2003), and CXCR3 (Nakajima et al., 2002). IFNγ can act with other inflammatory signals to further drive induction of chemokine expression, as shown by the example of synergy with TNFα to induce CXCL10 expression via p48/STAT-1α complexes and NFκB (Majumder et al., 1998). This emphasizes that during CNS inflammation IFNγ does not act in isolation, but rather in concert with a multitude of other inflammatory signals.

4.2.1 IFNγ in MS

IFNγ in the CNS has generally been regarded as playing a pro-pathogenic role in CNS inflammation. IFNγ is detectable in MS lesions (Traugott and Lebon, 1988), and in a clinical trial with MS patients, treatment with IFNγ was associated with an increased relapse rate (Panitch et al., 1987). The IFNγ-inducible chemokines CXCL9 and CXCL10 and their receptor CXCR3 (Simpson et al., 2000b) are expressed in MS lesions. IFNγ is also present in the CNS during TMEV (Rubio and Torres, 1991; Kohanawa et al., 1993), along with several IFNγ-inducible chemokines (Hoffman et al., 1999; Murray et al., 2000; Theil et al., 2000; Ransohoff et al., 2002). In vitro, CXCL10 and CCL5 were induced in primary rat astrocytes infected with TMEV, and this result could be replicated by applying IFNγ to uninfected cultures (Palma and Kim, 2001).

4.2.2 IFNγ in EAE

IFNγ production in the CNS correlates with the severity of EAE, appearing prior to onset of clinical symptoms, and declining during remission (Renno et al., 1995; Juedes et al., 2000). Numerous reports have described expression of a multitude of chemokines during EAE (Ransohoff et al., 1993; Tani and Ransohoff, 1994; Godiska et al., 1995; Eng et al., 1996). Given the complex milieu of immune signals present during CNS inflammation, it is difficult to understand the contribution of any single component to the inflammatory process. Several studies have attempted to address the question of whether a given chemokine or receptor is redundant or essential for the inflammatory process using antibody blockade and knockout animals. These studies have sometimes led to contradictory and confusing results. For example, antibody blockade of CCL2 was shown to inhibit EAE, but CCL2 knockout mice were just as susceptible to disease as wild types (Karpus et al., 1995; Tran et al., 2000a). Nevertheless, despite the wealth of studies associating IFNγ and IFNγ-induced chemokines with inflammatory processes, there is evidence that IFNγ may play a regulatory, disease-limiting role in CNS inflammation. Greater numbers of IFNγ-secreting T cells are present in the CNS of TNFα-receptor-I knockout mice with EAE, compared to wild types, despite the fact that disease in the knockout is more mild, with delayed onset (Wheeler et al., 2006). Antibody depletion of IFNγ exacerbated EAE (Billiau et al., 1988; Duong et al., 1992). Antibody depletion of IFNγ also rendered resistant C57Bl/6 mice susceptible to TMEV infection and exacerbated disease in SJL mice (Pullen et al., 1994; Rodriguez et al., 1995). TMEV-resistant 129Sv mice lacking the IFNγ receptor developed severe demyelination and neurological signs (Fiette et al., 1995).

4.2.3   Studies of EAE in Mice Lacking IFNγ or its Receptor

Striking evidence of an immunoregulatory role for IFNγ has come from studies of EAE in mice lacking IFNγ or its receptor. IFNγ–/– (GKO) mice are more susceptible to EAE than WT mice, and develop a severe lethal disease, with disseminated immune cell infiltration dominated by neutrophils (Ferber et al., 1996; Krakowski and Owens, 1996; Tran et al., 2000b). The characteristic phenotype of EAE in the GKO mouse is associated with differences in chemokine expression. The T cell-attracting chemokines CXCL10, CCL5 and CXCL11 were expressed at low levels or were undetectable in CNS of GKO mice with EAE, despite being strongly induced in EAE in the wild-type (Glabinski et al., 1999; Tran et al., 2000b; Hamilton et al., 2002). Conversely, neutrophil-attracting CXCL2 and CCL1 were substantially upregulated in CNS of these mice during disease (Tran et al., 2000b; Hamilton et al., 2002). A similarly severe disease phenotype, with a similar shift in chemokine expression is seen during EAE in mice lacking the IFNγ receptor (Willenborg et al., 1996; Tran et al., 2000b). These findings suggest that expression of certain chemokines during CNS inflammation is dependent on IFNγ signalling. The fact that this result persists despite differences in the background strain used in different studies argues for a central role for IFNγ in shaping chemokine responses.

Recent work from our lab has involved the administration of IFNγ to the CNS with the use of a viral vector administered intrathecally via the cisterna magna, to examine the influence of IFNγ on chemokine expression without the confounding effects of the multitude of other signals present during inflammation. Using this approach we have demonstrated that IFNγ can induce expression of selected chemokines (CXCL10 and CCL5) in meningeal and ependymal cells as well as astrocytes and microglia in the CNS parenchyma [Millward et al., 2007]. This IFNγ-induced expression of chemokines in the CNS was not associated with inflammation. When an infectious stimulus, pertussis toxin, was administered intraperitoneally, there was a significant increase in influx of immune cells to the CNS, as detected by flow cytometry, but this increase was not associated with histologically-detectable foci of inflammation. This argues for synergy between central IFNγ-directed chemokine expression and peripheral stimuli to promote entry of immune cells into the CNS.

4.3   Transgenic Models

Another elegant example of a peripheral infectious stimulus synergizing with central production of a chemokine to promote immune cell entry into the CNS is offered by transgenic mice expressing CCL2 in the CNS. Expression of CCL2 (a chemokine which can be induced by IFNγ (Zhou et al., 2001)) in oligodendrocytes under the control of the MBP promoter led to a dramatic accumulation of mononuclear cells, confined to the perivascular regions of the CNS, and not penetrating into the parenchyma, which could be augmented by intraperitoneal injection of LPS (Fuentes et al., 1995). Mice expressing CCL2 in astrocytes under the control of the GFAP promoter showed dramatic leukocyte infiltration, but only upon peripheral administration of pertussis toxin and complete Freund's adjuvant (CFA) (Huang et al., 2002). A similar theme is echoed in the example of transgenic expression of a another IFNγ-inducible chemokine, CXCL10, in astrocytes. GFAP promoter-driven expression of CXCL10 was associated with perivascular, meningeal and periventricular infiltration of predominantly neutrophils, but without overt pathology or clinical impairment (Boztug et al., 2002). When CFA and pertussis toxin were administered peripherally, there was a marked increase in the extent of immune infiltration, and a transient increase in the expression of cytokine and chemokine genes (Boztug et al., 2002). These results suggest that other factors may be required to fully activate the immune cells recruited to the CNS by CXCL10, and that the presence of CXCL10 in the CNS does not per se have a direct negative consequence on CNS cells.

Transgenic expression of another IFNγ-inducible chemokine in the CNS also led to neutrophil infiltration of the CNS. MBP promoter-driven expression of KC/CXCL1 in oligodendrocytes led to perivascular, meningeal and parenchymal infiltration of neutrophils (Tani et al., 1996). In this case mice demonstrated neurological signs, including postural instability and rigidity, ataxia, terminal wasting and ultimately decreased lifespan (Tani et al., 1996). Histopathology revealed microglial activation and disruption of the blood–brain barrier, but no evidence of axon disruption or dysmyelination (Tani et al., 1996). By contrast, transgenic expression of CCL21 in the CNS had profound consequences on myelin. Expression of CCL21 in oligodendrocytes led to severe neurological disease and death by 4 weeks of age (Chen et al., 2002). Histopathology revealed scattered leukocyte infiltration consisting mainly of neutrophils and eosinophils (but not lymphocytes), along with astrocyte and microglial activation. These mice showed hypomyelination and evidence of myelin breakdown producing spongiform myelinopathy (Chen et al., 2002). The severity of these effects and their relatively early onset reveal that chemokines can play a substantial role in facilitating normal developmental processes such as myelination, as well as their roles in directing leukocyte traffic during inflammation. The fact that transgenic expression of CCL19, a chemokine very similar to CCL21, did not produce any results underscores the exquisite specificity of the chemokine system (Chen et al., 2002).

Results from the transgenic studies emphasize that chemokines play a role in promoting the migration of immune cells into the CNS. In some cases this can lead to pathological consequences (CCL21, CXCL1/KC) but for other chemokines, additional immune stimuli such as pertussis toxin or CFA are required to produce pathological consequences. Similarly, virally-driven expression of IFNγ in the CNS can promote chemokine expression, and can synergize with pertussis toxin to drive immune cell entry into the CNS, but this also does not necessarily lead to inflammatory pathology. Data obtained from studies of EAE in the absence of IFNγ or its receptor indicate that IFNγ can have a disease-limiting role, likely due at least in part to its regulation of chemokine expression, and subsequent influence on leukocyte trafficking into the CNS.

5 Chemokines in Myelination

There is evidence that chemokines and their receptors are evolutionarily conserved, and that their chemotactic actions play a vital role in developmental processes (Huising et al., 2003; DeVries et al., 2006). In particular, chemokines are involved in helping to orchestrate the migration, proliferation and differentiation of oligodendrocyte precursor cells (OPCs) to myelinate axons in the developing CNS. CXCL1/KC enhanced proliferation of OPCs in culture, and spatial and temporal correlation of CXCL1/KC expression with developing myelin was observed in the ventral and dorsal postnatal spinal cord (Robinson and Franic, 2001). It is proposed that CXCL1/KC produced in a targeted manner by astrocytes and neurons acts synergistically with the principle OPC mitogen platelet-derived growth factor, which is present throughout (Robinson and Franic, 2001). CXCL1/KC was shown in vitro to inhibit spinal cord OPC migration and enhance cell-cell interaction, and this effect could be suppressed by antibody blockade of the receptor for CXCL1/KC, CXCR2 (Tsai et al., 2002). CXCL1/KC inhibited embryonic OPC migration when applied to rat E14 slice cultures, and wild-type OPCs microinjected into postnatal slice cultures arrested migration, while OPCs from CXCR2 knockouts (or wild-type cells in the presence of anti-CXCR2 blocking antibody) dispersed throughout the tissue (Tsai et al., 2002). Spinal cords of CXCR2 knockouts had reduced numbers of mature oligodendrocytes, which were abnormally distributed in the peripheral regions, accompanied by reduced and abnormal distribution of MBP-positive myelin (Tsai et al., 2002).

Other chemokines are also reported to influence OPCs. CXCL1/KC, CXCL8/IL-8, CXCL12/SDF-1α, and CCL5/RANTES (but not CCL2-MCP-1) increased proliferation of a murine OPC-like cell line Oli-neu, as well as mixed primary cortical cultures (Kadi et al., 2006). CXCL1/KC, CXCL8/IL-8 and CXCL12/SDF-1α also promoted MBP synthesis in these cultures in a dose dependent manner (Kadi et al., 2006). Nevertheless, some apparently contradictory reports have yet to be resolved. Maysami et al. (2006) reported that rat primary OPCs express CXCR4, and that CXCL12 blocked migratration, and while Dziembowska et al. (2005) showed CXCR4 expression on mouse OPCs, they reported that CXCL12 promoted migration.

Understanding the involvement of chemokines in regulating normal myelin development may lead to greater understanding of how to promote remyelination in the context of disease. For example, rapid demyelination can be induced experimentally by injection of lysophosphatidylcholine into mouse spinal cord. In response to this intervention, there was rapid transient induction of CCL2/MCP-1 and CCL3/MIP-1α, as well as GM-CSF and TNFα (Ousman and David, 2001). Administration of function-blocking antibodies to these molecules suppressed recruitment of T cells, neutrophils, monocytes, and phagocytic macrophages, and reduced the extent of demyelination after lysophosphatidylcholine injection (Ousman and David, 2001).

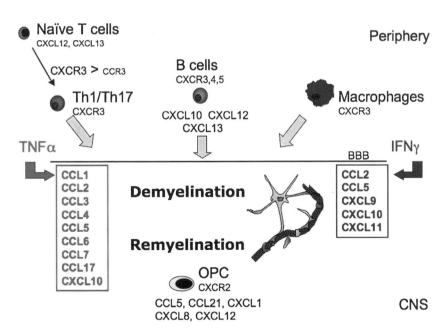

Fig. 9.1 Schematic to summarize the major points covered in this chapter. Demyelinating disease results primarily from entry of leukocytes to the CNS, which is directed by chemokines acting on receptors specifically expressed by leukocyte subsets such as T cells, B cells and macrophages. Neutrophils are not shown in the figure, for greater simplicity. The cytokines TNFα and IFNγ induce selected chemokines, these are indicated in boxes for each cytokine. Remyelination is portrayed as an intra-CNS event, the role of chemokines being to direct migration of oligodendrocyte precursor cells (OPC) (*See also color plates*).

6   Conclusions

One of the potential pitfalls of reviewing a biological field as complex as the world
of chemokines, is that whatever aspect one sets out to cover, there are multiple others
that also play an important role. The more one attempts to isolate chemokines from
the myriad other players in immune infiltration, neuroimmune demyelination and
neurobiological myelination and remyelination, the less completely one can treat
the topic. Furthermore, we are aware that we could not cite all of the relevant literature
for reasons of space and we apologize to those colleagues whose work we were not
able to include. The issues we hope to have shone some light on in this chapter
relate to the role played by chemokines in directing cellular migration and activation, not just of immune cells but also of cells resident in the CNS, and the role of
cytokines in directing the expression of those chemokines and their receptors. The
major points that we have covered are summarized in Fig. 9.1. Chemokines clearly
play a pivotal role in the CNS, represent an obvious target for therapeutic interventions and are likely to generate as much if not more interest in the coming decade
as in the last.

References

Akassoglou K, Probert L, Kontogeorgos G, Kollias G (1997) Astrocyte-specific but not neuron-
    specific transmembrane TNF triggers inflammation and degeneration in the central nervous
    system of transgenic mice. *J Immunol* 158:438–445.
Alt C, Laschinger M, Engelhardt B (2002) Functional expression of the lymphoid chemokines
    CCL19 (ELC) and CCL 21 (SLC) at the blood–brain barrier suggests their involvement in
    G-protein-dependent lymphocyte recruitment into the central nervous system during experi-
    mental autoimmune encephalomyelitis. *Eur J Immunol* 32:2133–2144.
Babbe H, Roers A, Waisman A, Lassmann H, Goebels N, Hohlfeld R, Friese M, Schroder R,
    Deckert M, Schmidt S, Ravid R, Rajewsky K (2000) Clonal expansions of CD8(+) T cells
    dominate the T cell infiltrate in active multiple sclerosis lesions as shown by micromanipula-
    tion and single cell polymerase chain reaction. *J Exp Med* 192:393–404.
Baggiolini M (1998) Chemokines and leukocyte traffic. *Nature* 392:565–568.
Balabanian K, Couderc J, Bouchet-Delbos L, Amara A, Berrebi D, Foussat A, Baleux F, Portier A,
    Durand-Gasselin I, Coffman RL, Galanaud P, Peuchmaur M, Emilie D (2003) Role of the
    chemokine stromal cell-derived factor 1 in autoantibody production and nephritis in murine
    lupus. *J Immunol* 170:3392–3400.
Bartosik-Psujek H, Stelmasiak Z (2005) The levels of chemokines CXCL8, CCL2 and CCL5 in
    multiple sclerosis patients are linked to the activity of the disease. *Eur J Neurol* 12:49–54.
Becher B, Prat A, Antel JP (2000) Brain-immune connection: immuno-regulatory properties of
    CNS-resident cells. *Glia* 29:293–304.
Begolka WS, Haynes LM, Olson JK, Padilla J, Neville KL, Dal Canto M, Palma J, Kim BS, Miller
    SD (2001) CD8-deficient SJL mice display enhanced susceptibility to Theiler's virus infection
    and increased demyelinating pathology. *J Neurovirol* 7:409–420.
Bettelli E, Pagany M, Weiner HL, Linington C, Sobel RA, Kuchroo VK (2003) Myelin oli-
    godendrocyte glycoprotein-specific T cell receptor transgenic mice develop spontaneous
    autoimmune optic neuritis. *J Exp Med* 197:1073–1081.

Biernacki K, Prat A, Blain M, Antel JP (2004) Regulation of cellular and molecular trafficking across human brain endothelial cells by Th1- and Th2-polarized lymphocytes. *J Neuropathol Exp Neurol* 63:223–232.

Billiau A, Heremans H, Vandekerckhove F, Dijkmans R, Sobis H, Meulepas E, Carton H (1988) Enhancement of experimental allergic encephalomyelitis in mice by antibodies against IFN-gamma. *J Immunol* 140:1506–1510.

Bitsch A, da Costa C, Bunkowski S, Weber F, Rieckmann P, Bruck W (1998) Identification of macrophage populations expressing tumor necrosis factor-alpha mRNA in acute multiple sclerosis. *Acta Neuropathol (Berl)* 95:373–377.

Bitsch A, Kuhlmann T, Da Costa C, Bunkowski S, Polak T, Bruck W (2000) Tumour necrosis factor alpha mRNA expression in early multiple sclerosis lesions: correlation with demyelinating activity and oligodendrocyte pathology. *Glia* 29:366–375.

Blakemore WF, Smith PM, Franklin RJ (2000) Remyelinating the demyelinated CNS. *Novartis Found Symp* 231:289–298; discussion 298–306.

Boehm U, Klamp T, Groot M, Howard JC (1997) Cellular responses to interferon-gamma. *Annu Rev Immunol* 15:749–795.

Bonilla S, Alarcon P, Villaverde R, Aparicio P, Silva A, Martinez S (2002) Haematopoietic progenitor cells from adult bone marrow differentiate into cells that express oligodendroglial antigens in the neonatal mouse brain. *Eur J Neurosci* 15:575–582.

Boztug K, Carson MJ, Pham-Mitchell N, Asensio VC, DeMartino J, Campbell IL (2002) Leukocyte infiltration, but not neurodegeneration, in the CNS of transgenic mice with astrocyte production of the CXC chemokine ligand 10. *J Immunol* 169:1505–1515.

Brisebois M, Zehntner SP, Estrada J, Owens T, Fournier S (2006) A pathogenic role for CD8 + T cells in a spontaneous model of demyelinating disease. *J Immunol* 177:2403–2411.

Brosnan CF, Bornstein MB, Bloom BR (1981) The effects of macrophage depletion on the clinical and pathologic expression of experimental allergic encephalomyelitis. *J Immunol* 126:614–620.

Calderon TM, Eugenin EA, Lopez L, Kumar SS, Hesselgesser J, Raine CS, Berman JW (2006) A role for CXCL12 (SDF-1alpha) in the pathogenesis of multiple sclerosis: regulation of CXCL12 expression in astrocytes by soluble myelin basic protein. *J Neuroimmunol* 177:27–39.

Chang JR, Zaczynska E, Katsetos CD, Platsoucas CD, Oleszak EL (2000) Differential expression of TGF-beta, IL-2, and other cytokines in the CNS of Theiler's murine encephalomyelitis virus-infected susceptible and resistant strains of mice. *Virology* 278:346–360.

Chang A, Tourtellotte WW, Rudick R, Trapp BD (2002) Premyelinating oligodendrocytes in chronic lesions of multiple sclerosis. *N Engl J Med* 346:165–173.

Chen SC, Leach MW, Chen Y, Cai XY, Sullivan L, Wiekowski M, Dovey-Hartman BJ, Zlotnik A, Lira SA (2002) Central nervous system inflammation and neurological disease in transgenic mice expressing the CC chemokine CCL21 in oligodendrocytes. *J Immunol* 168:1009–1017.

Cole KE, Strick CA, Paradis TJ, Ogborne KT, Loetscher M, Gladue RP, Lin W, Boyd JG, Moser B, Wood DE, Sahagan BG, Neote K (1998) Interferon-inducible T cell alpha chemoattractant (I-TAC): a novel non-ELR CXC chemokine with potent activity on activated T cells through selective high affinity binding to CXCR3. *J Exp Med* 187:2009–2021.

Columba-Cabezas S, Serafini B, Ambrosini E, Aloisi F (2003) Lymphoid chemokines CCL19 and CCL21 are expressed in the central nervous system during experimental autoimmune encephalomyelitis: implications for the maintenance of chronic neuroinflammation. *Brain Pathol* 13:38–51.

Compston A, Coles A (2002) Multiple sclerosis. *Lancet* 359:1221–1231.

Corcione A, Casazza S, Ferretti E, Giunti D, Zappia E, Pistorio A, Gambini C, Mancardi GL, Uccelli A, Pistoia V (2004) Recapitulation of B cell differentiation in the central nervous system of patients with multiple sclerosis. *Proc Natl Acad Sci U S A* 101:11064–11069.

Croitoru-Lamoury J, Guillemin GJ, Boussin FD, Mognetti B, Gigout LI, Cheret A, Vaslin B, Le Grand R, Brew BJ, Dormont D (2003) Expression of chemokines and their receptors in human and simian astrocytes: evidence for a central role of TNF alpha and IFN gamma in CXCR4 and CCR5 modulation. *Glia* 41:354–370.

Cyster JG (1999) Chemokines and cell migration in secondary lymphoid organs. *Science* 286:2098–2102.

DeVries ME, Kelvin AA, Xu L, Ran L, Robinson J, Kelvin DJ (2006) Defining the origins and evolution of the chemokine/chemokine receptor system. *J Immunol* 176:401–415.

Drulovic J, Mostarica-Stojkovic M, Levic Z, Stojsavljevic N, Pravica V, Mesaros S (1997) Interleukin-12 and tumor necrosis factor-alpha levels in cerebrospinal fluid of multiple sclerosis patients. *J Neurol Sci* 147:145–150.

Dubois-Dalcq M, Ffrench-Constant C, Franklin RJ (2005) Enhancing central nervous system remyelination in multiple sclerosis. *Neuron* 48:9–12.

Duong TT, St. Louis J, Gilbert JJ, Finkelman FD, Strejan GH (1992) Effect of anti-interferon-gamma and anti-interleukin-2 monoclonal antibody treatment on the development of actively and passively induced experimental allergic encephalomyelitis in the SJL/J mouse. *J Neuroimmunol* 36:105–115.

Dyment DA, Ebers GC, Sadovnick AD (2004) Genetics of multiple sclerosis. *Lancet Neurol* 3:104–110.

Dziembowska M, Tham TN, Lau P, Vitry S, Lazarini F, Dubois-Dalcq M (2005) A role for CXCR4 signaling in survival and migration of neural and oligodendrocyte precursors. *Glia* 50:258–269.

Eng LF, Ghirnikar RS, Lee YL (1996) Inflammation in EAE: role of chemokine/cytokine expression by resident and infiltrating cells. *Neurochem Res* 21:511–525.

Engelhardt B, Ransohoff RM (2005) The ins and outs of T-lymphocyte trafficking to the CNS: anatomical sites and molecular mechanisms. *Trends Immunol* 26:485–495.

Ferber IA, Brocke S, Taylor-Edwards C, Ridgway W, Dinisco C, Steinman L, Dalton D, Fathman CG (1996) Mice with a disrupted IFN-gamma gene are susceptible to the induction of experimental autoimmune encephalomyelitis (EAE). *J Immunol* 156:5–7.

Fiette L, Aubert C, Muller U, Huang S, Aguet M, Brahic M, Bureau JF (1995) Theiler's virus infection of 129Sv mice that lack the interferon alpha/beta or interferon gamma receptors. *J Exp Med* 181:2069–2076.

Friese MA, Fugger L (2005) Autoreactive CD8 + T cells in multiple sclerosis: a new target for therapy? *Brain* 128:1747–1763.

Friese MA, Montalban X, Willcox N, Bell JI, Martin R, Fugger L (2006) The value of animal models for drug development in multiple sclerosis. *Brain* 129:1940–1952.

Fuentes ME, Durham SK, Swerdel MR, Lewin AC, Barton DS, Megill JR, Bravo R, Lira SA (1995) Controlled recruitment of monocytes and macrophages to specific organs through transgenic expression of monocyte chemoattractant protein-1. *J Immunol* 155:5769–5776.

Furlan R, Rovaris M, Martinelli Boneschi F, Khademi M, Bergami A, Gironi M, Deleidi M, Agosta F, Franciotta D, Scarpini E, Uccelli A, Zaffaroni M, Kurne A, Comi G, Olsson T, Filippi M, Martino G (2005) Immunological patterns identifying disease course and evolution in multiple sclerosis patients. *J Neuroimmunol* 165:192–200.

Gage FH, Kempermann G, Palmer TD, Peterson DA, Ray J (1998) Multipotent progenitor cells in the adult dentate gyrus. *J Neurobiol* 36:249–266.

Gimenez MA, Sim J, Archambault AS, Klein RS, Russell JH (2006) A tumor necrosis factor receptor 1-dependent conversation between central nervous system-specific T cells and the central nervous system is required for inflammatory infiltration of the spinal cord. *Am J Pathol* 168:1200–1209.

Glabinski AR, Krakowski M, Han Y, Owens T, Ransohoff RM (1999) Chemokine expression in GKO mice (lacking interferon-gamma) with experimental autoimmune encephalomyelitis. J Neurovirol 5:95–101.

Glabinski AR, Bielecki B, O'Bryant S, Selmaj K, Ransohoff RM (2002) Experimental autoimmune encephalomyelitis: CC chemokine receptor expression by trafficking cells. *J Autoimmun* 19:175–181.

Glabinski AR, Bielecki B, Ransohoff RM (2003a) Chemokine upregulation follows cytokine expression in chronic relapsing experimental autoimmune encephalomyelitis. *Scand J Immunol* 58:81–88.

Glabinski AR, Bielecki B, Kolodziejski P, Han Y, Selmaj K, Ransohoff RM (2003b) TNF-alpha microinjection upregulates chemokines and chemokine receptors in the central nervous system without inducing leukocyte infiltration. *J Interferon Cytokine Res* 23:457–466.

Glabinski AR, Bielecki B, Kawczak JA, Tuohy VK, Selmaj K, Ransohoff RM (2004) Treatment with soluble tumor necrosis factor receptor (sTNFR):Fc/p80 fusion protein ameliorates relapsing-remitting experimental autoimmune encephalomyelitis and decreases chemokine expression. *Autoimmunity* 37:465–471.

Godiska R, Chantry D, Dietsch GN, Gray PW (1995) Chemokine expression in murine experimental allergic encephalomyelitis. *J Neuroimmunol* 58:167–176.

Goebeler M, Yoshimura T, Toksoy A, Ritter U, Brocker EB, Gillitzer R (1997) The chemokine repertoire of human dermal microvascular endothelial cells and its regulation by inflammatory cytokines. *J Invest Dermatol* 108:445–451.

Gold R, Linington C, Lassmann H (2006) Understanding pathogenesis and therapy of multiple sclerosis via animal models: 70 years of merits and culprits in experimental autoimmune encephalomyelitis research. *Brain* 129:1953–1971.

Goverman J, Woods A, Larson L, Weiner LP, Hood L, Zaller DM (1993) Transgenic mice that express a myelin basic protein-specific T cell receptor develop spontaneous autoimmunity. *Cell* 72:551–560.

Grell M, Douni E, Wajant H, Lohden M, Clauss M, Maxeiner B, Georgopoulos S, Lesslauer W, Kollias G, Pfizenmaier K, Scheurich P (1995) The transmembrane form of tumor necrosis factor is the prime activating ligand of the 80 kDa tumor necrosis factor receptor. *Cell* 83:793–802.

Guo H, Jin YX, Ishikawa M, Huang YM, van der Meide PH, Link H, Xiao BG (1998) Regulation of beta-chemokine mRNA expression in adult rat astrocytes by lipopolysaccharide, proinflammatory and immunoregulatory cytokines. *Scand J Immunol* 48:502–508.

Hamilton NH, Banyer JL, Hapel AJ, Mahalingam S, Ramsay AJ, Ramshaw IA, Thomson SA (2002) IFN-gamma regulates murine interferon-inducible T cell alpha chemokine (I-TAC) expression in dendritic cell lines and during experimental autoimmune encephalomyelitis (EAE). *Scand J Immunol* 55:171–177.

Hoffman LM, Fife BT, Begolka WS, Miller SD, Karpus WJ (1999) Central nervous system chemokine expression during Theiler's virus-induced demyelinating disease. *J Neurovirol* 5:635–642.

Huang D, Han Y, Rani MR, Glabinski A, Trebst C, Sorensen T, Tani M, Wang J, Chien P, O'Bryan S, Bielecki B, Zhou ZL, Majumder S, Ransohoff RM (2000) Chemokines and chemokine receptors in inflammation of the nervous system: manifold roles and exquisite regulation. *Immunol Rev* 177:52–67.

Huang D, Tani M, Wang J, Han Y, He TT, Weaver J, Charo IF, Tuohy VK, Rollins BJ, Ransohoff RM (2002) Pertussis toxin-induced reversible encephalopathy dependent on monocyte chemoattractant protein-1 overexpression in mice. *J Neurosci* 22:10633–10642.

Huising MO, Stet RJ, Kruiswijk CP, Savelkoul HF, Lidy Verburg-van Kemenade BM (2003) Molecular evolution of CXC chemokines: extant CXC chemokines originate from the CNS. *Trends Immunol* 24:307–313.

Inoue A, Koh CS, Yahikozawa H, Yanagisawa N, Yagita H, Ishihara Y, Kim BS (1996) The level of tumor necrosis factor-alpha producing cells in the spinal cord correlates with the degree of Theiler's murine encephalomyelitis virus-induced demyelinating disease. *Int Immunol* 8:1001–1008.

Janatpour MJ, Hudak S, Sathe M, Sedgwick JD, McEvoy LM (2001) Tumor necrosis factor-dependent segmental control of MIG expression by high endothelial venules in inflamed lymph nodes regulates monocyte recruitment. *J Exp Med* 194:1375–1384.

Juedes AE, Hjelmstrom P, Bergman CM, Neild AL, Ruddle NH (2000) Kinetics and cellular origin of cytokines in the central nervous system: insight into mechanisms of myelin oligodendrocyte glycoprotein-induced experimental autoimmune encephalomyelitis. *J Immunol* 164:419–426.

Kadi L, Selvaraju R, de Lys P, Proudfoot AE, Wells TN, Boschert U (2006) Differential effects of chemokines on oligodendrocyte precursor proliferation and myelin formation in vitro. *J Neuroimmunol* 174:133–146.

Karpus WJ, Lukacs NW, McRae BL, Strieter RM, Kunkel SL, Miller SD (1995) An important role for the chemokine macrophage inflammatory protein-1 alpha in the pathogenesis of the T cell-mediated autoimmune disease, experimental autoimmune encephalomyelitis. *J Immunol* 155:5003–5010.

Kassiotis G, Kollias G (2001) Uncoupling the proinflammatory from the immunosuppressive properties of tumor necrosis factor (TNF) at the p55 TNF receptor level. Implications for pathogenesis and therapy of autoimmune demyelination. *J Exp Med* 193:427–434.

Kassiotis G, Bauer J, Akassoglou K, Lassmann H, Kollias G, Probert L (1999) A tumor necrosis factor-induced model of human primary demyelinating diseases develops in immunodeficient mice. *Eur J Immunol* 29:912–917.

Keegan BM, Noseworthy JH (2002) Multiple sclerosis. *Annu Rev Med* 53:285–302.

Keirstead HS, Nistor G, Bernal G, Totoiu M, Cloutier F, Sharp K, Steward O (2005) Human embryonic stem cell-derived oligodendrocyte progenitor cell transplants remyelinate and restore locomotion after spinal cord injury. *J Neurosci* 25:4694–4705.

Kiefer R, Kieseier BC, Hartung HP (2002) Immune-mediated neuropathies. *Adv Neurol* 88:111–131.

Kivisakk P, Mahad DJ, Callahan MK, Trebst C, Tucky B, Wei T, Wu L, Baekkevold ES, Lassmann H, Staugaitis SM, Campbell JJ, Ransohoff RM (2003) Human cerebrospinal fluid central memory CD4 + T cells: evidence for trafficking through choroid plexus and meninges via P-selectin. *Proc Natl Acad Sci U S A* 100:8389–8394.

Kivisakk P, Mahad DJ, Callahan MK, Sikora K, Trebst C, Tucky B, Wujek J, Ravid R, Staugaitis SM, Lassmann H, Ransohoff RM (2004) Expression of CCR7 in multiple sclerosis: implications for CNS immunity. *Ann Neurol* 55:627–638.

Knopf PM, Harling-Berg CJ, Cserr HF, Basu D, Sirulnick EJ, Nolan SC, Park JT, Keir G, Thompson EJ, Hickey WF (1998) Antigen-dependent intrathecal antibody synthesis in the normal rat brain: tissue entry and local retention of antigen-specific B cells. *J Immunol* 161:692–701.

Kohanawa M, Nakane A, Minagawa T (1993) Endogenous gamma interferon produced in central nervous system by systemic infection with Theiler's virus in mice. *J Neuroimmunol* 48:205–211.

Kollias G, Douni E, Kassiotis G, Kontoyiannis D (1999) On the role of tumor necrosis factor and receptors in models of multiorgan failure, rheumatoid arthritis, multiple sclerosis and inflammatory bowel disease. *Immunol Rev* 169:175–194.

Korner H, Goodsall AL, Lemckert FA, Scallon BJ, Ghrayeb J, Ford AL, Sedgwick JD (1995) Unimpaired autoreactive T-cell traffic within the central nervous system during tumor necrosis factor receptor-mediated inhibition of experimental autoimmune encephalomyelitis. *Proc Natl Acad Sci U S A* 92:11066–11070.

Korner H, Riminton DS, Strickland DH, Lemckert FA, Pollard JD, Sedgwick JD (1997a) Critical points of tumor necrosis factor action in central nervous system autoimmune inflammation defined by gene targeting. *J Exp Med* 186:1585–1590.

Korner H, Lemckert FA, Chaudhri G, Etteldorf S, Sedgwick JD (1997b) Tumor necrosis factor blockade in actively induced experimental autoimmune encephalomyelitis prevents clinical disease despite activated T cell infiltration to the central nervous system. *Eur J Immunol* 27:1973–1981.

Krakowski M, Owens T (1996) Interferon-gamma confers resistance to experimental allergic encephalomyelitis. *Eur J Immunol* 26:1641–1646.

Krumbholz M, Theil D, Derfuss T, Rosenwald A, Schrader F, Monoranu CM, Kalled SL, Hess DM, Serafini B, Aloisi F, Weckerle H, Hohlfeld R, Meinl E (2005) BAFF is produced by astrocytes and up-regulated in multiple sclerosis lesions and primary central nervous system lymphoma. *J Exp Med* 201:195–200.

Krumbholz M, Theil D, Cepok S, Hemmer B, Kivisakk P, Ransohoff RM, Hofbauer M, Farina C, Derfuss T, Hartle C, Newcombe J, Hohlfeld R, Meinl E (2006) Chemokines in multiple sclerosis: CXCL12 and CXCL13 up-regulation is differentially linked to CNS immune cell recruitment. *Brain* 129:200–211.

Lassmann H, Bruck W, Lucchinetti C, Rodriguez M (1997) Remyelination in multiple sclerosis. *Mult Scler* 3:133–136.

Lennon VA, Kryzer TJ, Pittock SJ, Verkman AS, Hinson SR (2005) IgG marker of optic-spinal multiple sclerosis binds to the aquaporin-4 water channel. *J Exp Med* 202:473–477.

Linington C, Engelhardt B, Kapocs G, Lassman H (1992) Induction of persistently demyelinated lesions in the rat following the repeated adoptive transfer of encephalitogenic T cells and demyelinating antibody. *J Neuroimmunol* 40:219–224.

Litzenburger T, Bluthmann H, Morales P, Pham-Dinh D, Dautigny A, Wekerle H, Iglesias A (2000) Development of myelin oligodendrocyte glycoprotein autoreactive transgenic B lymphocytes: receptor editing In vivo after encounter of a self-antigen distinct from myelin oligodendrocyte glycoprotein. *J Immunol* 165:5360–5366.

Lucchinetti C, Bruck W, Parisi J, Scheithauer B, Rodriguez M, Lassmann H (2000) Heterogeneity of multiple sclerosis lesions: implications for the pathogenesis of demyelination. *Ann Neurol* 47:707–717.

Ludwin SK (1981) Pathology of demyelination and remyelination. *Adv Neurol* 31:123–168.

MacEwan DJ (2002) TNF receptor subtype signalling: differences and cellular consequences. *Cell Signal* 14:477–492.

Madsen LS, Andersson EC, Jansson L, Krogsgaard M, Andersen CB, Engberg J, Strominger JL, Svejgaard A, Hjorth JP, Holmdahl R, Wucherpfennig KW, Fugger L (1999) A humanized model for multiple sclerosis using HLA-DR2 and a human T-cell receptor. *Nat Genet* 23:343–347.

Magliozzi R, Columba-Cabezas S, Serafini B, Aloisi F (2004) Intracerebral expression of CXCL13 and BAFF is accompanied by formation of lymphoid follicle-like structures in the meninges of mice with relapsing experimental autoimmune encephalomyelitis. *J Neuroimmunol* 148:11–23.

Mahad D, Callahan MK, Williams KA, Ubogu EE, Kivisakk P, Tucky B, Kidd G, Kingsbury GA, Chang A, Fox RJ, Mack M, Sniderman MB, Ravid R, Staugaitis SM, Stins MF, Ransohoff RM (2006) Modulating CCR2 and CCL2 at the blood–brain barrier: relevance for multiple sclerosis pathogenesis. *Brain* 129:212–223.

Mahad DJ, Howell SJ, Woodroofe MN (2002) Expression of chemokines in the CSF and correlation with clinical disease activity in patients with multiple sclerosis. *J Neurol Neurosurg Psychiatry* 72:498–502.

Majumder S, Zhou LZ, Chaturvedi P, Babcock G, Aras S, Ransohoff RM (1998) p48/STAT-1alpha-containing complexes play a predominant role in induction of IFN-gamma-inducible protein, 10kDa (IP-10) by IFN-gamma alone or in synergy with TNF-alpha. *J Immunol* 161:4736–4744.

Marshall CA, Suzuki SO, Goldman JE (2003) Gliogenic and neurogenic progenitors of the subventricular zone: who are they, where did they come from, and where are they going? *Glia* 43:52–61.

Mastronardi FG, Ackerley CA, Arsenault L, Roots BI, Moscarello MA (1993) Demyelination in a transgenic mouse: a model for multiple sclerosis. *J Neurosci Res* 36:315–324.

Matejuk A, Dwyer J, Ito A, Bruender Z, Vandenbark AA, Offner H (2002) Effects of cytokine deficiency on chemokine expression in CNS of mice with EAE. *J Neurosci Res* 67:680–688.

Matthys P, Vermeire K, Billiau A (2001) Mac-1(+) myelopoiesis induced by CFA: a clue to the paradoxical effects of IFN-gamma in autoimmune disease models. *Trends Immunol* 22:367–371.

Maysami S, Nguyen D, Zobel F, Pitz C, Heine S, Hopfner M, Stangel M (2006) Modulation of rat oligodendrocyte precursor cells by the chemokine CXCL12. *Neuroreport* 17:1187–1190.

McManus C, Berman JW, Brett FM, Staunton H, Farrell M, Brosnan CF (1998a) MCP-1, MCP-2 and MCP-3 expression in multiple sclerosis lesions: an immunohistochemical and in situ hybridization study. *J Neuroimmunol* 86:20–29.

McManus CM, Brosnan CF, Berman JW (1998b) Cytokine induction of MIP-1 alpha and MIP-1 beta in human fetal microglia. *J Immunol* 160:1449–1455.

Meinl E, Krumbholz M, Hohlfeld R (2006) B lineage cells in the inflammatory central nervous system environment: migration, maintenance, local antibody production, and therapeutic modulation. *Ann Neurol* 59:880–892.

Millward JM, Caruso M, Campbell IL, Gauldie J, Owens T (2007) IFN-gamma-induced chemokines synergize with pertussis toxin to promote T cell entry to the central nervous system. J Immunol 178:8175–8182.

Moser B, Wolf M, Walz A, Loetscher P (2004) Chemokines: multiple levels of leukocyte migration control. *Trends Immunol* 25:75–84.

Muller DM, Pender MP, Greer JM (2004) Chemokines and chemokine receptors: potential therapeutic targets in multiple sclerosis. *Curr Drug Targets Inflamm Allergy* 3:279–290.

Murphy CA, Hoek RM, Wiekowski MT, Lira SA, Sedgwick JD (2002) Interactions between hemopoietically derived TNF and central nervous system-resident glial chemokines underlie initiation of autoimmune inflammation in the brain. *J Immunol* 169:7054–7062.

Murray PD, Krivacic K, Chernosky A, Wei T, Ransohoff RM, Rodriguez M (2000) Biphasic and regionally-restricted chemokine expression in the central nervous system in the Theiler's virus model of multiple sclerosis. *J Neurovirol* 6 Suppl 1:S44–S52.

Nakajima C, Mukai T, Yamaguchi N, Morimoto Y, Park WR, Iwasaki M, Gao P, Ono S, Fujiwara H, Hamaoka T (2002) Induction of the chemokine receptor CXCR3 on TCR-stimulated T cells: dependence on the release from persistent TCR-triggering and requirement for IFN-gamma stimulation. *Eur J Immunol* 32:1792–1801.

Ngo VN, Korner H, Gunn MD, Schmidt KN, Riminton DS, Cooper MD, Browning JL, Sedgwick JD, Cyster JG (1999) Lymphotoxin alpha/beta and tumor necrosis factor are required for stromal cell expression of homing chemokines in B and T cell areas of the spleen. *J Exp Med* 189:403–412.

Noseworthy JH, Lucchinetti C, Rodriguez M, Weinshenker BG (2000) Multiple sclerosis. *N Engl J Med* 343:938–952.

O'Riordan JI, Gallagher HL, Thompson AJ, Howard RS, Kingsley DP, Thompson EJ, McDonald WI, Miller DH (1996) Clinical, CSF, and MRI findings in Devic's neuromyelitis optica. *J Neurol Neurosurg Psychiatry* 60:382–387.

Ohmori Y, Hamilton TA (1993) Cooperative interaction between interferon (IFN) stimulus response element and kappa B sequence motifs controls IFN gamma- and lipopolysaccharide-stimulated transcription from the murine IP-10 promoter. *J Biol Chem* 268:6677–6688.

Ohmori Y, Schreiber RD, Hamilton TA (1997) Synergy between interferon-gamma and tumor necrosis factor-alpha in transcriptional activation is mediated by cooperation between signal transducer and activator of transcription 1 and nuclear factor kappaB. *J Biol Chem* 272:14899–14907.

Oleszak EL, Chang JR, Friedman H, Katsetos CD, Platsoucas CD (2004) Theiler's virus infection: a model for multiple sclerosis. *Clin Microbiol Rev* 17:174–207.

Oliver AR, Lyon GM, Ruddle NH (2003) Rat and human myelin oligodendrocyte glycoproteins induce experimental autoimmune encephalomyelitis by different mechanisms in C57BL/6 mice. *J Immunol* 171:462–468.

Olson JK, Ludovic Croxford J, Miller SD (2004) Innate and adaptive immune requirements for induction of autoimmune demyelinating disease by molecular mimicry. *Mol Immunol* 40:1103–1108.

Ousman SS, David S (2001) MIP-1alpha, MCP-1, GM-CSF, and TNF-alpha control the immune cell response that mediates rapid phagocytosis of myelin from the adult mouse spinal cord. *J Neurosci* 21:4649–4656.

Owens T, Wekerle H, Antel J (2001) Genetic models for CNS inflammation. *Nat Med* 7:161–166.

Pajusto M, Ihalainen N, Pelkonen J, Tarkkanen J, Mattila PS (2004) Human in vivo-activated CD45R0(+) CD4(+) T cells are susceptible to spontaneous apoptosis that can be inhibited by the chemokine CXCL12 and IL-2, -6, -7, and -15. *Eur J Immunol* 34:2771–2780.

Palma JP, Kim BS (2001) Induction of selected chemokines in glial cells infected with Theiler's virus. *J Neuroimmunol* 117:166–170.

Panitch HS, Hirsch RL, Haley AS, Johnson KP (1987) Exacerbations of multiple sclerosis in patients treated with gamma interferon. *Lancet* 1:893–895.

Pashenkov M, Soderstrom M, Link H (2003) Secondary lymphoid organ chemokines are elevated in the cerebrospinal fluid during central nervous system inflammation. *J Neuroimmunol* 135:154–160.

Pluchino S, Quattrini A, Brambilla E, Gritti A, Salani G, Dina G, Galli R, Del Carro U, Amadio S, Bergami A, Furlan R, Comi G, Vescovi AL, Martino G (2003) Injection of adult neurospheres induces recovery in a chronic model of multiple sclerosis. *Nature* 422:688–694.

Prat A, Biernacki K, Lavoie JF, Poirier J, Duquette P, Antel JP (2002) Migration of multiple sclerosis lymphocytes through brain endothelium. *Arch Neurol* 59:391–397.

Probert L, Akassoglou K, Pasparakis M, Kontogeorgos G, Kollias G (1995) Spontaneous inflammatory demyelinating disease in transgenic mice showing central nervous system-specific expression of tumor necrosis factor alpha. *Proc Natl Acad Sci U S A* 92:11294–11298.

Pullen LC, Miller SD, Dal Canto MC, Van der Meide PH, Kim BS (1994) Alteration in the level of interferon-gamma results in acceleration of Theiler's virus-induced demyelinating disease. *J Neuroimmunol* 55:143–152.

Raivich G, Banati R (2004) Brain microglia and blood-derived macrophages: molecular profiles and functional roles in multiple sclerosis and animal models of autoimmune demyelinating disease. *Brain Res Brain Res Rev* 46:261–281.

Ransohoff RM, Tani M (1998) Do chemokines mediate leukocyte recruitment in post-traumatic CNS inflammation? *Trends Neurosci* 21:154–159.

Ransohoff RM, Hamilton TA, Tani M, Stoler MH, Shick HE, Major JA, Estes ML, Thomas DM, Tuohy VK (1993) Astrocyte expression of mRNA encoding cytokines IP-10 and JE/MCP-1 in experimental autoimmune encephalomyelitis. *FASEB J* 7:592–600.

Ransohoff RM, Wei T, Pavelko KD, Lee JC, Murray PD, Rodriguez M (2002) Chemokine expression in the central nervous system of mice with a viral disease resembling multiple sclerosis: roles of CD4(+) and CD8(+) T cells and viral persistence. *J Virol* 76:2217–2224.

Ransohoff RM, Kivisakk P, Kidd G (2003) Three or more routes for leukocyte migration into the central nervous system. *Nat Rev Immunol* 3:569–581.

Renno T, Krakowski M, Piccirillo C, Lin JY, Owens T (1995) TNF-alpha expression by resident microglia and infiltrating leukocytes in the central nervous system of mice with experimental allergic encephalomyelitis. Regulation by Th1 cytokines. *J Immunol* 154:944–953.

Riminton SD, Korner H, Strickland DH, Lemckert FA, Pollard JD, Sedgwick JD (1998) Challenging cytokine redundancy: inflammatory cell movement and clinical course of experimental autoimmune encephalomyelitis are normal in lymphotoxin-deficient, but not tumor necrosis factor-deficient, mice. *J Exp Med* 187:1517–1528.

Robinson S, Franic LA (2001) Chemokine GRO1 and the spatial and temporal regulation of oligodendrocyte precursor proliferation. *Dev Neurosci* 23:338–345.

Rodriguez M, Pavelko K, Coffman RL (1995) Gamma interferon is critical for resistance to Theiler's virus-induced demyelination. *J Virol* 69:7286–7290.

Rossi D, Zlotnik A (2000) The biology of chemokines and their receptors. *Annu Rev Immunol* 18:217–242.

Rossi CP, Delcroix M, Huitinga I, McAllister A, van Rooijen N, Claassen E, Brahic M (1997) Role of macrophages during Theiler's virus infection. *J Virol* 71:3336–3340.

Rottman JB, Slavin AJ, Silva R, Weiner HL, Gerard CG, Hancock WW (2000) Leukocyte recruitment during onset of experimental allergic encephalomyelitis is CCR1 dependent. *Eur J Immunol* 30:2372–2377.

Rubio N, Torres C (1991) IL-1, IL-2 and IFN-gamma production by Theiler's virus-induced encephalomyelitic SJL/J mice. *Immunology* 74:284–289.

Rubio N, Sanz-Rodriguez F, Lipton HL (2006) Theiler's virus induces the MIP-2 chemokine (CXCL2) in astrocytes from genetically susceptible but not from resistant mouse strains. *Cell Immunol* 239:31–40.

Ruuls SR, Hoek RM, Ngo VN, McNeil T, Lucian LA, Janatpour MJ, Korner H, Scheerens H, Hessel EM, Cyster JG, McEvoy LM, Sedgwick JD (2001) Membrane-bound tnf supports secondary lymphoid organ structure but is subservient to secreted tnf in driving autoimmune inflammation. *Immunity* 15:533–543.

Sato S, Reiner SL, Jensen MA, Roos RP (1997) Central nervous system cytokine mRNA expression following Theiler's murine encephalomyelitis virus infection. *J Neuroimmunol* 76:213–223.

Seguin R, Biernacki K, Rotondo RL, Prat A, Antel JP (2003) Regulation and functional effects of monocyte migration across human brain-derived endothelial cells. *J Neuropathol Exp Neurol* 62:412–419.

Serafini B, Rosicarelli B, Magliozzi R, Stigliano E, Aloisi F (2004) Detection of ectopic B-cell follicles with germinal centers in the meninges of patients with secondary progressive multiple sclerosis. *Brain Pathol* 14:164–174.

Sharief MK, Hentges R (1991) Association between tumor necrosis factor-alpha and disease progression in patients with multiple sclerosis. *N Engl J Med* 325:467–472.

Sierra A, Rubio N (1993) Theiler's murine encephalomyelitis virus induces tumour necrosis factor-alpha in murine astrocyte cell cultures. *Immunology* 78:399–404.

Simpson JE, Newcombe J, Cuzner ML, Woodroofe MN (1998) Expression of monocyte chemoattractant protein-1 and other beta-chemokines by resident glia and inflammatory cells in multiple sclerosis lesions. *J Neuroimmunol* 84:238–249.

Simpson J, Rezaie P, Newcombe J, Cuzner ML, Male D, Woodroofe MN (2000a) Expression of the beta-chemokine receptors CCR2, CCR3 and CCR5 in multiple sclerosis central nervous system tissue. *J Neuroimmunol* 108:192–200.

Simpson JE, Newcombe J, Cuzner ML, Woodroofe MN (2000b) Expression of the interferon-gamma-inducible chemokines IP-10 and Mig and their receptor, CXCR3, in multiple sclerosis lesions. *Neuropathol Appl Neurobiol* 26:133–142.

Sunnemark D, Eltayeb S, Wallstrom E, Appelsved L, Malmberg A, Lassmann H, Ericsson-Dahlstrand A, Piehl F, Olsson T (2003) Differential expression of the chemokine receptors CX3CR1 and CCR1 by microglia and macrophages in myelin-oligodendrocyte-glycoprotein-induced experimental autoimmune encephalomyelitis. *Brain Pathol* 13:617–629.

Tani M, Ransohoff RM (1994) Do chemokines mediate inflammatory cell invasion of the central nervous system parenchyma? *Brain Pathol* 4:135–143.

Tani M, Fuentes ME, Peterson JW, Trapp BD, Durham SK, Loy JK, Bravo R, Ransohoff RM, Lira SA (1996) Neutrophil infiltration, glial reaction, and neurological disease in transgenic mice expressing the chemokine N51/KC in oligodendrocytes. *J Clin Invest* 98:529–539.

Taupin V, Renno T, Bourbonniere L, Peterson AC, Rodriguez M, Owens T (1997) Increased severity of experimental autoimmune encephalomyelitis, chronic macrophage/microglial reactivity, and demyelination in transgenic mice producing tumor necrosis factor-alpha in the central nervous system. *Eur J Immunol* 27:905–913.

Theil DJ, Tsunoda I, Libbey JE, Derfuss TJ, Fujinami RS (2000) Alterations in cytokine but not chemokine mRNA expression during three distinct Theiler's virus infections. *J Neuroimmunol* 104:22–30.

Tran EH, Hoekstra K, van Rooijen N, Dijkstra CD, Owens T (1998) Immune invasion of the central nervous system parenchyma and experimental allergic encephalomyelitis, but not leukocyte extravasation from blood, are prevented in macrophage-depleted mice. *J Immunol* 161:3767–3775.

Tran EH, Kuziel WA, Owens T (2000a) Induction of experimental autoimmune encephalomyelitis in C57BL/6 mice deficient in either the chemokine macrophage inflammatory protein-1alpha or its CCR5 receptor. *Eur J Immunol* 30:1410–1415.

Tran EH, Prince EN, Owens T (2000b) IFN-gamma shapes immune invasion of the central nervous system via regulation of chemokines. *J Immunol* 164:2759–2768.

Traugott U, Lebon P (1988) Multiple sclerosis: involvement of interferons in lesion pathogenesis. *Ann Neurol* 24:243–251.

Traugott U, Reinherz EL, Raine CS (1983) Multiple sclerosis: distribution of T cell subsets within active chronic lesions. *Science* 219:308–310.

Tsai HH, Frost E, To V, Robinson S, Ffrench-Constant C, Geertman R, Ransohoff RM, Miller RH (2002) The chemokine receptor CXCR2 controls positioning of oligodendrocyte precursors in developing spinal cord by arresting their migration. *Cell* 110:373–383.

van Oosten BW, Barkhof F, Truyen L, Boringa JB, Bertelsmann FW, von Blomberg BM, Woody JN, Hartung HP, Polman CH (1996) Increased MRI activity and immune activation in two multiple sclerosis patients treated with the monoclonal anti-tumor necrosis factor antibody cA2. *Neurology* 47:1531–1534.

von Andrian UH, Mackay CR (2000) T-cell function and migration. Two sides of the same coin. *N Engl J Med* 343:1020–1034.

Warrington AE, Asakura K, Bieber AJ, Ciric B, Van Keulen V, Kaveri SV, Kyle RA, Pease LR, Rodriguez M (2000) Human monoclonal antibodies reactive to oligodendrocytes promote remyelination in a model of multiple sclerosis. *Proc Natl Acad Sci U S A* 97:6820–6825.

Wheeler RD, Zehntner SP, Kelly LM, Bourbonniere L, Owens T (2006) Elevated interferon gamma expression in the central nervous system of tumour necrosis factor receptor 1-deficient mice with experimental autoimmune encephalomyelitis. *Immunology* 118:527–538.

Willenborg DO, Fordham S, Bernard CC, Cowden WB, Ramshaw IA (1996) IFN-gamma plays a critical down-regulatory role in the induction and effector phase of myelin oligodendrocyte glycoprotein-induced autoimmune encephalomyelitis. *J Immunol* 157:3223–3227.

Windrem MS, Nunes MC, Rashbaum WK, Schwartz TH, Goodman RA, McKhann G, 2nd, Roy NS, Goldman SA (2004) Fetal and adult human oligodendrocyte progenitor cell isolates myelinate the congenitally dysmyelinated brain. *Nat Med* 10:93–97.

Zehntner SP, Brisebois M, Tran E, Owens T, Fournier S (2003) Constitutive expression of a costimulatory ligand on antigen-presenting cells in the nervous system drives demyelinating disease. *FASEB J* 17:1910–1912.

Zehntner SP, Brickman C, Bourbonniere L, Remington L, Caruso M, Owens T (2005) Neutrophils that infiltrate the central nervous system regulate T cell responses. *J Immunol* 174:5124–5131.

Zhao C, Fancy SP, Magy L, Urwin JE, Franklin RJ (2005) Stem cells, progenitors and myelin repair. *J Anat* 207:251–258.

Zhou ZH, Han Y, Wei T, Aras S, Chaturvedi P, Tyler S, Rani MR, Ransohoff RM (2001) Regulation of monocyte chemoattractant protein (MCP)-1 transcription by interferon-gamma (IFN-gamma) in human astrocytoma cells: postinduction refractory state of the gene, governed by its upstream elements. *FASEB J* 15:383–392.

10
Chemokine Actions in the CNS: Insights from Transgenic Mice

Marcus Müller and Iain L. Campbell

1 Introduction

Historically the central nervous system (CNS) has been viewed as a relatively immune sheltered tissue. Under physiological conditions the CNS is devoid of leukocytes, including professional antigen presenting cells (APC), is deficient in key immune accessory molecules such as major histocompatibility molecules (MHC) and is protected by an effective blood brain barrier. Significantly, however, in numerous pathological states including infectious diseases and autoimmune disorders (e.g. multiple sclerosis) immune cells are effectively recruited to and infiltrate in the CNS. This immune response can be a two-edged sword required on the one hand to control infection and facilitate tissue repair and regeneration but on the other causing tissue injury that can result in life threatening complications. Therefore, understanding the mechanisms that control the trafficking of leukocytes to the CNS and the subsequent interactions between these cells that contribute to tissue injury has significant implications.

Key to our understanding of the mechanisms that govern leukocyte trafficking has been the relatively recent discovery of a large super-family of mostly small proteins (termed chemokines) that are leukocyte chemoattractant molecules (for reviews see (Rot and von Andrian, 2004; Charo and Ransohoff, 2006). The chemokines are grouped in four subfamilies, CXC, CC, C, and CX3C, based on the arrangement of NH2-terminal cysteine amino acids. It is now clear that chemokines have major roles in regulating leukocyte migration to the brain in inflammatory disease states (Engelhardt, 2006; Rebenko-Moll et al., 2006). The chemokine gene super-family can be further divided into two broad groups based on their gene expression properties. The first group are the constitutively expressed chemokines of which the CXC family chemokine CXCL12 (SDF-1) and the CX3C family member CX3CL1 (fractalkine) are the best characterized. In contrast to the constitutively expressed chemokines, a second larger group of chemokines consisting of many members of both the CXC and CC families, are not detectable under normal conditions in most non-immune tissues including the brain. However, abundant CNS production of chemokines belonging to this second group is found following a variety of insults including trauma, ischemia, infection and inflammation. The induction of these

T.E. Lane et al. (eds.), *Central Nervous System Diseases and Inflammation.*
© Springer 2008

chemokines following such insults can be mediated by a number of factors including microbial products (e.g., LPS) and host defense molecules such as cytokines (e.g., IL-1, IFN-γ and TNF). Cells intrinsic to the nervous system, including neurons, the macroglia and microglia, all have the ability to produce different chemokines (reviewed in Asensio and Campbell (1999) and Ubogu et al. (2006)). Moreover, the surfaces of these same cells are decorated with a variety of different chemokine receptors. Thus, the CNS seemingly has its own chemokine ligand/receptor network, which is consistent with the growing awareness that the function of chemokines could extend well beyond the regulation of leukocyte trafficking in the brain (Asensio and Campbell, 1999; Mennicken et al., 1999).

So it can be seen that the CNS biology of the chemokines is likely to extend well beyond the singular function of leukocyte chemotaxis. A number of important questions are therefore raised concerning the functions of the chemokines in the CNS. Some key questions include: (1) Do the presence of chemokines in different disordered states reflect their primary involvement or is an epiphenomenon? (2) Is there a cause and effect relationship between the production of a particular chemokine (or chemokines) and the development of specific molecular and cellular neuropathological alterations? (3) What are the mechanisms that underlie chemokine-mediated actions in the CNS? (4) Could manipulation of the chemokines or their receptors be used as a rational means for therapeutic intervention? To begin to address these questions experimental approaches are required that are non-invasive and organism-based. Of particular significance in this regard has been the application of genetic engineering technology that has permitted the germline transmission of foreign genes (transgenes) in the mouse. This is accomplished via a fusion gene construct (transgene) in which the gene of interest is placed under the transcriptional control of a cell-specific promoter. The transgene is then introduced (most commonly by microinjection) into the pronucleus of fertilized eggs, which are then implanted in the oviduct of pseudopregnant recipient mice. The transgenic approach has a number of advantages over other approaches such as intracerebral infusion or in vitro cell culture. Transgenic modeling allows the prolonged, reproducible delivery of a specific pure protein to specific cells in the intact CNS. The actions of the transgene product are exerted and can be assessed in a milieu where the anatomic and physiologic interactions of the CNS are preserved. Moreover, individual lines of mice in which the transgene is stably integrated into the genome can be developed and provide an unrestricted source of animals identical for the introduced genetic and molecular alteration, hence permitting systematic multi-level analysis of pathological, electrophysiological, neuroendocrinological and behavioral manifestations. For more detailed reviews of these techniques see (Galli-Taliadoros et al., 1995; Campbell and Gold, 1996; Campbell et al., 1998).

The application of genetically-manipulated animal models to study the CNS biology of the chemokines and their receptors has emerged as an important experimental approach. In this chapter, we will discuss findings from studies based on the use of CNS-chemokine transgenic models. In particular, we will focus on the contribution that these models have made to our understanding of the basic functions and mechanisms of actions of chemokines in the CNS.

2   Transgenic Production of Chemokines in the CNS

As mentioned in the preceding, a great many chemokines belonging in particular to the CXC and CC families are known to be induced in the brain during a variety of different pathologic states (Asensio and Campbell, 1999; Ransohoff, 2002; Rebenko-Moll et al., 2006; Ubogu et al., 2006). The consequences of chemokine expression in the unmanipulated CNS and in some cases in stimulus-evoked disease models have been examined in transgenic mice with CNS-targeted expression of specific chemokines. A summary of the salient features of transgenic models in which CNS-specific promoter constructs were used to drive expression of a specific chemokine gene is given in Table 10.1 and will be outlined in more detail in the following commentary.

2.1   CXCL1(GRO-α/KC/N51)

The chemokine CXCL1 (also known as GRO-α/KC/N51) contains an N-terminal ELR-amino acid motif and belongs to the CXC-chemokine family. CXCL1 binds to the CXCR2 receptor found predominantly on the surface of neutrophils. Not surprisingly, CXCL1 is a potent neutrophil chemoattractant. That this chemokine has a wider function in the CNS is supported by numerous studies. Cells intrinsic to the CNS including neurons (Horuk et al., 1997), astrocytes (Danik et al., 2003; Flynn et al., 2003), microglia (Filipovic et al., 2003; Flynn et al., 2003), oligodendrocyte precursor cells (Nguyen and Stangel, 2001) and mature oligodendrocytes (Omari et al., 2006) have been reported to have the CXCR2 receptor. A role for CXCR2/CXCL1 interaction is documented in rodents in the development and maintenance of the oligodendrocyte lineage, myelination, and white matter in the neonatal and adult CNS (Padovani-Claudio et al., 2006).

The CNS function of this chemokine was examined in transgenic mice by placing the CXCL1 gene under the transcriptional control of the myelin basic protein (MBP) promoter (Tani et al., 1996). In the resulting transgenic mice (termed MBP-KC) transgene-encoded mRNA levels were coincident with the developmentally-regulated pattern of endogenous MBP, peaking at 2–3 week of age, before declining to low levels characteristically seen in adult brain. During the period of maximal transgene expression, the brain of MBP-KC mice was heavily infiltrated with neutrophils that accumulated in perivascular, meningeal and parenchymal sites. At this time, despite loss of blood brain barrier (BBB) integrity and focal gliosis, there was no evidence of brain tissue destruction nor did affected animals display abnormal neurological signs. These results underscore the ability of CXCL1 to promote neutrophil migration to and extravasation in the CNS. However, the absence of any tissue destruction suggests that this chemokine does not activate the neutrophils.

Curiously, in one high transgene expressing line, from >40 days of age, mice exhibited a syndrome of progressive neurological dysfunction consisting of slowing

Table 10.1 Studies in transgenic mice with specific CNS-targeted chemokine

Chemokine	Promoter/targeted cell	Spontaneous phenotype	Induced disease phenotype	References
CXCL1	MBP/oligodendrocytes	Infiltration of neutrophils	Not examined	(Tani et al., 1996)
CCL2	GFAP/astrocytes	Infiltration of monocytes, long-term: delayed encephalopathy and impaired microglial function	Reversible encephalopathy after systemic injection of pertussis toxin	(Huang et al., 2002) (Huang et al., 2005) (Elhofy et al., 2005)
			Decreased severity of PLP-induced EAE	
			Accelerated β-Amyloid deposition in APP overexpressing mice	(Yamamoto et al., 2005)
CCL2	MBP/oligodendrocytes	Perivascular accumulation of monocytes and macrophages	Increase of perivascular accumulations of monocytes after systemic injection of lipopolysaccharide (LPS)	(Fuentes et al., 1995)
			Encephalopathy after systemic injection of pertussis toxin	(Toft-Hansen et al., 2006)
CCL19	MBP/oligodendrocytes	None	Not examined	
CCL21	MBP/oligodendrocytes	Early onset encephalopathy with myelin breakdown and gliosis, accumulation of neutrophils and eosinophils	Not examined	
CXCL10	GFAP/astrocytes	Subarachnoidal and parenchymal (minor) infiltration of mixed leukocytes	Amplification of leukocyte accumulation after non-specific immunization with complete Freund's adjuvants	(Boztug et al., 2002)

of the righting reflex, clumsiness, increase of rigidity of the hindlimbs and tail and profound truncal instability. The major neuropathological findings were florid microglial activation and BBB disruption without dysmyelination. Despite the severity of this neurological disorder (which resulted in premature death), there was no histological evidence of damage to neurons, myelin or axons. Although the basis for the neurological dysfunction developing in the MBP-KC mice is unknown, Tani et al. (1996) speculated that this might be due to the glial and BBB perturbations. It is also possible as the authors suggested that chronic exposure of neurons to the chemokine may directly compromise their function. In support of such a mechanism, CXCL1 can enhance both stimulus-evoked and spontaneous postsynaptic currents (Ragozzino et al., 1998), and increase neurotransmitter release and reduce long-term depression (Giovannelli et al., 1998), in Purkinje neurons. On the other hand, despite accumulating evidence that CXCL1 influences oligodendrocyte precursor cell and differentiated oligodendrocyte function both in vitro and in vivo, surprisingly, MBP-KC mice were reported to exhibit normal CNS myelination both in the acute and chronic disorders (Tani et al., 1996).

2.2 CXCL10 (IP-10)

CXCL10 (also known as IP-10) belongs to a sub-family of chemokines that include CXCL9 (MIG) and CXCL11 (ITAC) within the CXC family. Unlike CXCL1, the chemokines in this sub-family lack the N-terminal ELR-amino acid motif. Expression of the CXCL9, CXCL10 and CXCL11 genes is induced by the cytokine IFN-γ and all three chemokines bind to a common receptor, CXCR3. High levels of CXCR3 can be found principally on activated CD4[+] Th1 and CD8[+] T-cells and NK-cells (Taub et al., 1993, 1996a; Hopkins and Rothwell, 1995; Liao et al., 1995; Loetscher et al., 1996; Cole et al., 1998). In keeping with this cellular receptor distribution, CXC9, CXCL10 and CXCL11 all promote the trafficking of these immune cells in vitro and are strongly implicated in the generation of type I immune responses associated with anti-viral host defense, transplant rejection and autoimmunity (Liu et al., 2005). Evidence from rodent studies indicate that a functional CXCR3 receptor exists on microglia (Rappert et al., 2004) and neurons (Nelson and Gruol, 2004). Descriptive studies confirm the presence of high levels of CXCL10 RNA and protein, particularly in astrocytes and neurons, in a variety of experimental and human neurological diseases including viral infection and autoimmune diseases (Klein, 2004; Liu et al., 2005).

A transgenic mouse model was developed in which the GFAP promoter was used to accomplish astrocyte-targeted production of CXCL10 (see Fig. 10.1) (Boztug et al., 2002). Astrocyte-localized production of the GFAP-transgene encoded CXCL10 RNA was confirmed while production of CXCL10 at the protein level was demonstrated by immunoblotting. The levels of these transgene-encoded products were shown to be similar to levels of the endogenous CXCL10 induced following viral infection of the mouse brain confirming the

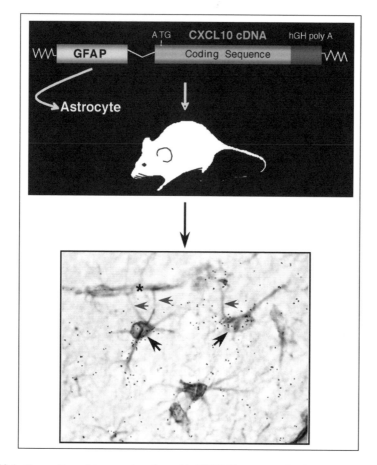

Fig. 10.1 Generation of transgenic mice with CXCL10 gene production targeted to astro-cytes. A cDNA encoding murine CXCL10 was inserted downstream of the murine GFAP pro-moter and upstream of a human growth hormone (hGH) polyadenylation signal sequence. This transgene construct was microinjected into the germ line of mice to generate a stable transgenic line. The presence of CXCL10 gene expression in astrocytes (lower panel) was confirmed by in situ hybridization for CXCL10 RNA combined with immunohistochemistry for GFAP protein. Note that some CXCL10 RNA positive astrocytes (black arrows) extend processes (blue arrows) to a blood vessel (asterisk) that is close by (*See also color plates*).

pathophysiological relevance of this model. These GFAP-CXCL10 transgenic mice bred normally and appeared physically unimpaired. Histological evaluation of the brain revealed a predominantly subarachnoidal and meningeal accumulation of mixed leukocytes with only minor parenchymal infiltration (see Fig. 10.2). The accumulation of leukocytes in the brain of these transgenic mice was mark-edly amplified by systemic immune challenge after immunization with complete Freund's adjuvant and pertussis toxin. Immunophenotypic characterization of these infiltrates revealed surprisingly, that the majority of infiltrating leukocytes

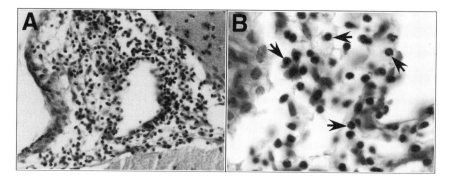

Fig. 10.2 Histological evaluation of the brain from GFAP-CXCL10 transgenic mice. A. appearance of a perivascular leukocyte infiltrate in the meninges (haematoxylin and eosin stained section; original magnification ×100). B. numerous neutrophils were observed (arrows) in the leukocyte infiltrates, (haematoxylin & eosin stained section; original magnification ×1,000) (*See also color plates*).

were neutrophils and macrophages with a minority being represented by CD+ T-cells. These observations suggest that constitutive astrocyte production of CXCL10 in the CNS does not provide an effective signal for the recruitment of T-cells from the periphery. This conclusion is in keeping with the findings from more recent reports that found that mice deficient for CXCL10 exhibit similar clinical and pathological outcomes as wild type controls for experimental autoimmune encephalomyelitis (EAE-a CD4 + T-cell-mediated disease) (Klein et al., 2004) as well as murine hepatitis virus encephalitis (Stiles et al., 2006).

The finding that neutrophils were preferentially recruited to the CNS of the GFAP-CXCL10 transgenic mice was unexpected since these cells were not known to have CXCR3. This point was confirmed by the studies of Boztug et al. (2002) who used fluorescence activated cell sorting analysis to show that the neutrophil population present in the CNS of the GFAP-CXCL10 transgenic mice were CXCR3 negative. Moreover, the results of a survey of the cerebral expression of a large number of other chemokines including ELR-CXC chemokines known to have neutrophil chemoattractant properties proved unremarkable. Therefore, the basis for the preferential accumulation of neutrophils in the brain of the transgenic mice is unknown. In all, these findings argue for a CXCR3-independent chemoattractant mechanism mediated by CXCL10 for the recruitment of neutrophils to the brain. It is interesting to note that an alternative receptor for CXCL10 has been reported on endothelial and epithelial cells (Soejima and Rollins, 2001). However, the identity of this novel CXCL10 binding moiety and its functional signature is currently unknown.

Of further note in these GFAP-CXCL10 transgenic mice was the lack of any gliosis or degenerative features that would be expected if there existed a destructive immunoinflammatory response in the CNS. So while constitutive, astrocyte-targeted production of CXCL10 can promote the recruitment of leukocytes to the CNS, this chemokine lacks the ability to further influence these cells, in particular, to drive a functional immune response. Moreover, although there are reports that a functional CXCR3 exists on neurons (Nelson and Gruol, 2004) and microglia (Rappert et al., 2004) in the rodent CNS, there would seem to be little effect of CXCL10 on these cells in this transgenic model. A possible explanation for this lack of effect of CXCL10 is that there is desensitization to the chemokine through CXCR3 downregulation. In support of this idea, hippocampal slices from wild type but not GFAP-CXCL10 transgenic mice in response to acute exposure to recombinant CXCL10 protein were shown to exhibit altered synaptic plasticity that was CXCR3-dependent (Vlkolinsky et al., 2004).

Interestingly, the role of CXCL10 as an effective T-cell chemoattractant may depend on the tissue in which the chemokine is produced. In contrast to the CNS, transgenic production of CXCL10 targeted to the beta cells of the islets of Langerhans induced spontaneous infiltration of large numbers of CD4 + and CD8 + T-cells around the islets (Rhode et al., 2005). Moreover, islet CXCL10 production markedly accelerated a stimulus-evoked autoimmune process by enhancing the migration of antigen-specific T-cells to the islets resulting in the development type I diabetes.

2.3   CCL2 (MCP-1)

The chemokine CCL2 (also known as MCP-1 and JE) was first identified as a monocyte attracting chemokine (Rollins, 1991; Rollins et al., 1991) but also has been shown to attract a wide variety of other leukocytes, including, activated T-cells, dendritic cells, mast cells and basophils (Taub et al., 1996b; Gunn et al., 1997; Siveke and Hamann, 1998). Although CCL2 binds to a number of CC receptors the primary receptor for this chemokine is CCR2. Cells within the CNS including astrocytes and microglia have been shown to be positive for CCR2 (Banisadr et al., 2002), while the ligand is produced by astrocytes (Ransohoff et al., 1993; Glabinski et al., 1996). CCL2 is implicated in the pathogenesis of a wide variety of experimental and human neurodegenerative and immune-mediated disorders (Cartier et al., 2005).

The first transgenic model developed to achieve CCL2 production in the CNS of mice employed the MBP-promoter to drive expression of the CCL2 gene in oligodendrocytes (Fuentes et al., 1995). Temporal expression of transgene-encoded CCL2 RNA followed that of endogenous MBP peaking between the second and third post-natal week before declining. In addition, CCL2 protein was demonstrated to be present in the brain of the transgenic but not wild type mice during peak transgene gene expression. In concordance with the temporal profile of transgene-encoded CCL2 production, discrete monocytic cell infiltrates were observed throughout the brain predominantly in perivascular sites in the meninges, choroid plexus, and brain parenchymal white matter. The intensity of these infiltrates could be increased as well as further parenchymal infiltration stimulated, following systemic

LPS administration to the MBP-MCP1 transgenic mice. Importantly, there was no evidence of either spontaneous or stimulus-evoked CNS injury, particularly in white matter, in these transgenic mice. The observations in this model confirmed the potent monocyte chemoattractant properties of CCL2 and further indicate that while CCL2 can provide a signal for the recruitment of these cells to the CNS, the chemokine does not directly stimulate these cells to engage in an active inflammatory process.

However, a more recent study using these transgenic mice revealed that a systemic injection of pertussis toxin can induce a reversible encephalopathy with clinical signs, including tremor, inactivity, limb clasping and weight loss (Toft-Hansen et al., 2006). Encephalopathic signs were correlated with a more diffuse distribution of monocytes, which were less restricted to the perivascular space suggesting a role for the inflammatory process in the clinical symptoms. However the pathological basis for the physical impairments in these mice was not determined and it therefore remains unclear whether tissue injury was also involved. Increased levels of metalloproteinases were found in pertussis toxin treated MBP-CCL2 transgenic mice, suggesting a critical need for metalloproteinases for the parenchymal infiltration of the monocytes. Weight loss and parenchymal infiltration, but not perivascular accumulation, were markedly resolved by the broad-spectrum metalloproteinase inhibitor BB-94/Batimastat. The findings suggest that a critical event in inducing disease in this model is metalloproteinase-dependent leukocyte migration across the astroglial basement membrane of the blood–brain barrier, which is induced by pertussis toxin.

As noted above, astrocytes but not oligodendrocytes have been found to be a major source of CCL2 in various CNS pathologies. The production of CCL2 under the control of the MBP-promoter follows a similar temporal pattern as the endogenous gene and produces a spike of CCL2 activity. The delivery of maximal transgene-derived CCL2 in this model occurs during a developmentally sensitive period before subsiding making this model less relevant for studying the effects of CCL2 on the adult brain. In order to address these issues, Huang and co-workers generated transgenic mice with astrocyte-targeted expression of the CCL2 gene driven by the human GFAP promoter (Huang et al., 2002, 2005). These GFAP-CCL2 transgenic mice produced CCL2 in the CNS at levels similar to those produced during EAE, yet, these animals exhibited little spontaneous inflammation. This contrasts with the MBP-CCL2 transgenic mouse model in which widespread and marked perivascular accumulation of monocytes was seen. The explanation for this difference between these models is not clear but it may reflect the chronic and more generalized production of high levels of CCL2 in the case of the GFAP-CCL2 transgenic mice. However, systemic immunization with complete Freund's adjuvant and injection of pertussis toxin induces a transient encephalopathy with elevated levels of IFN-γ and IL-2 in the GFAP-CCL2 transgenic mice. Interestingly, the encephalopathy was less severe in CCL2 transgenic mice on a T-cell deficient background, indicating the involvement of T-cells and not only monocytes in the observed encephalopathy (Huang et al., 2002). Moreover, this phenotype was shown to be dependent on CCR2 indicating that under these conditions CCR2 is the principal receptor for CCL2 in the murine CNS. Although a number of studies have suggested that CCR2 activation in the periphery downregulates Th1-T-cell development and can augment Th2-T-cell development (Daly and Rollins, 2003), the findings in the GFAP-CCL2 transgenic mouse indicate that in the CNS, CCL2 may recruit Th1 T-cells or direct T-cell polarization toward type 1 cytokine production. This concept is also supported by the finding that mice deficient for the CCL2 receptor, CCR2, are resistant to the development of the CD4$^+$ T-cell-mediated autoimmune disease, EAE (Fife et al., 2000). However, in findings that are difficult to reconcile with these conclusions, it has been reported that GFAP-CCL2 transgenic mice developed significantly milder EAE disease than littermate controls (Elhofy et al., 2005). Antigen-specific T-cells recovered from the GFAP-CCL2 transgenic mice showed decrease proliferative response to autoantigen and secreted less IFN-γ and had lower levels of IL-12 receptor RNA. These studies suggest that rather than augmenting Th1 cells, that chronic CCL2 production in the CNS impairs the function of these cells.

Although it has been possible to produce stimulus-evoked encephalopathy in MBP- and GFAP-CCL2 transgenic mice by systemic innate immune challenge these disorders are largely transient and therefore do not model the increased levels of CNS CCL2 that have been reported to occur in chronic neurological diseases such as human immunodeficiency virus type 1-associated dementia, amyotrophic lateral sclerosis, and multiple sclerosis. However, Huang et al. (2005) report from a more recent study that GFAP-CCL2 transgenic mice develop a spontaneous, delayed encephalopathy from >7 months of age. These mice presented with loss of weight, a hunched posture, slow response to tactile stimuli and a flaccid tail. Some animals even developed a hind-limb paralysis at >12 months of age. The brain exhibited perivascular infiltrates made up of activated macrophages, microglia with morphological activation, and impaired BBB. Surprisingly, no evidence was found for demyelination or of reduced neurons, axons, or synapses of neural components. Microglia in the GFAP-CCL2 transgenic mice, despite the appearance of activation and increased CD45 immunoreactivity, were found to be defective in their response to environmental stimuli in vitro. The Authors propose that chronic elevation of CCL2 in the CNS leads to desensitization of CCR2 and microglial dysfunction.

Further support for the notion that chronic CCL2 leads to microglial dysfunction comes from a study by Yamamoto et al. who generated bigenic GFAP-CCL2/tg 2576 mice that, in addition to chronic CCL2, also overproduce a mutant amyloid precursor protein (APP) in the CNS (Yamamoto et al., 2005). These workers found a significant increase in amyloid deposition in CCL2/APP bigenic mice compared with APP singly transgenic mice. Despite being accompanied by an increase in lesion-associated macrophages/microglia, evidence is provided suggesting that these cells show reduced phagocytic function leading to accelerated amyloid deposition. These findings underline the importance of considering the participation of chemokines and chemokine driven monocyte accumulation and function in the pathogenesis of degenerative CNS diseases.

2.4   CCL19 (MIP-3β/ELC/exodus-3) and CCL21 (6Ckine/.SLC/exodus-2)

The chemokines CCL19 and CCL21 have been implicated in the migration of lymphocytes and are produced in secondary lymphoid organs such as the lymph nodes. Both these chemokines bind to CCR7 and CCR11. Mice lacking CCR7 have impaired migration of lymphocytes into secondary lymphoid organs (Forster et al., 1999). CCL21 but not CCL19 has also been reported to bind to CXCR3 (Rappert et al., 2002). CCL19 and CCL21 are localized in venules surrounded by CCR7 positive inflammatory cells in the brain and spinal cord of mice with EAE and functional studies suggest that CCL19 and CCL21 are involved in T lymphocyte migration into the central nervous system during EAE (Alt et al., 2002). However, recent studies in CCR7 deficient mice suggest that CCL19 and CCL21 function may be dispensable in EAE (Pahuja et al., 2006). This interpretation is complicated in the case of CCL21 since this chemokine could still play a role in EAE in CCR7 deficient mice via CXCR3 on encephalitogenic T-cells. Moreover, CCL21 was reported to be produced by neurons in the murine brain following ischemic insult, highlighting the existence of an intrinsic source of this chemokine in the CNS that could interact with microglia via CXCR3 (Biber et al., 2002).

To study the biological role of the chemokine ligands CCL19 and CCL21, transgenic mice were developed that expressed either CCL19 or CCL21 in oligodendrocytes of the CNS using an MBP-promoter construct (Chen et al., 2002a). As expected, transgene-encoded CCL19 and CCL20 mRNA was developmentally regulated and peaked at 2–3 weeks of age. Expression of CCL19 or CCL21 protein in the CNS of the transgenic mice was confirmed by immunohistochemical staining. CCL19 producing transgenic mice did not show any clinical or histological phenotype, which led to the conclusion that CCL19, at least from diffuse cerebral expression, is not able to attract leukocytes into the brain or cause a CNS pathology. In contrast, transgenic mice with MBP-promoter driven production of CCL21 developed a rapid onset encephalopathy with tremor, ataxia and weight loss. Most of the CCL21 producing animals died within 4 weeks. Studies in vitro documented a chemotactic effect of CCL21 for thymocytes, naïve T cells, dendritic cells and B cells and not for macrophages and neutrophils (Hedrick and Zlotnik, 1997; Nagira et al., 1997; Kellermann et al., 1999). However, cerebral infiltrates in CCL21 producing transgenic mice consisted almost exclusively of neutrophils and eosinophils. These infiltrates were associated with gliosis, microglial activation and demyelination. Infiltrating lymphocytes were not observed-even in areas in which the BBB was disrupted. The lack of lymphocyte infiltration is a notable finding because transgenic expression of CCL21 in the pancreas induced lymphocyte influx and led to the formation of lymph node like structures (Fan et al., 2000; Chen et al., 2002b). It would thus appear that tissue-specific factors that might include for example, specific adhesion molecules and other cofactors, are required for CCL21 to induce lymphocyte recruitment.

As pointed out by the authors the phenotypic differences observed between MBP-CCL21 and MBP-cCL19 transgenic mice is very interesting given that these chemokines are functionally related and bind to the same cell surface receptors, CCR7 and CCR11. However, and as noted above CCL21 but not CCL19 also binds to CXCR3, which in the case of murine microglia produces functional alterations in these cells (Rappert et al., 2002). Interestingly, the same workers have reported that neurons in the ischemic brain produce CCL21 (Biber et al., 2001). These observations have led to the suggestion that CCL21 and CXCR3 may mediate neuron-glia interactions during disease conditions. It is speculated that CCL21 derived from transgenic oligodendrocytes in the brain of the MBP-CCL21 mice may have interacted with CXCR3 on microglia inducing an inflammatory response that led to the influx of inflammatory cells into the CNS. While further studies, such as crossing the MBP-CCL21 transgenic mice with CXCR3 deficient animals, would help to clarify this mechanism, as noted in the preceding discussion, transgenic mice with astrocyte-targeted production of the primary CXCR3 ligand CXCL10 do not share the severe neurological phenotype of the MBP-CCL2 mice.

3 Concluding Discussion

An important constraint with the transgenic approach that needs to be considered relates to the spatial and temporal control of transgene expression. The production of chemokines in pathologic states such as in EAE is regulated in a temporal fashion that precedes and then overlaps with the severity of the disease process. Moreover, the spatial pattern of expression of chemokines such as CXCL10 is focally restricted to the immunoinflammatory lesions. The production of chemokines achieved in pathologic states therefore would favor the establishment of highly localized chemoattractant gradients that then guide the trafficking of leukocytes. In contrast, in the transgenic model the chemokine is produced more chronically and more diffusely throughout the CNS. In addition to the potential for receptor down-regulation and desensitization of the ligand response, it is likely that any chemoattractant gradient in this milieu would be much weaker in strength and more widely distributed. Technically achieving more acute and focal expression of the chemokine to better replicate pathologic states in the adult brain using transgenic modeling is clearly desirable but difficult to achieve at this time. Although constructs have been engineered that permit the temporal control of transgene expression in the CNS, more precise spatial targeting that would allow for highly localized production of the chemokine in neural cells such as astrocytes is not yet possible.

Despite the drawbacks noted in the preceding discussion, it is clear that development of molecular genetic methods that allow for the targeted manipulation of gene expression in an intact organism has been important in contributing to our understanding of the actions of chemokines in the CNS in physiologic and pathologic states (see Table 10.1). Promoter-driven expression of different biologically relevant chemokines in the CNS of transgenic mice highlight the ability of some but not all of these molecules to stimulate the recruitment but commonly not the functional activation of specific subsets of leukocytes to the CNS. With certain chemokines such as CXCL10 and CCL21, their transgenic production in the CNS led to the accumulation of unexpected leukocyte phenotypes that could not have been predicted based on either in vitro chemotaxis assays or the known cellular distribution of the receptor. This to some extent reflects the complexity of the actions and interactions between chemokines and their receptors many of which we know to be promiscuous for the ligands they bind. Moreover, differences in the immunophenotype of leukocytes recruited to the CNS versus peripheral tissues following organ-specific transgene driven chemokine production indicates that tissue-specific influences can clearly determine how a chemokine will behave in that milieu. There are still considerable gaps in our understanding of chemokine biology and neurobiology. However, ongoing and future studies employing approaches such as transgenic modeling no doubt will begin to fill these gaps.

Acknowledgements The authors studies referred to in this article were funded by USPHS grants MH 62231 and NS044905.

References

Alt, C., Laschinger, M. and Engelhardt, B. (2002). Functional expression of the lymphoid chemokines CCL19 (ELC) and CCL 21 (SLC) at the blood–brain barrier suggests their involvement in G-protein-dependent lymphocyte recruitment into the central nervous system during experimental autoimmune encephalomyelitis. *Eur J Immunol*, 32, 2133–2144.

Asensio, V. C. and Campbell, I. L. (1999). Chemokines in the CNS: plurifunctional mediators in diverse states. *Trends Neurosci*, 22, 504–512.

Banisadr, G., Queraud-Lesaux, F., Boutterin, M. C., Pelaprat, D., Zalc, B., Rostene, W., Haour, F. and Parsadaniantz, S. M. (2002). Distribution, cellular localization and functional role of CCR2 chemokine receptors in adult rat brain. *J Neurochem*, 81, 257–269.

Biber, K., Sauter, A., Brouwer, N., Copray, S. C. and Boddeke, H. W. (2001). Ischemia-induced neuronal expression of the microglia attracting chemokine secondary lymphoid-tissue chemokine (SLC). *Glia*, 34, 121–133.

Biber, K., Dijkstra, I., Trebst, C., De Groot, C. J., Ransohoff, R. M. and Boddeke, H. W. (2002). Functional expression of CXCR3 in cultured mouse and human astrocytes and microglia. *Neuroscience*, 112, 487–497.

Boztug, K., Carson, M. J., Pham-Mitchell, N., Asensio, V. C., DeMartino, J. and Campbell, I. L. (2002). Leukocyte infiltration, but not neurodegeneration, in the CNS of transgenic mice with astrocyte production of the CXC chemokine ligand 10. *J Immunol*, 169, 1505–1515.

Campbell, I. L. and Gold, L. H. (1996). Transgenic modeling of neuropsychiatric disorders. *Mol Psychiatry*, 1, 105–120.

Campbell, I. L., Stalder, A. K., Akwa, Y., Pagenstecher, A. and Asensio, V. C. (1998). Transgenic models to study the actions of cytokines in the central nervous system. *Neuroimmunomodulation*, 5, 126–135.

Cartier, L., Hartley, O., Dubois-Dauphin, M. and Krause, K. H. (2005). Chemokine receptors in the central nervous system: role in brain inflammation and neurodegenerative diseases. *Brain Res Brain Res Rev*, 48, 16–42.

Charo, I. F. and Ransohoff, R. M. (2006). The many roles of chemokines and chemokine receptors in inflammation. *N Engl J Med*, 354, 610–621.

Chen, S. C., Leach, M. W., Chen, Y., Cai, X. Y., Sullivan, L., Wiekowski, M., Dovey-Hartman, B. J., Zlotnik, A. and Lira, S. A. (2002a). Central nervous system inflammation and neurological disease in transgenic mice expressing the CC chemokine CCL21 in oligodendrocytes. *J Immunol*, 168, 1009–1017.

Chen, S. C., Vassileva, G., Kinsley, D., Holzmann, S., Manfra, D., Wiekowski, M. T., Romani, N. and Lira, S. A. (2002b). Ectopic expression of the murine chemokines CCL21a and CCL21b induces the formation of lymph node-like structures in pancreas, but not skin, of transgenic mice. *J Immunol*, 168, 1001–1008.

Cole, K. E., Strick, C. A., Paradis, T. J., Ogborne, K. T., Loetscher, M., Gladue, R. P., Lin, W., Boyd, J. G., Moser, B., Wood, D. E., Sahagan, B. G. and Neote, K. (1998). Interferon-inducible T cell alpha chemoattractant (I-TAC): a novel non-ELR CXC chemokine with potent activity on activated T cells through selective high affinity binding to CXCR3. *J Exp Med*, 187, 2009–2021.

Daly, C. and Rollins, B. J. (2003). Monocyte chemoattractant protein-1 (CCL2) in inflammatory disease and adaptive immunity: therapeutic opportunities and controversies. *Microcirculation*, 10, 247–257.

Danik, M., Puma, C., Quirion, R. and Williams, S. (2003). Widely expressed transcripts for chemokine receptor CXCR1 in identified glutamatergic, gamma-aminobutyric acidergic, and cholinergic neurons and astrocytes of the rat brain: a single-cell reverse transcription-multiplex polymerase chain reaction study. *J Neurosci Res*, 74, 286–295.

Elhofy, A., Wang, J., Tani, M., Fife, B. T., Kennedy, K. J., Bennett, J., Huang, D., Ransohoff, R. M. and Karpus, W. J. (2005). Transgenic expression of CCL2 in the central nervous system prevents experimental autoimmune encephalomyelitis. *J Leukoc Biol*, 77, 229–237.

Engelhardt, B. (2006). Molecular mechanisms involved in T cell migration across the blood–brain barrier. *J Neural Transm,* 113, 477–485.

Fan, L., Reilly, C. R., Luo, Y., Dorf, M. E. and Lo, D. (2000). Cutting edge: ectopic expression of the chemokine TCA4/SLC is sufficient to trigger lymphoid neogenesis. *J Immunol*, 164, 3955–3959.

Fife, B. T., Huffnagle, G. B., Kuziel, W. A. and Karpus, W. J. (2000). CC chemokine receptor 2 is critical for induction of experimental autoimmune encephalomyelitis. *J Exp Med*, 192, 899–905.

Filipovic, R., Jakovcevski, I. and Zecevic, N. (2003). GRO-alpha and CXCR2 in the human fetal brain and multiple sclerosis lesions. *Dev Neurosci*, 25, 279–290.

Flynn, G., Maru, S., Loughlin, J., Romero, I. A. and Male, D. (2003). Regulation of chemokine receptor expression in human microglia and astrocytes. *J Neuroimmunol*, 136, 84–93.

Forster, R., Schubel, A., Breitfeld, D., Kremmer, E., Renner-Muller, I., Wolf, E. and Lipp, M. (1999). CCR7 coordinates the primary immune response by establishing functional microenvironments in secondary lymphoid organs. *Cell*, 99, 23–33.

Fuentes, M. E., Durham, S. K., Swerdel, M. R., Lewin, A. C., Barton, D. S., Megill, J. R., Bravo, R. and Lira, S. A. (1995). Controlled recruitment of monocytes and macrophages to specific organs through transgenic expression of monocyte chemoattractant protein-1. *J Immunol*, 155, 5769–5776.

Galli-Taliadoros, L. A., Sedgwick, J. D., Wood, S. A. and Korner, H. (1995). Gene knock-out technology: a methodological overview for the interested novice. *J Immunol Methods*, 181, 1–15.

Giovannelli, A., Limatola, C., Ragozzino, D., Mileo, A. M., Ruggieri, A., Ciotti, M. T., Mercanti, D., Santoni, A. and Eusebi, F. (1998). CXC chemokines interleukin-8 (IL-8) and growth-related gene product alpha (GROalpha) modulate Purkinje neuron activity in mouse cerebellum. *J Neuroimmunol*, 92, 122–132.

Glabinski, A. R., Balasingam, V., Tani, M., Kunkel, S. L., Strieter, R. M., Yong, V. W. and Ransohoff, R. M. (1996). Chemokine monocyte chemoattractant protein-1 is expressed by astrocytes after mechanical injury to the brain. *J Immunol*, 156, 4363–4368.

Gunn, M. D., Nelken, N. A., Liao, X. and Williams, L. T. (1997). Monocyte chemoattractant protein-1 is sufficient for the chemotaxis of monocytes and lymphocytes in transgenic mice but requires an additional stimulus for inflammatory activation. *J Immunol*, 158, 376–383.

Hedrick, J. A. and Zlotnik, A. (1997). Identification and characterization of a novel beta chemokine containing six conserved cysteines. *J Immunol*, 159, 1589–1593.

Hopkins, S. J. and Rothwell, N. J. (1995). Cytokines and the nervous system. I: expression and recognition [see comments]. *Trends Neurosci*, 18, 83–88.

Horuk, R., Martin, A. W., Wang, Z., Schweitzer, L., Gerassimides, A., Guo, H., Lu, Z., Hesselgesser, J., Perez, H. D., Kim, J., Parker, J., Hadley, T. J. and Peiper, S. C. (1997). Expression of chemokine receptors by subsets of neurons in the central nervous system. *J Immunol*, 158, 2882–2890.

Huang, D., Tani, M., Wang, J., Han, Y., He, T. T., Weaver, J., Charo, I. F., Tuohy, V. K., Rollins, B. J. and Ransohoff, R. M. (2002). Pertussis toxin-induced reversible encephalopathy dependent on monocyte chemoattractant protein-1 overexpression in mice. *J Neurosci*, 22, 10633–10642.

Huang, D., Wujek, J., Kidd, G., He, T. T., Cardona, A., Sasse, M. E., Stein, E. J., Kish, J., Tani, M., Charo, I. F., Proudfoot, A. E., Rollins, B. J., Handel, T. and Ransohoff, R. M. (2005). Chronic expression of monocyte chemoattractant protein-1 in the central nervous system causes delayed encephalopathy and impaired microglial function in mice. *FASEB J*, 19, 761–772.

Kellermann, S. A., Hudak, S., Oldham, E. R., Liu, Y. J. and McEvoy, L. M. (1999). The CC chemokine receptor-7 ligands 6Ckine and macrophage inflammatory protein-3 beta are potent chemoattractants for in vitro- and in vivo-derived dendritic cells. *J Immunol*, 162, 3859–3864.

Klein, R. S. (2004). Regulation of neuroinflammation: the role of CXCL10 in lymphocyte infiltration during autoimmune encephalomyelitis. *J Cell Biochem,* 92, 213–222.

Klein, R. S., Izikson, L., Means, T., Gibson, H. D., Lin, E., Sobel, R. A., Weiner, H. L. and Luster, A. D. (2004). IFN-inducible protein 10/CXC chemokine ligand 10-independent induction of experimental autoimmune encephalomyelitis. *J Immunol*, 172, 550–559.

Liao, F., Rabin, R. L., Yannelli, J. R., Koniaris, L. G., Vanguri, P. and Farber, J. M. (1995). Human mig chemokine: biochemical and functional characterization. *J Exp Med*, 182, 1301–1314.

Liu, L., Callahan, M. K., Huang, D. and Ransohoff, R. M. (2005). Chemokine receptor CXCR3: an unexpected enigma. *Curr Top Dev Biol*, 68, 149–181.

Loetscher, M., Gerber, B., Loetscher, P., Jones, S. A., Piali, L., Clark-Lewis, I., Baggiolini, M. and Moser, B. (1996). Chemokine receptor specific for IP10 and mig: structure, function, and expression in activated T-lymphocytes. *J Exp Med*, 184, 963–969.

Mennicken, F., Maki, R., de Souza, E. B. and Quirion, R. (1999). Chemokines and chemokine receptors in the CNS: a possible role in neuroinflammation and patterning. *Trends Pharmacol Sci*, 20, 73–78.

Nagira, M., Imai, T., Hieshima, K., Kusuda, J., Ridanpaa, M., Takagi, S., Nishimura, M., Kakizaki, M., Nomiyama, H. and Yoshie, O. (1997). Molecular cloning of a novel human CC chemokine secondary lymphoid-tissue chemokine that is a potent chemoattractant for lymphocytes and mapped to chromosome 9p13. *J Biol Chem*, 272, 19518–19524.

Nelson, T. E. and Gruol, D. L. (2004). The chemokine CXCL10 modulates excitatory activity and intracellular calcium signaling in cultured hippocampal neurons. *J Neuroimmunol*, 156, 74–87.

Nguyen, D. and Stangel, M. (2001). Expression of the chemokine receptors CXCR1 and CXCR2 in rat oligodendroglial cells. *Brain Res Dev Brain Res*, 128, 77–81.

Omari, K. M., John, G., Lango, R. and Raine, C. S. (2006). Role for CXCR2 and CXCL1 on glia in multiple sclerosis. *Glia*, 53, 24–31.

Padovani-Claudio, D. A., Liu, L., Ransohoff, R. M. and Miller, R. H. (2006). Alterations in the oligodendrocyte lineage, myelin, and white matter in adult mice lacking the chemokine receptor CXCR2. *Glia*, 54, 471–483.

Pahuja, A., Maki, R. A., Hevezi, P. A., Chen, A., Verge, G. M., Lechner, S. M., Roth, R. B., Zlotnik, A. and Alleva, D. G. (2006). Experimental autoimmune encephalomyelitis develops in CC chemokine receptor 7-deficient mice with altered T-cell responses. *Scand J Immunol*, 64, 361–369.

Ragozzino, D., Giovannelli, A., Mileo, A. M., Limatola, C., Santoni, A. and Eusebi, F. (1998). Modulation of the neurotransmitter release in rat cerebellar neurons by GRO beta. *Neuroreport*, 9, 3601–3606.

Ransohoff, R. M. (2002). The chemokine system in neuroinflammation: an update. *J Infect Dis*, 186 Suppl 2, S152–S156.

Ransohoff, R. M., Hamilton, T. A., Tani, M., Stoler, M. H., Shick, H. E., Major, J. A., Estes, M. L., Thomas, D. M. and Tuohy, V. K. (1993). Astrocyte expression of mRNA encoding cytokines IP-10 and JE/MCP-1 in experimental autoimmune encephalomyelitis. *FASEB J*, 7, 592–600.

Rappert, A., Biber, K., Nolte, C., Lipp, M., Schubel, A., Lu, B., Gerard, N. P., Gerard, C., Boddeke, H. W. and Kettenmann, H. (2002). Secondary lymphoid tissue chemokine (CCL21) activates CXCR3 to trigger a Cl- current and chemotaxis in murine microglia. *J Immunol*, 168, 3221–3226.

Rappert, A., Bechmann, I., Pivneva, T., Mahlo, J., Biber, K., Nolte, C., Kovac, A. D., Gerard, C., Boddeke, H. W., Nitsch, R. and Kettenmann, H. (2004). CXCR3-dependent microglial recruitment is essential for dendrite loss after brain lesion. *J Neurosci*, 24, 8500–8509.

Rebenko-Moll, N. M., Liu, L., Cardona, A. and Ransohoff, R. M. (2006). Chemokines, mononuclear cells and the nervous system: heaven (or hell) is in the details. *Curr Opin Immunol*, 18, 683–689.

Rhode, A., Pauza, M. E., Barral, A. M., Rodrigo, E., Oldstone, M. B., von Herrath, M. G. and Christen, U. (2005). Islet-specific expression of CXCL10 causes spontaneous islet infiltration and accelerates diabetes development. *J Immunol*, 175, 3516–3524.

Rollins, B. J. (1991). JE/MCP-1: an early-response gene encodes a monocyte-specific cytokine. *Cancer Cells*, 3, 517–524.

Rollins, B. J., Walz, A. and Baggiolini, M. (1991). Recombinant human MCP-1/JE induces chemotaxis, calcium flux, and the respiratory burst in human monocytes. *Blood*, 78, 1112–1116.

Rot, A. and von Andrian, U. H. (2004). Chemokines in innate and adaptive host defense: basic chemokinese grammar for immune cells. *Annu Rev Immunol*, 22, 891–928.

Siveke, J. T. and Hamann, A. (1998). T helper 1 and T helper 2 cells respond differentially to chemokines. *J Immunol*, 160, 550–554.

Soejima, K. and Rollins, B. J. (2001). A functional IFN-gamma-inducible protein-10/CXCL10-specific receptor expressed by epithelial and endothelial cells that is neither CXCR3 nor glycosaminoglycan. *J Immunol*, 167, 6576–6582.

Stiles, L. N., Hardison, J. L., Schaumburg, C. S., Whitman, L. M. and Lane, T. E. (2006). T cell antiviral effector function is not dependent on CXCL10 following murine coronavirus infection. *J Immunol*, 177, 8372–8380.

Tani, M., Fuentes, M. E., Peterson, J. W., Trapp, B. D., Durham, S. K., Loy, J. K., Bravo, R., Ransohoff, R. M. and Lira, S. A. (1996). Neutrophil infiltration, glial reaction, and neurological disease in transgenic mice expressing the chemokine N51/KC in oligodendrocytes. *J Clin Invest*, 98, 529–539.

Taub, D. D., Lloyd, A. R., Wang, J. M., Oppenheim, J. J. and Kelvin, D. J. (1993). The effects of human recombinant MIP-1 alpha, MIP-1 beta, and RANTES on the chemotaxis and adhesion of T cell subsets. *Adv Exp Med Biol*, 351, 139–146.

Taub, D. D., Longo, D. L. and Murphy, W. J. (1996a). Human interferon-inducible protein-10 induces mononuclear cell infiltration in mice and promotes the migration of human T lymphocytes into the peripheral tissues and human peripheral blood lymphocytes-SCID mice. *Blood*, 87, 1423–1431.

Taub, D. D., Ortaldo, J. R., Turcovski-Corrales, S. M., Key, M. L., Longo, D. L. and Murphy, W. J. (1996b). Beta chemokines costimulate lymphocyte cytolysis, proliferation, and lymphokine production. *J Leukoc Biol*, 59, 81–89.

Toft-Hansen, H., Buist, R., Sun, X. J., Schellenberg, A., Peeling, J. and Owens, T. (2006). Metalloproteinases control brain inflammation induced by pertussis toxin in mice overexpressing the chemokine CCL2 in the central nervous system. *J Immunol*, 177, 7242–7249.

Ubogu, E. E., Cossoy, M. B. and Ransohoff, R. M. (2006). The expression and function of chemokines involved in CNS inflammation. *Trends Pharmacol Sci*, 27, 48–55.

Vlkolinsky, R., Siggins, G. R., Campbell, I. L. and Krucker, T. (2004). Acute exposure to CXC chemokine ligand 10, but not its chronic astroglial production, alters synaptic plasticity in mouse hippocampal slices. *J Neuroimmunol*, 150, 37–47.

Yamamoto, M., Horiba, M., Buescher, J. L., Huang, D., Gendelman, H. E., Ransohoff, R. M. and Ikezu, T. (2005). Overexpression of monocyte chemotactic protein-1/CCL2 in beta-amyloid precursor protein transgenic mice show accelerated diffuse beta-amyloid deposition. *Am J Pathol*, 166, 1475–1485.

11
Chemokines and Spinal Cord Injury

Maya N. Hatch and Hans S. Keirstead

Abstract   The immune system plays a critical role in CNS disorders and spinal cord injury (SCI). Primary trauma to the adult mammalian spinal cord is immediately followed by secondary degeneration in which the inflammatory response is thought to be detrimental. This inflammatory response is mediated by small, chemotropic cytokines, called chemokines, which are secreted by a variety of cell types in the CNS including neurons, glia, and vascular cells. Here, we review studies which provide insight into the functional role of chemokines in neuroinflammation and disease, with an emphasis on SCI. More specifically, this review emphasizes studies which indicate that ablation of the T cell chemotactic CXC chemokine ligand 10 (CXCL10) results in diminished neuropathology associated with decreased immune cell infiltration into the CNS. Importantly, these findings reveal that targeting chemokines such as CXCL10 may offer a powerful therapeutic approach for the treatment of neuroinflammatory diseases.

1   Spinal Cord Injury

Traumatic injury to the spinal cord is a combination of primary and secondary phases resulting from some form of mechanical insult. The direct tissue damage caused by the injury, known as the primary phase, initiates a cascade of events that lead to a detrimental secondary phase. Secondary degeneration is defined by a cascade of chemical and physiological events which include activation of voltage (dependent or agonist) gated channels, ion leaks, activation of calcium-dependent enzymes such as proteases, lipases, and nucleases, mitochondrial dysfunction, vascular changes and cellular infiltration and activation (Hausmann, 2003; Ramer et al., 2005). These processes lead to cell death, progressive tissue loss and cystic cavitation evolving away from the initial trauma site (Schwab and Bartholdi, 1996; Beattie et al., 2002) which can ultimately result in the functional deficits seen after SCI. Primary tissue damage is often irreversible, whereas secondary degeneration is amenable to pharmacological treatment. Although it is unclear which particular component of secondary degeneration is most significant to SCI severity, it is widely accepted that attenuating secondary degeneration would be beneficial.

T.E. Lane et al. (eds.), *Central Nervous System Diseases and Inflammation.*
© Springer 2008

A number of different strategies for treating SCI have been attempted over the years including rehabilitation, the use of exercise to enhance axonal regeneration (Neeper et al., 1995; Gomez-Pinilla et al., 2001; Cotman and Engesser-Cesar, 2002; Cotman and Berchtold, 2002; McDonald et al., 2002; Van Meeteren et al., 2003), and surgical decompression and/or fusion (Fehlings and Tator, 1999; Fehlings et al., 2001; Tator, 2006). These treatments are often costly, labor intensive and induce minimal improvement (Tator, 2006). On the other hand, neuroprotective agents targeting secondary injury mechanisms offer greater promise and have been more widely used. Two of the most widely used agents are high-dose methylpred-nisolone (Bracken et al., 1998; Ramer et al., 2005) and GM1 gangliosides (Geisler, 1993; Geisler et al., 2001; Ramer et al., 2005). GM1 gangliosides have been shown to enhance recovery rate from SCI, but did not improve overall functional outcome, and methylprednisolone has been shown to have limited effect on the outcome of CNS trauma in clinical trials (Hall, 2001; Bracken et al., 1992; Ramer et al., 2005). However, to date no pharmacological treatment targeting secondary degeneration has been shown to fully prevent the complicated pathological cascade triggered by SCI, and the efficacy of methylprednisolone, currently the standard of care for spinal cord injured patients, has been questioned (Hurlbert, 2000; Hugenholtz, 2003; Kwon et al., 2004; Ramer et al. 2005; Tator 2006). Clearly, more research is needed to identify a better suited therapeutic target.

Animal models are invaluable for SCI research because they allow for precise manipulations and measurements, as well as a means to evaluate potential thera-peutic interventions. Important aspects of animal models include clinical relevance, reliability, sensitivity, and feasibility of use. Several different animal models for SCI have been developed, but most fall into one of two major injury categories, contusion or laceration. Typically, the characteristics of the nervous system architecture, biochemistry, and physiology are conserved across species, allowing for reliable and valid comparisons. However, it should be noted that some differ-ences may exist that can produce varying results, such as a reduced sensitivity to excitotoxic damage in some strains of mice, and the atypical absence of gross cavitation following SCI in mice (Sroga et al., 2003). Strain differences in rats may not affect histological damage, but can influence the behavioral recovery (Mills, Hains et al., 2001). Such interspecies differences are critical when considering translation of animal findings to human treatment. There are significant genetic differences between rodents and humans, which underlie the species specificity of antibodies, probes, and drugs. Thus, in some instances it is not possible to extrapolate directly from rodent to human.

2   Inflammation Following Spinal Cord Injury

Studies of the etiology of MS and related animal models, in which pro-inflammatory cells are considered central to the development of disease, indicate that the immune response can be detrimental to a compromised CNS (Olsson, 1995; Schulze-Koops et al., 1995; Lane et al., 2000). Likewise, immune cell infiltration to SCI sites is a major contributor to secondary degeneration (Blight, 1985; Dusart and Schwab, 1994; Blight et al., 1995; Popovich et al., 1997; Bethea and Dietrich, 2002; Gonzalez et al., 2003; Glaser et al., 2004). After primary insult to the spinal cord, there is immediate blood brain barrier disruption and cell death of local neurons, astrocytes, oligodendrocytes, and endothelial cells (Blight, 1985; Bethea and Dietrich, 2002; Hagg and Oudega, 2006). During this acute phase of injury signals are released that lead to massive neutrophil influx (Schnell et al., 1999; Saville et al., 2004; Hagg and Oudega, 2006), vascular damage and subsequent inflammatory responses involving T lymphocyte and macrophage/microglia infiltration, which have been shown to produce neurotoxic and pro-inflammatory compounds that lead to further cell death and secondary disease pathology (Giulian et al., 1993; Martiney et al., 1998; Popovich, 2000; Hausmann, 2003; Hagg and Oudega, 2006).

Several studies have indicated that attenuation of the pro-inflammatory response to CNS injury reduces the incidence and/or severity of disease, underscoring the pro-inflammatory response as a key target for therapeutic intervention (Lane et al., 2000; Fife et al., 2001; Moalem et al., 1999; Popovich et al., 1999; Hauben et al., 2000; Ghirnikar et al., 2001; Gris et al., 2004). For example, selective depletion of peripheral macrophages in spinal cord injured rats resulted in marked improvements in overground locomotion (Popovich et al., 1999), and blockade of the CD11d/CD18 integrin with monoclonal antibodies decreased neutrophil influx to the injured spinal cord and delayed the entry of hematogenous macrophages, improving behavioral function (Gris et al., 2004). Along these lines, administration of antibodies against CD81 (Dijkstra et al., 2006) and IL-6 receptor (Okada et al., 2004) have induced improved functional recovery and reduced astrogliosis as well as influx of inflammatory cells. In a slightly different approach, administration of minocycline, a second generation tetracycline, induced a reduction in secondary degeneration and improved locomotor function by decreasing microglia/macrophage activation and diminishing oligodendrocyte and microglial/macrophage death (Stirling et al., 2004). Although all of the aforementioned treatments seem effective in reducing inflammation, they may have too widespread effects for SCI clinical use. Therefore, a more complete understanding of the molecular and cellular mechanisms governing leukocyte trafficking and accumulation within the CNS is needed.

3 Chemokines and Inflammation

Chemokines are small 8–10kda heparin-binding proteins that regulate leukocyte trafficking during injury or inflammation. They were originally studied due to their role in cell regulation and localization during development and are now also known to be critical players in inflammation and adaptive immune responses during a variety of CNS pathologies. Due to their relatively specific actions as well as their malleability, chemokine and chemokine receptor targeting make an attractive therapeutic approach for neuroinflammatory diseases.

Four families of chemokines have been identified based on relative positions of their cysteine residues and functional characteristics. The two major subfamilies of chemokines include the alpha chemokines, or CXC chemokines, and the beta chemokines, or CC chemokines. The CC chemokines, named due to two adjacent cysteine residues, are generally associated with promoting both NK cell and mononuclear cell migration like microglia/macrophages during inflammation. The CXC chemokines, denoted due to two cysteine residues separated by any amino acid, can attract multiple cell types and are further subdivided into ELR (Glu-Leu-Arg) containing and non-ELR containing motifs (Baggiolini et al., 1997; Clark-Lewis et al., 1991). ELR containing CXC chemokines induce the directional migration of neutrophils, whereas the non-ELR CXC chemokines are able to attract various cells types like natural killer cells as well as activated T cells (Th1 subset) by signaling through the CXCR3 receptor (Farber, 1997; Hildebrant et al., 2004).

Chemokines are activated by binding to their respective CXC or CC receptors thereby eliciting intracellular signaling cascades that regulate the movement of the cell. These receptors are G-protein coupled with seven transmembrane regions and are often capable of binding more than one CC or CXC chemokine, respectively. A variety of cell types express chemokine receptors, including endothelial cells, lymphocytes, macrophages, and resident CNS cells such as neurons, astrocytes and microglia (Taub and Oppenheim, 1994; Luster, 1998).

Chemokine secretion has been detected in a number of different diseases and their presence caused the activation and accumulation of leukocytes (Baggiolini et al., 1994). This initial phase of leukocyte invasion is thought to be especially destructive, and is preceded by the expression of chemokines (Ghirnikar et al., 1998; Luster, 1998; McTigue et al., 1998; Ransohoff and Tani, 1998; Liu et al., 2001; Gonzalez et al., 2003). Indeed, the upregulation of various chemokines has also been associated with many neuroinflammatory pathologies (Galimberti et al., 2004; Banisor et al., 2005; Bartosik-Psujek and Stelmasiak, 2005; Charo and Ransohoff, 2006; Ubogu et al., 2006). While the expression patterns of many chemokines during CNS inflammatory diseases have not yet been thoroughly investigated, a few non-ELR CXC chemokines are the current focus of several studies. More specifically, the chemokines Mig/CXCL9 (monokine inducible by interferon gamma), CXCL10/IP-10 and CXCL11 are related interferon (IFN)-γ-inducible chemokines that play crucial roles in Th1 inflammatory mediated diseases such as allograft rejection and multiple sclerosis (Gerard and Rollins, 2001; Liu et al., 2001). They all act by binding to the CXCR3 receptor on CD8 + T lymphocytes, activated CD4 + T lymphocytes, and natural killer cells (Baggiolini et al., 1997; Bonecchi et al., 1998; Lane et al., 2000; Ogasawara et al., 2002; Liu et al., 2001). Although all three of these chemokines are induced by IFN-γ, and bind the CXCR3 receptor, they appear to have distinct biological roles in vivo (Liu et al., 2005).

The T lymphocyte chemoattractant chemokine CXCL10 is the most widely studied of the three CXCR3 binding chemokines and is known to be expressed during neuroinflammatory disorders (Luster et al, 1995; McTigue et al., 1998; Liu et al., 2001; Sui et al., 2004). In addition to its role as a chemoattractant, CXCL10 has been shown to enhance IFN-γ release and is involved in the development of antigen-driven T-cell responses, suggesting that it plays an important role in the overall evolution of Th1 immune responses (Dufour et al., 2002; O'Garra et al., 1998; Tsunoda et al., 2004).

The role of CXCL10 in lymphocyte trafficking into the CNS has been well documented utilizing the rodent animal model of MS, EAE and MHV (Dufour et al., 2002; Liu et al., 2001; Tsunoda et al., 2004). Studies have found elevated levels of CXCL10 in the cerebral spinal fluid (CSF) of MS patients during periods of clinical attack (Sorensen et al., 1999) as well as in astrocytes within active plaque lesions in the CNS (Balashov et al., 1999; Sorensen et al., 2002). Additionally, the majority of T cells and macrophages infiltrating the CNS of MS patients express the receptor for CXCL10, CXCR3 (Sorensen et al., 2002; Simpson et al., 2000). Indeed, it has been demonstrated that when CXCL10 levels are decreased, there is a corresponding reduction in inflammation and disease severity (Balashov et al., 1999). Furthermore, in Liu et al (2001) antibody neutralization of the chemokine CXCL10 reduced inflammatory cell invasion and demyelination, and improved neurological function in a viral model of MS (Liu et al., 2001). Together, these studies demonstrate an important role for CXCL10 in the pathogenesis of neuroinflammatory disorders such as MS and they support chemokine targeting as a means for therapeutic intervention (Sorensen et al., 1999, 2002; Balashov et al., 1999; Liu et al., 2001; Fife et al. 2001).

Aside from its role in neuroinflammation, CXCL10 is also recognized to have potent angiostatic activity in vivo and in vitro (Belperio et al., 2000; Strieter et al., 2002; Angiolillo et al., 1995; Glaser et al., 2004) where it inhibits angiogenesis by tumor-associated vasculature and promotes tumor necrosis (Maione et al., 1990; Arenberg et al., 1996; Belperio et al., 2000; Strieter et al., 2002). CXCL10 exerts these angiostatic effects by inhibiting endothelial cell growth, proliferation and chemotaxis as well as inhibiting angiogenic activities such as basic fibroblast growth factor-induced neovascularization (Angiolillo et al., 1995).

The link between CXCL10's dual activating (neuroinflammatory) and inhibiting (angiostatic) activities had not been elucidated until 2003 when Lasagni et al. (2003) found a different, unrecognized CXCR3 receptor, named CXCR3B. They found that the CXCR3 receptor is alternatively spliced into two variants, CXCR3A (the classically known receptor) and CXCR3B (newly found receptor). Binding of CXCL10 to CXCR3A causes the activation of proliferation and chemotaxis to the target cell, as shown in its role in neuroinflammation, while binding of CXCL10 to CXCR3B has inhibitory, angiogenic effects (Lasagni et al., 2003).

All of the above studies show that CXCL10 is strongly involved in neuroinflammatory diseases (Liu et al., 2000; Dufour et al., 2002), and is implicated in angiogenesis as well (Belperio et al., 2000; Strieter et al., 2002; Angiolillo et al., 1995; Glaser et al., 2004). In addition to this, CXCL10 has also been found to play an important role in other CNS disorders including demyelinating neuropathies (Kieseier et al., 2002) and human immunodeficiency virus dementia (Sui et al., 2004) where it has a prominent role in apoptosis of neurons in the brain (Sui et al., 2004, 2006). Collectively, all of these studies show CXCL10's crucial role in triggering T lymphocyte infiltration and pathology within the CNS.

4   Chemokines and Spinal Cord Injury

Although the role of chemokines in various CNS diseases has been well documented (Liu et al., 2000; Dufour et al., 2002; Kieseier et al., 2002; Sui et al., 2004,
2006; Bendall, 2005; Charo and Ransohoff, 2006; Ubogu et al., 2006), their role
in SCI specifically, has not. The few existing studies presented below clearly illustrate the important role of chemokines in spinal cord pathology and the potential
for chemokine-directed therapies in treating SCI.

A few studies have delineated the expression pattern of specific chemokines after
SCI. In 1997, Bartholdi and Schwab documented for the first time the upregulation
of the cytokines TNF-α, and IL-1 and the chemokines macrophage inflammatory
protein-1α (MIP-1α) and MIP-1β 1h after SCI (Bartholdi and Schwab, 1997).
Following this study two other groups showed that in addition to MIP-1, other
chemokines such as GRO-α and LIX (neutrophil chemoattractants), monocyte
chemoattractant protein-1 (MCP-1), MCP-5 and inducible protein 10 (IP-10) were
also upregulated at varying times after SCI (McTigue et al., 1998; Lee et al., 2000).
Further implicating a role for chemokines in SCI, Ghirnikar et al. (2000) demonstrated that infusion of a broad chemokine receptor antagonist, vMIPII, suppressed
gliotic reactions, reduced neuronal damage and dramatically halted inflammatory
cellular infiltration following SCI. These changes were also accompanied by
increased levels of an endogenous apoptosis inhibitor, thereby reducing neuronal
apoptosis (Ghirnikar et al., 2001). Notably in a follow up study, infusion of vMIPII
also lead to lower hematogenous infiltration, reduced axonal degeneration and
increased neuronal survival (Ghirnikar et al., 2001).

In a more specified targeting of chemokines following SCI, a group from Ohio
found increased production of MCP-1 and MCP-3 1 day following SCI; MCP-1
receptor CCR2 knockout led to reduced macrophage infiltration and myelin accumulation at the lesion epicenter 7 days post injury (Ma et al., 2002). In addition,
these animals also showed reduced expression of the chemokine receptors CCR1
and CCR5, both receptors for MIP-1 and RANTES (CCL5), supporting the hypothesis that MCP-1 interactions affect other chemokine expression patterns that further
lead to SCI damage. In addition, Jones et al. (2005) demonstrated that intraspinal
T cell accumulation was accompanied by dramatic increases in the chemokines
CXCL10 (IP-10) and CCL5 (RANTES) mRNA expression (Jones et al., 2005).
These chemokines were expressed 2- to 17-fold higher in transgenic mice with
myelin basic protein (MBP)-reactive T cells, and were associated with faster T cell
influx and enhanced macrophage activation within the spinal cord.

Elevated levels of the chemokine CXCL10 have been found in the spinal cords of
various SCI models and have been shown to be upregulated for as long as 14 days
following SCI (Gonzalez et al., 2003; Jones et al., 2005; Lee et al., 2000; McTigue
et al., 1998). We have demonstrated that the chemokine CXCL10 is upregulated by
6h following SCI and remains above basal levels 14 days post-injury (Gonzalez et al.,
2003). Moreover, the upregulation of CXCL10 also correlated with an exacerbated
immune cell infiltration into the damaged spinal cord (Jones et al., 2005). Further
supporting a role for CXCL10 in inflammation are studies that show the upregulated

expression of CXCL10 in the cerebral spinal fluid (CSF) of patients with MS (Sorensen et al., 1999), the CSF of patients with Guillen-Barre syndrome and demyelinating polyradiculoneuropathy (CIDP) (Kieseier et al., 2002), and within the CNS of spinal cord injured animals (Jones et al., 2005). Pathogenic Th1 T cells preferentially express the receptor for CXCL10, CXCR3 and these CXCR3 expressing T cells predominate in MS, EAE and SCI alike (Balashov et al., 1999; Sorensen et al., 1999; Jones et al., 2005). Therefore, selective manipulation of the chemokine CXCL10, which recruits/activates T cells, may be an attractive therapeutic target for secondary degeneration following SCI.

Work from our laboratory has further demonstrated that administration of a neutralizing antibody against CXCL10 (anti-CXCL10 antibody) into mice with SCI resulted in a pronounced reduction in neuroinflammation (Gonzalez et al., 2003). In addition, anti-CXCL10 antibody treatment also decreased the infiltrating activated macrophage response. These findings are particularly important to our understanding of SCI, as activated leukocytes are known to release a wide variety of reactive oxygen and nitrogen species, and activated mononuclear phagocytes are known to release neurotoxins, all of which contribute to the detrimental secondary degenerative cascade of SCI (Giulian et al., 1993; Martiney et al., 1998; Satake et al., 2000; Bao et al. 2004, 2005). Although microglia have been shown by others to express the CXCL10 receptor CXCR3 in vitro and in vivo, hematogenous macrophages have not (Van Der Meer et al., 2001; Biber et al., 2002). Therefore, it is likely that the decrease in infiltrating hematogenous macrophages that results from anti-CXCL10 treatment is a downstream effect of the decrease of infiltrating lymphocytes, as activated CD4 + T lymphocytes secrete RANTES, a chemoattractant for hematogenous macrophages (Fransen et al., 2000).

Most importantly, in addition to the reduction in neuroinflammation, anti-CXCL10 antibody treatment has been shown to significantly ameliorate locomotor deficits (Gonzalez et al., 2003). Although the mechanisms of action remains elusive, the treatment did lead to significant increases in tissue sparing (Glaser et al., 2004) as well as a reduction in neuronal loss and apoptosis, a key component in the secondary degenerative cascade that is ultimately linked to locomotion (Glaser et al., 2006). Apoptosis has been shown to contribute to cell death following SCI, a fact that has resulted in the development of many therapeutic approaches for SCI (Li et al., 2000; Demjen et al., 2004; Casha et al., 2005). Likewise, other studies in HIV dementia have demonstrated that CXCL10 plays a neurotoxic role in the disease (van Marle et al., 2004), and that neuronal apoptosis is mediated by CXCL10 over-expression (Sui et al., 2004). These observations support the notion that CXCL10 is detrimental in the CNS. Thus, glial and neuronal cell survival may be adversely affected by CXCL10 recruited/activated T cells, yet it remains to be determined if CXCL10 has a direct neurotoxic effect on neurons following SCI.

CXCL10 has also been shown to inhibit angiogenesis in vitro and in vivo (Strieter et al., 1995; Angiolillo et al., 1995; Luster et al., 1995) as mentioned in the previous section. Data from our laboratory has demonstrated that anti-CXCL10 treatment significantly enhanced blood vessel formation after SCI in vivo (Glaser et al., 2004). In this study, anti-CXCL10 antibody treatment led to a significant increase (182%) in the number of blood vessels at the injury epicenter following SCI. Increased blood flow restores oxygen and nutrients to tissue, creating a supportive environment for cell survival and growth. Indeed, angiogenesis has been shown to precede nerve regeneration (Zhang and Guth, 1997; Weidner et al., 2001; Kwon et al., 2002). Nerve regeneration does not necessarily imply that functional synaptic connections are being made, and may not contribute to behavioral recovery. Rather, it is an indicator that the environment is growth permissive. Thus, angiogenesis is believed to be crucial for wound healing and may be an important factor in establishing the proper environment for nerve regeneration. This study clearly demonstrated that CXCL10 plays a critical role in vasculature remodeling following SCI, and that angiogenesis is enhanced following anti-CXCL10 treatment.

Further supporting the theory that anti-CXCL10 treatment establishes a growth permissive environment following SCI, is a recent report from our laboratory demonstrating that neutralization of CXCL10 in spinal cord injured animals led to significant changes in genes involved in nervous system development and regenerative responses (Glaser et al., 2006). Moreover, quantification of biotinylated dextran amines (BDA) labeled corticospinal axons indicated a significant increase in the amount of axons caudal to the injury site in mice treated with anti-CXCL10 as compared to untreated mice (Glaser et al., 2006). Since revascularization occurs prior to nerve regrowth (Zhang and Guth, 1997; Weidner et al., 2001; Kwon et al., 2002), and anti-CXCL10 antibody treatment promotes angiogenesis (Glaser et al., 2006), these data clearly support the idea that our treatment renders a growth permissive environment that facilitates CNS repair.

Collectively, all of the studies above document that immune cell infiltration can be extremely detrimental to CNS disorders and to SCI specifically. Immune mediated damage is guided in part by chemokines and their specific expression and action can dictate the level of disease pathology. Therefore, targeting chemokine action following SCI represents a viable therapeutic strategy.

Acknowledgments The authors thank Rafael Gonzalez and Janette Glaser for discussion and advice.

References

Angiolillo, A. L., C. Sgadari, et al. (1995). Human interferon-inducible protein 10 is a potent inhibitor of angiogenesis in vivo. *J Exp Med* **182**(1): 155–162.

Arenberg, D. A., S. L. Kunkel, et al. (1996). Interferon-gamma-inducible protein 10 (IP-10) is an angiostatic factor that inhibits human non-small cell lung cancer (NSCLC) tumorigenesis and spontaneous metastases. *J Exp Med* **184**(3): 981–992.

Baggiolini, M., B. Moser, et al. (1994). Interleukin-8 and related chemotactic cytokines. The Giles Filley Lecture. *Chest* **105**(3 Suppl): 95S–98S.

Baggiolini, M., B. Dewald, et al. (1997). Human chemokines: an update. *Annu Rev Immunol* **15**: 675–705.

Balashov, K. E., J. B. Rottman, et al. (1999). CCR5(+) and CXCR3(+) T cells are increased in multiple sclerosis and their ligands MIP-1alpha and IP-10 are expressed in demyelinating brain lesions. *Proc Natl Acad Sci U S A* **96**(12): 6873–6878.

Banisor, I., T. P. Leist, et al. (2005). Involvement of beta-chemokines in the development of inflammatory demyelination. *J Neuroinflammation* **2**(1): 7.

Bao, F., Y. Chen, et al. (2004). An anti-CD11d integrin antibody reduces cyclooxygenase-2 expression and protein and DNA oxidation after spinal cord injury in rats. *J Neurochem* **90**(5): 1194–1204.

Bao, F., G. A. Dekaban, et al. (2005). Anti-CD11d antibody treatment reduces free radical formation and cell death in the injured spinal cord of rats. *J Neurochem* **94**(5): 1361–1373.

Bartholdi, D. and M. E. Schwab (1997). Expression of pro-inflammatory cytokine and chemokine mRNA upon experimental spinal cord injury in mouse: an in situ hybridization study. *Eur J Neurosci* **9**(7): 1422–1438.

Bartosik-Psujek, H. and Z. Stelmasiak (2005). The levels of chemokines CXCL8, CCL2 and CCL5 in multiple sclerosis patients are linked to the activity of the disease. *Eur J Neurol* **12**(1): 49–54.

Beattie, M. S., G. E. Hermann, et al. (2002). Cell death in models of spinal cord injury. *Prog Brain Res* **137**: 37–47.

Belperio, J. A., M. P. Keane, et al. (2000). CXC chemokines in angiogenesis. *J Leukoc Biol* **68**(1): 1–8.

Bendall, L. (2005). Chemokines and their receptors in disease. *Histol Histopathol* **20**(3): 907–926.

Bethea, J. R. and W. D. Dietrich (2002). Targeting the host inflammatory response in traumatic spinal cord injury. *Curr Opin Neurol* **15**(3): 355–360.

Biber, K., I. Dijkstra, et al. (2002). Functional expression of CXCR3 in cultured mouse and human astrocytes and microglia. *Neuroscience* **112**(3): 487–497.

Blight, A. R. (1985). Delayed demyelination and macrophage invasion: a candidate for secondary cell damage in spinal cord injury. *Cent Nerv Syst Trauma* **2**(4): 299–315.

Blight, A. R., T. I. Cohen, et al. (1995). Quinolinic acid accumulation and functional deficits following experimental spinal cord injury. *Brain* **118** (Pt 3): 735–752.

Bonecchi, R., G. Bianchi, et al. (1998). Differential expression of chemokine receptors and chemotactic responsiveness of type 1 T helper cells (Th1s) and Th2s. *J Exp Med* **187**(1): 129–34.

Bracken, M. B., M. J. Shepard, et al. (1992). Methylprednisolone or naloxone treatment after acute spinal cord injury: 1-year follow-up data. Results of the second National Acute Spinal Cord Injury Study. *J Neurosurg* **76**(1): 23–31.

Bracken, M. B., M. J. Shepard, et al. (1998). Methylprednisolone or tirilazad mesylate administration after acute spinal cord injury: 1-year follow up. Results of the third National Acute Spinal Cord Injury randomized controlled trial. *J Neurosurg* **89**(5): 699–706.

Casha, S., W. R. Yu, et al. (2005). FAS deficiency reduces apoptosis, spares axons and improves function after spinal cord injury. *Exp Neurol* **196**(2): 390–400.

Charo, I. F. and R. M. Ransohoff (2006). The many roles of chemokines and chemokine receptors in inflammation. *N Engl J Med* **354**(6): 610–621.

Clark-Lewis, I., C. Schumacher, et al. (1991). Structure-activity relationships of interleukin-8 determined using chemically synthesized analogs. Critical role of NH2-terminal residues and evidence for uncoupling of neutrophil chemotaxis, exocytosis, and receptor binding activities. *J Biol Chem* **266**(34): 23128–23134.

Cotman, C. and C. Engesser-Cesar (2002). Exercise enhances and protects brain function. *Exerc sport sci rev* **30**: 75–79.

Cotman, C. W. and N. C. Berchtold (2002). Exercise: a behavioral intervention to enhance brain health and plasticity. *Trends Neurosci* **25**(6): 295–301.

Demjen, D., S. Klussmann, et al. (2004). Neutralization of CD95 ligand promotes regeneration and functional recovery after spinal cord injury. *Nat Med* **10**(4): 389–395.

Dijkstra, S., S. Duis, et al. (2006). Intraspinal administration of an antibody against CD81 enhances functional recovery and tissue sparing after experimental spinal cord injury. *Exp Neurol* **202**(1): 57–66.

Dufour, J. H., M. Dziejman, et al. (2002). IFN-gamma-inducible protein 10 (IP-10; CXCL10)-deficient mice reveal a role for IP-10 in effector T cell generation and trafficking. *J Immunol* **168**(7): 3195–204.

Dusart, I. and M. E. Schwab (1994). Secondary cell death and the inflammatory reaction after dorsal hemisection of the rat spinal cord. *Eur J Neurosci* **6**(5): 712–724.

Farber, J. M. (1997). Mig and IP-10: CXC chemokines that target lymphocytes. *J Leukoc Biol* **61**(3): 246–57.

Fehlings, M. G. and C. H. Tator (1999). An evidence-based review of decompressive surgery in acute spinal cord injury: rationale, indications, and timing based on experimental and clinical studies. *J Neurosurg* **91**(1 Suppl): 1–11.

Fehlings, M. G., L. H. Sekhon, et al. (2001). The role and timing of decompression in acute spinal cord injury: what do we know? What should we do? *Spine* **26**(24 Suppl): S101–S110.

Fife, B. T., K. J. Kennedy, et al. (2001). CXCL10 (IFN-gamma-inducible protein-10) control of encephalitogenic CD4 + T cell accumulation in the central nervous system during experimental autoimmune encephalomyelitis. *J Immunol* **166**(12): 7617–7624.

Fransen, S., K. F. Copeland, et al. (2000). RANTES production by T cells and CD8-mediated inhibition of human immunodeficiency virus gene expression before initiation of potent antiretroviral therapy predict sustained suppression of viral replication. *J Infect Dis* **181**(2): 505–512.

Galimberti, D., N. Bresolin, et al. (2004). Chemokine network in multiple sclerosis: role in pathogenesis and targeting for future treatments. *Expert Rev Neurother* **4**(3): 439–453.

Geisler, F. H. (1993). GM-1 ganglioside and motor recovery following human spinal cord injury. *J Emerg Med* **11**(1 Suppl): 49–55.

Geisler, F. H., W. P. Coleman, et al. (2001). The Sygen multicenter acute spinal cord injury study. *Spine* **26**(24 Suppl): S87–S98.

Gerard, C. and B. J. Rollins (2001). Chemokines and disease. *Nat Immunol* **2**(2): 108–15.

Ghirnikar, R. S., Y. L. Lee, et al. (1998). Chemokine inhibition in rat stab wound brain injury using antisense oligodeoxynucleotides. *Neurosci Lett* **247**(1): 21–24.

Ghirnikar, R. S. and Y. L. Lee, et al. (2000). Chemokine antagonist infusion attenuates cellular infiltration following spinal cord contusion injury in rat. *J Neurosci Res* **59**(1): 63–73.

Ghirnikar, R. S., Y. L. Lee, et al. (2001). Chemokine antagonist infusion promotes axonal sparing after spinal cord contusion injury in rat. *J Neurosci Res* **64**(6): 582–589.

Giulian, D., M. Corpuz, et al. (1993). Reactive mononuclear phagocytes release neurotoxins after ischemic and traumatic injury to the central nervous system. *J Neurosci Res* **36**(6): 681–693.

Glaser, J., R. Gonzalez, et al. (2004). Neutralization of the chemokine CXCL10 enhances tissue sparing and angiogenesis following spinal cord injury. *J Neurosci Res* **77**(5): 701–708.

Glaser, J., R. Gonzalez, et al. (2006). Neutralization of the chemokine CXCL10 reduces apoptosis and increases axon sprouting after spinal cord injury. *J Neurosci Res* **84**(4): 724–734.

Gomez-Pinilla, F., Z. Ying, et al. (2001). Differential regulation by exercise of BDNF and NT-3 in rat spinal cord and skeletal muscle. *Eur J Neurosci* **13**(6): 1078–1084.

Gonzalez, R., J. Glaser, et al. (2003). Reducing inflammation decreases secondary degeneration and functional deficit after spinal cord injury. *Exp Neurol* **184**(1): 456–463.

Gris, D., D. R. Marsh, et al. (2004). Transient blockade of the CD11d/CD18 integrin reduces secondary damage after spinal cord injury, improving sensory, autonomic, and motor function. *J Neurosci* **24**(16): 4043–4051.

Hagg, T. and M. Oudega (2006). Degenerative and spontaneous regenerative processes after spinal cord injury. *J Neurotrauma* **23**(3–4): 264–280.

Hall, E. (2001). Pharmacological treatment of acute spinal cord injury: how do we build on past success? *Journal of spinal cord medicine* **24**(3): 142–146.

Hauben, E., U. Nevo, et al. (2000). Autoimmune T cells as potential neuroprotective therapy for spinal cord injury. *Lancet* **355**(9200): 286–287.

Hausmann, O. N. (2003). Post-traumatic inflammation following spinal cord injury. *Spinal Cord* **41**(7): 369–378.

Hildebrandt, G. C., L. A. Corrion, et al. (2004). Blockade of CXCR3 receptor:ligand interactions reduces leukocyte recruitment to the lung and the severity of experimental idiopathic pneumonia syndrome. *J Immunol* **173**(3): 2050–9.

Hugenholtz, H. (2003). Methylprednisolone for acute spinal cord injury: not a standard of care. *CMAJ* **168**: 1145–1146.

Hurlbert, R. J. (2000). Methylprednisolone for acute spinal cord injury: an inappropriate standard of care. *J Neurosurg* **93**(1 Suppl): 1–7.

Jones, T. B., R. P. Hart, et al. (2005). Molecular control of physiological and pathological T-cell recruitment after mouse spinal cord injury. *J Neurosci* **25**(28): 6576–6583.

Kieseier, B. C., M. Tani, et al. (2002). Chemokines and chemokine receptors in inflammatory demyelinating neuropathies: a central role for IP-10. *Brain* **125** (Pt 4): 823–834.

Kwon, B. K., J. F. Borisoff, et al. (2002). Molecular targets for therapeutic intervention after spinal cord injury. *Mol Interv* **2**(4): 244–258.

Kwon, B. K., W. Tetzlaff, et al. (2004). Pathophysiology and pharmacologic treatment of acute spinal cord injury. *Spine J* **4**(4): 451–464.

Lane, T. E., M. T. Liu, et al. (2000). A central role for CD4(+) T cells and RANTES in virus-induced central nervous system inflammation and demyelination. *J Virol* **74**(3): 1415–1424.

Lasagni, L., M. Francalanci, et al. (2003). An alternatively spliced variant of CXCR3 mediates the inhibition of endothelial cell growth induced by IP-10, Mig, and I-TAC, and acts as functional receptor for platelet factor 4. *J Exp Med* **197**(11): 1537–49.

Lee, Y. L., K. Shih, et al. (2000). Cytokine chemokine expression in contused rat spinal cord. *Neurochem Int* **36**(4–5): 417–425.

Li, M., V. O. Ona, et al. (2000). Functional role and therapeutic implications of neuronal caspase-1 and -3 in a mouse model of traumatic spinal cord injury. *Neuroscience* **99**(2): 333–342.

Liu, M. T., H. S. Keirstead, et al. (2001). Neutralization of the chemokine cxcl10 reduces inflammatory cell invasion and demyelination and improves neurological function in a viral model of multiple sclerosis. *J Immunol* **167**(7): 4091–4097.

Luster, A. D. (1998). Chemokines–chemotactic cytokines that mediate inflammation. *N Engl J Med* **338**(7): 436–445.

Luster, A. D., S. M. Greenberg, et al. (1995). The IP-10 chemokine binds to a specific cell surface heparan sulfate site shared with platelet factor 4 and inhibits endothelial cell proliferation. *J Exp Med* **182**(1): 219–231.

Ma, M., T. Wei, et al. (2002). Monocyte recruitment and myelin removal are delayed following spinal cord injury in mice with CCR2 chemokine receptor deletion. *J Neurosci Res* **68**(6): 691–702.

Maione, T. E., G. S. Gray, et al. (1990). Inhibition of angiogenesis by recombinant human platelet factor-4 and related peptides. *Science* **247**(4938): 77–79.

Martiney, J. A., C. Cuff, et al. (1998). Cytokine-induced inflammation in the central nervous system revisited. *Neurochem Res* **23**(3): 349–359.

McDonald, J. W., D. Becker, et al. (2002). Late recovery following spinal cord injury. Case report and review of the literature. *J Neurosurg* **97**(2 Suppl): 252–265.

McTigue, D. M., M. Tani, et al. (1998). Selective chemokine mRNA accumulation in the rat spinal cord after contusion injury. *J Neurosci Res* **53**(3): 368–376.

Mills, C.D., B. Hains et al. (2001). "Strain and model differences in behavioral outcomes after spinal cord injury in rats." *J neurotrauma* **18**: 743–65.

Moalem, G., R. Leibowitz-Amit, et al. (1999). Autoimmune T cells protect neurons from secondary degeneration after central nervous system axotomy. *Nat Med* **5**(1): 49–55.

Neeper, S., F. Gomez-Pinilla, et al. (1995). Exercise and brain neurotrophins. *Nature* **373**: 109.

O'Garra, A., L. M. McEvoy, et al. (1998). T-cell subsets: chemokine receptors guide the way. *Curr Biol* **8**(18): R646–R649.

Ogasawara, K., S. K. Yoshinaga, et al. (2002). Inducible costimulator costimulates cytotoxic activity and IFN-gamma production in activated murine NK cells. *J Immunol* **169**(7): 3676–85.

Okada, S., M. Nakamura, et al. (2004). Blockade of interleukin-6 receptor suppresses reactive astrogliosis and ameliorates functional recovery in experimental spinal cord injury. *J Neurosci Res* **76**(2): 265–276.

Olsson, T. (1995). Cytokine-producing cells in experimental autoimmune encephalomyelitis and multiple sclerosis. *Neurology* **45**(6 Suppl 6): S11–S15.

Popovich, P. G. (2000). Immunological regulation of neuronal degeneration and regeneration in the injured spinal cord. *Prog Brain Res* **128**: 43–58.

Popovich, P. G., P. Wei, et al. (1997). Cellular inflammatory response after spinal cord injury in Sprague-Dawley and Lewis rats. *J Comp Neurol* **377**(3): 443–464.

Popovich, P. G., Z. Guan, et al. (1999). Depletion of hematogenous macrophages promotes partial hindlimb recovery and neuroanatomical repair after experimental spinal cord injury. *Exp Neurol* **158**(2): 351–365.

Ramer, L., M. Ramer, et al. (2005). Setting the stage for functional repair of spinal cord injuries: a cast of thousands. *Spinal cord* **43**: 134–161.

Ransohoff, R. M. and M. Tani (1998). Do chemokines mediate leukocyte recruitment in posttraumatic CNS inflammation? *Trends Neurosci* **21**(4): 154–159.

Satake, K., Y. Matsuyama, et al. (2000). Nitric oxide via macrophage iNOS induces apoptosis following traumatic spinal cord injury. *Brain Res Mol Brain Res* **85**(1–2): 114–122.

Saville, L. R., C. H. Pospisil, et al. (2004). A monoclonal antibody to CD11d reduces the inflammatory infiltrate into the injured spinal cord: a potential neuroprotective treatment. *J Neuroimmunol* **156**(1–2): 42–57.

Schnell, L., S. Fearn, et al. (1999). Acute inflammatory responses to mechanical lesions in the CNS: differences between brain and spinal cord. *Eur J Neurosci* **11**(10): 3648–3658.

Schulze-Koops, H., P. E. Lipsky, et al. (1995). Elevated Th1- or Th0-like cytokine mRNA in peripheral circulation of patients with rheumatoid arthritis. Modulation by treatment with anti-ICAM-1 correlates with clinical benefit. *J Immunol* **155**(10): 5029–5037.

Schwab, M. E. and D. Bartholdi (1996). Degeneration and regeneration of axons in the lesioned spinal cord. *Physiol Rev* **76**(2): 319–70.

Simpson, J. E., J. Newcombe, et al. (2000). Expression of the interferon-gamma-inducible chemokines IP-10 and Mig and their receptor, CXCR3, in multiple sclerosis lesions. *Neuropathol Appl Neurobiol* **26**(2): 133–142.

Sorensen, T. L., M. Tani, et al. (1999). Expression of specific chemokines and chemokine receptors in the central nervous system of multiple sclerosis patients. *J Clin Invest* **103**(6): 807–815.

Sorensen, T. L., C. Trebst, et al. (2002). Multiple sclerosis: a study of CXCL10 and CXCR3 co-localization in the inflamed central nervous system. *J Neuroimmunol* **127**(1–2): 59–68.

Sroga, J. M., T. B. Jones, et al. (2003). Rats and mice exhibit distinct inflammatory reactions after spinal cord injury. *J Comp Neurol* **462**(2): 223–240.

Stirling, D. P., K. Khodarahmi, et al. (2004). Minocycline treatment reduces delayed oligodendrocyte death, attenuates axonal dieback, and improves functional outcome after spinal cord injury. *J Neurosci* **24**(9): 2182–2190.

Strieter, R. M., S. L. Kunkel, et al. (1995). Interferon gamma-inducible protein 10 (IP-10), a member of the C-X-C chemokine family, is an inhibitor of angiogenesis. *Biochem Biophys Res Commun* **210**(1): 51–57.

Strieter, R. M., J. A. Belperio, et al. (2002). CXC chemokines in angiogenesis related to pulmonary fibrosis. *Chest* **122**(6 Suppl): 298S–301S.

Sui, Y., R. Potula, et al. (2004). Neuronal apoptosis is mediated by CXCL10 overexpression in simian human immunodeficiency virus encephalitis. *Am J Pathol* **164**(5): 1557–1566.

Sui, Y., L. Stehno-Bittel, et al. (2006). CXCL10-induced cell death in neurons: role of calcium dysregulation. *Eur J Neurosci* **23**(4): 957–64.

Tator, C. H. (2006). Review of treatment trials in human spinal cord injury: issues, difficulties, and recommendations. *Neurosurgery* **59**(5): 957–982; discussion 982–987.

Taub, D. D. and J. J. Oppenheim (1994). Chemokines, inflammation and the immune system. *Ther Immunol* **1**(4): 229–46.

Tsunoda, I., T. E. Lane, et al. (2004). Distinct roles for IP-10/CXCL10 in three animal models, Theiler's virus infection, EAE, and MHV infection, for multiple sclerosis: implication of differing roles for IP-10. *Mult Scler* **10**(1): 26–34.

Ubogu, E. E., M. B. Cossoy, et al. (2006). The expression and function of chemokines involved in CNS inflammation. *Trends Pharmacol Sci* **27**(1): 48–55.

Van Der Meer, P., S. H. Goldberg, et al. (2001). Expression pattern of CXCR3, CXCR4, and CCR3 chemokine receptors in the developing human brain. *J Neuropathol Exp Neurol* **60**(1): 25–32.

van Marle, G., S. Henry, et al. (2004). Human immunodeficiency virus type 1 Nef protein mediates neural cell death: a neurotoxic role for IP-10. *Virology* **329**(2): 302–318.

Van Meeteren, N. L., R. Eggers, et al. (2003). Locomotor recovery after spinal cord contusion injury in rats is improved by spontaneous exercise. *J Neurotrauma* **20**(10): 1029–1037.

Weidner, N., A. Ner, et al. (2001). Spontaneous corticospinal axonal plasticity and functional recovery after adult central nervous system injury. *Proc Natl Acad Sci U S A* **98**(6): 3513–3518.

Zhang, Z. and L. Guth (1997). Experimental spinal cord injury: Wallerian degeneration in the dorsal column is followed by revascularization, glial proliferation, and nerve regeneration. *Exp Neurol* **147**(1): 159–171.

12

The Usual Suspects: Chemokines and Microbial Infection of the Central Nervous System

Michelle J. Hickey, Linda N. Stiles, Chris S. Schaumburg, and Thomas E. Lane

1 Introduction

For many years, the central nervous system (CNS) was considered an "immunologically privileged site" – a perspective based on limited immune surveillance when compared to peripheral tissue, muted expression of MHC molecules in the context of an apparent lack of professional antigen presenting cells, and the absence of a classical lymphatic drainage system. Together, these observations supported the notion that the CNS was unable to mount and/or support an immune response. However, over time this view evolved and it is now clear that CNS tissue is neither immunologically inert nor privileged, rather, its immune response is exquisitely sensitive to antigenic challenge. Indeed, overwhelming evidence now indicates that upon microbial infection of the CNS there is often a dynamic and orchestrated localized immune response that culminates with infiltration of antigen-specific lymphocytes, usually resulting in control and elimination of the invading pathogen. It is important to note that not all effective immune responses originating in the CNS are completely beneficial to the host; alternatively, there are instances where immune cell infiltration following infection is associated with severe neuropathology resulting in death or chronic neurodegenerative disease.

The signaling events governing leukocyte infiltration into the CNS in response to infection are complex and depend on many factors including type of pathogen, e.g. intracellular or extracellular, the route of infection, cellular tropism (neuron and/or glial cells), and genetic background of the host. Nevertheless, it is now apparent that leukocyte trafficking is partially dependent on a class of small (7–17 kDa) chemotactic cytokines known as chemokines (*chemo*tactic cyto*kine*). Chemokines represent a family of over 40 proteins, which for the most part are secreted into the environment and function by binding to chemokine receptors in the form of G protein-coupled receptors (GPCRs) that are expressed on numerous cell types. Four sub-families of chemokines have been identified and defined based on structural criteria relating to the location of conserved cysteine residues within the amino-terminus of the protein. Chemokines were initially discovered close to 30 years ago and their strong association with various human inflammatory diseases led some researches to theorize, and later confirm, that they influence leukocyte recruitment into inflamed tissue. In fact, it is now accepted that this

T.E. Lane et al. (eds.), *Central Nervous System Diseases and Inflammation.*
© Springer 2008

superfamily of proteins plays an important role in numerous biological processes ranging from maintaining the organizational integrity of secondary lymphoid tissue to participating in various aspects of both innate and adaptive immune responses following microbial infection. For excellent reviews on chemokine and chemokine receptor signaling as well as a comprehensive overview of identified chemokine receptors and their ligands, please see (Charo and Ransohoff 2006; Le et al., 2004; Luster 1998).

This chapter will focus on highlighting recent insights into chemokine and chemokine receptor expression in both host defense and disease following either viral or bacterial infections of the CNS. In all cases, we have tried to provide information pertaining to regulation of chemokine expression and functional roles within the context of animal models of disease as well as clinical disease. We certainly acknowledge that other microbial pathogens e.g. fungi and parasites are capable of infecting and replicating within the CNS and chemokines/chemokine receptors have been suggested to participate in host responses. However, our focus was weighted by the fact that the overwhelming majority of clinical infections involving the CNS relate to viral or bacterial infection.

2 Viral Infection of the CNS

Evaluation of the immune response following viral infection of the CNS is important as a number of viruses are capable of infecting the CNS (Table 12.1). Indeed, exposure to a neurotropic virus at some point during the course of a normal human lifespan is almost inevitable. While the majority of these encounters will result in a benign and clinically silent course of infection often associated with life-long

Table 12.1 Diseases associated with Viral Infection of the CNS

Pathogen	Primary host cell(s)	Disease
Virus		
RNA		
LCMV	Neurons, astrocytes, glia	Meningitis, meningoencephalitis
Measles	Neurons	Meningitis, demyelination, encephalitis
West Nile	Neurons	Meningitis, encephalitis
Coronavirus	Glia; neurons	Encephalomyelitis, demyelination
TMEV	Microglia, macrophages	Encephalomyelitis, demyelination
DNA		
HSV	Neurons	Blindness, corneal scarring
CMV	Astrocytes	Encephalitis
Retrovirus		
HIV	Microglia	HAD, HIVE

persistence of virus, others can be quite severe and potentially life-threatening. A common pathological feature of many neurotropic viruses is acute meningitis and/ or encephalitis that occur as a result of neuroinflammation due to chemotactic signals derived within the CNS in response to infection. While immune cell infiltration may be beneficial by attracting antigen-specific T cells required for reducing viral load, in many of these cases, the amplified inflammatory responses can also be dangerous and lead to permanent, immune-mediated neurologic damage and/or death. It is now widely accepted that resident cells of the CNS are quite capable of secreting a number of different chemokines that undoubtedly contribute to directing distinct subsets of leukocytes into the CNS based upon surface expression of chemokine receptors (Lane et al., 2006). Therefore, defining chemokine/chemokine receptor expression profiles within the CNS of viral-infected hosts provides important information on potential targets for therapeutic intervention.

2.1   Lymphocytic Choriomeningitis Virus (LCMV)

LCMV is an ambisense RNA virus and a member of the *Arenaviridae* family (Emonet et al., 2006). It is a rodent-borne viral infectious disease in humans that presents as aseptic meningitis, encephalitis or meningoencephalitis (Jahrling and Peters 1992). Infection of the murine CNS with LCMV results in a well-established model of viral meningitis (Buchmeier et al., 1980; Buchmeier and Zajac 1999). LCMV infection of the CNS of adult, immunocompetent mice results in a monophasic disease characterized by leukocyte infiltration into distinct anatomic regions of the brain and ultimately death between 6–8 days post-infection (p.i.) (Buchmeier et al., 1980). Specifically, infiltrating CD8 + T cells are essential in contributing to cell damage and death in LCMV-infected mice. Instillation of LCMV into the CNS of immunocompetent mice reveals that expression of chemokine RNA in the CNS precedes the infiltration of immune cells (Asensio and Campbell 1997). Transcripts for CXCL10, CCL2, CCL4, CCL5, and CCL7 are apparent in the CNS of LCMV infected mice 3 days p.i. and increase by day 6, correlating with accumulation of activated T cells (Asensio and Campbell 1997). Astrocytes are, in part, responsible for expression of certain chemokines including CXCL10 following LCMV infection (Asensio et al., 1999). These early results suggested that chemokine expression in the CNS represents an early host response to intracranial (i.c.) infection with LCMV and drives subsequent immunopathology.

Characterization of chemokine receptors has revealed that CCR5 and CXCR3 are readily detected on infiltrating T cells present within the CNS of LCMV-infected mice (Christensen et al., 2004; Nansen et al., 2000, 2002). Studies using CCR5-deficient mice (CCR5−/− mice) suggested that CCR5 signaling on T cells is dispensable with regards to regulating T cell migration into the CNS, as LCMV-infection of CCR5−/− mice results in a lethal T cell-mediated meningitis with no change in the composition of the cellular infiltrate within the CNS of CCR5–/– mice compared to controls (Nansen et al., 2002). In contrast, LCMV infection of the CNS in CXCR3–/– mice results in a dramatic increase in survival that correlated with paucity in CD8 + T cells migrating into the parenchyma from the meninges; however, no alterations in the distribution of viral antigen between infected CXCR3–/– mice and wildtype control mice were observed (Christensen et al., 2004). Importantly, reconstitution of LCMV-infected CXCR3–/– mice with CXCR3 + / + CD8 + T cells restored susceptibility to viral-induced meningitis (Christensen et al., 2004). These data highlight the importance of CXCR3 in allowing positional migration of effector T cells to sites of infection during on-going viral-induced neuroinflammation.

Ligands capable of binding to CXCR3 include the non-ELR CXC chemokines CXCL9, -10, and -11. These chemokines lack the ELR motif that is a common feature in the amino terminus of the other members within the CXC subfamily. Subsequent studies indicated that CXCL10, but not CXCL9 or CXCL11, is the key signaling molecule responsible for recruiting effector T cells into the CNS of LCMV-infected mice (Christensen et al., 2006). However, given the overlapping chemokine expression profiles that exist within the CNS of LCMV-infected mice there is the possibility of functional redundancy as well as compensatory mechanisms that may be employed in the absence of specific chemokine signaling pathways. Indeed, LCMV-infection of mice deficient in both CXCR3 and CCR5 (CXCR3/CCR5-deficient mice) indicated that lack of both receptors does not impair generation of virus-specific T cells suggesting that signaling through these receptors is not critical in generating virus-specific T cells (de Lemos et al., 2005). Although T cell infiltration into the CNS of either CXCR3–/– or CXCR3/CCR5-deficient mice was reduced compared to wild-type mice, there were greater numbers of CD8 + T cells present in the neural parenchyma of double-deficient mice compared to CXCR3–/– mice, indicating that CCR5 may function to negatively regulate the antiviral CD8 + T cell response (de Lemos et al., 2005).

2.2 West Nile Virus (WNV)

West Nile Virus (WNV) is a flavivirus that cycles between primary and secondary hosts, including avian/mosquito vectors and humans/mammals, respectively (Sejvar and Marfin 2006). WNV was first isolated from an infected patient in Uganda in 1937 and has caused sporadic outbreaks in Africa and Asia. WNV represents a re-emerging viral pathogen as the virus was isolated from a flamingo in New York City in 1999. Subsequently, the virus spread west and, in the process, has had significant impact on specific populations of birds. In addition, humans are susceptible to infection, which can range from mild flu-like symptoms (West Nile Fever) to more serious neurological disease characterized by meningitis and encephalitis.

Peripheral infection of susceptible mice (C57BL/6) with neurotropic clinical isolates of WNV results in viral entry and disease that recapitulates many of the viral and pathological parameters of the human disease. Neurons are the primary targets of viral infection resulting in multifocal encephalitis characterized by infiltrating T cells and macrophages. Upon WNV infection, neurons secrete CXCL10 that presumably serves to attract CXCR3 + T cells into the CNS. In support of this are studies indicating that either antibody neutralization of CXCL10 or infection of CXCL10–/– mice results in diminished infiltration of CXCR3 + CD8 + T cells into the brain accompanied by an increase in virus and disease severity (Klein et al., 2005). These data support a protective role for CXCL10 in host defense against WNV infection by attracting virus-specific effector T cells into the CNS.

Studies by Murphy and colleagues suggest an alternative chemokine signaling pathway is important in host defense (Glass et al., 2005, 2006; Lim et al., 2006). WNV infection of CCR5–/– mice resulted in a rapid and uniformly fatal infection characterized by an impaired ability to clear virus from the brain and reduced numbers of T cells compared to infected CCR5 + / + mice (Glass et al., 2005). Adoptive transfer of WNV-immune splenocytes into infected CCR5–/– mice restored protection that correlated with increased T cell infiltration into the CNS. Supporting the importance of CCR5 signaling in host defense against WNV-induced neurologic disease are findings that indicate humans harboring the CCRDelta32 deletion have increased susceptibility to WNV-induced disease (Glass 2006). Indeed, CCR5Delta32 homozygosity was significantly associated with fatal outcome after WNV infection in one patient cohort examined highlighting the importance of this receptor in defense following viral infection of the CNS.

In addition, another mosquito-borne flavivirus, Japanese encephalitis virus (JEV), induces an acute encephalitis, in which the role of the host immune response with regards to either host defense and/or disease has not been well-characterized (Solomon and Winter 2004). However, animal models of JEV-induced encephalitis suggest that disease severity correlates with increased expression of proinflammatory genes (Chen et al., 2004, 2000; Suzuki et al., 2000; Winter et al., 2004). Chemokines, including CCL5 (a ligand for CCR5) are expressed within the CNS of JEV-infected humans suggesting a potential role in controlling leukocyte infiltration into the CNS (Chen et al., 2000; Suzuki et al., 2000). While levels of IgM and IgG were higher within the cerebral spinal fluid (CSF) of survivors of JEV infection, expression of chemokines CCL8 and CCL5 were elevated in nonsurvivors, implying a potential role in disease pathogenesis (Winter et al., 2004).

2.3   Herpes Simplex Viruses (HSV)

Herpes simplex viruses (HSV) are extremely prevalent human pathogens with seroconversion rates approaching 60% worldwide (Carr and Tomanek 2006; Looker and Garnett 2005). Infection by HSV type 1 (HSV-1) is the leading cause of blindness in the industrialized world due to corneal infection that leads to the disease herpetic corneal keratitis. The inflammatory response associated with these viruses strongly correlates with morbidity as a result of the development of lesions and subsequent scarring. Therefore, understanding the signals regulating inflammation in response to HSV-1 infection of the cornea may provide insight into relevant targets for immunotherapy.

The first thorough study to characterize the chemokine gene expression profile within the cornea of HSV-1-infected mice was performed in 1996 using PCR (Su et al., 1996). Chemokines detected included CXCL1, CXCL2, CXCL9, CXCL10 and CCL5 (Su et al., 1996). With the exception of CXCL10, expression of chemokines as well as proinflammatory cytokines is transient and occurs during the innate immune response. Recent investigations into molecular signaling pathways involved in initiating chemokine gene expression in response to HSV-1 infection of the CNS have revealed differential requirements for members of the TLR family with regards to chemokine gene expression. TLR2 is required for production of various proinflammatory cytokines as well as chemokines, including CXCL1, CXCL2, CXCL4, and CXCL5 (Aravalli et al., 2005). In contrast, increased production of CXCL9 and CXCL10 required both TLR9 as well as type I interferon signaling pathways (Wuest et al., 2006). HSV-1 infection of either TLR2−/− or TLR9−/− mice resulted in an attenuated inflammatory response highlighting a previously unappreciated role for TLR signaling pathways in regulating viral neuropathogenesis (Aravalli et al., 2005; Wuest et al., 2006). Additional studies by various groups employing either neutralizing antibodies or chemokine/chemokine receptor knock-out mice have also aided in defining the functional contributions of these molecules to disease. CXCL2 promotes neutrophil trafficking by signaling through the receptor CXCR2 expressed on the cell surface. Administration of anti-CXCL2 antibody to HSV-1-infected mice or infection of CXCR2−/− mice revealed limited neutrophil accumulation within the cornea (Maertzdorf et al., 2002; Yan et al., 1998). Blocking CXCL10 during the acute phase of HSV-1 ocular infection resulted in increased viral titers within the stroma and trigeminal ganglion, but limited corneal pathology, although spread of virus from the cornea stroma into the retina was markedly restricted in anti-CXCL10 treated mice (Carr et al., 2003). In addition, viral antigen was co-localized with infiltrating CD11b + cells suggesting that viral infection of inflammatory cells facilitates spread of the virus to other restricted anatomical regions of the eye (Carr et al., 2003). These findings led investigators to hypothesize that upon HSV-1 infection of the cornea, CXCL10 serves to orchestrate inflammatory responses that function to control viral replication, but also may serve to disseminate virus to other regions of the eye by enabling infection of inflammatory cells (Carr et al., 2003). HSV-1 infection of CXCR3−/− mice revealed somewhat surprising results as mice deficient in CXCR3 signaling exhibited an overall increase in survival rate compared to infected wildtype mice, which was associated with a 2-fold increase in the frequency of infiltrating T cells within the trigeminal ganglion (Wickham et al., 2005). Similarly, HSV-1 infection of mice lacking CCR5 allowed for characterization of the contributions of CCR5 ligands CCL3 and CCL5 to defense and/or disease. HSV-1 infection of CCR5−/− mice resulted in no difference in mortality despite deficiencies in controlling viral replication in the eye (Carr et al., 2006). This approach highlights the potential overlapping compensatory mechanisms that are inherent within the chemokine family.

2.4 Measles Virus (MV)

Measles virus is an enveloped negative-strand RNA virus, and a member of the paramyxovirus family (Griffin and Bellini 1996). Infection with MV typically results in an acute febrile illness and rash due to viral replication first in mono- and lymphocytic cells, and then in a variety of tissues (Katz 1995). MV infection is typically self-limiting due to the generation of the virus-specific immune response (Griffin and Bellini 1996). In the rare case, however, MV re-emerges in fully immunocompetent individuals in the CNS months to years following the resolution of the acute infection resulting in a fatal disease known as subacute sclerosing panencephalitis (SSPE) (ter Meulen et al., 1983). The pathological hallmarks of SSPE include demyelination, gliosis, and T lymphocyte infiltration into the CNS accompanied by expression of chemokines and cytokines (Anlar et al., 2001; Hofman et al., 1991; Nagano et al., 1994; Schnorr et al., 1997). As infiltration of activated immune cells into the CNS is important in disease pathogenesis in SSPE, understanding the signals regulating leukocyte invasion is critical to potential disease intervention. Infection of human astrocyte cultures with MV (either live virus or ultraviolet-inactivated) results in production of mRNA transcripts for CCL2, CCL3, CCL4, and CCL5 (Noe et al., 1999; Xiao et al., 1998). Addition of cytokines IFN-γ and TNF-α to viral-infected cultures did not result in a synergistic effect with regards to chemokine transcript expression, but rather selectively inhibited CCL2 and CCL4 transcript levels (Xiao et al., 1998). These intriguing results suggest localized expression of cytokines may regulate chemokine gene expression and modulate disease by altering the composition of the cellular infiltrate. Analysis of the cerebral spinal fluid (CSF) of SSPE patients contained elevated levels of chemokines, in which CXCL10 was predominantly expressed (Saruhan-Direskeneli et al., 2005).

The receptor for MV, CD46, is expressed at low levels on neurons, oligodendrocytes, and astrocytes in normal brains (McQuaid and Cosby 2002). Mouse models for MV pathogenesis have been hampered by the fact that mice do not express CD46 and are normally not infected by MV. However, a transgenic mouse expressing human CD46 on neurons under the control of the neuron-specific enolase promoter (NSE) is susceptible to infection and replication of human MV strains (Rall et al., 1997). Intracranial infection of transgenic NSE-CD46 neonate (day 1) mice with MV results in widespread viral replication in neurons that leads to acute encephalitis, seizures, weight loss and ataxia, with death occurring by day 15–20 p.i. (Rall et al., 1997). T and B lymphocytes enter the CNS and are found in association with MV-antigen expressing neurons following MV infection in NSE-CD46 mice (Manchester et al., 1999). The contribution of these cells to host defense was determined using adoptive transfer of immune cell subsets into transgenic CD46-expressing mice. These studies revealed that CD4 + T lymphocytes in combination with CD8 + T lymphocytes or B cells are necessary for protection during MV infection (Tishon et al., 2006). Additionally, neurons contributed to chemokine expression in the CNS following MV infection of NSE-CD46 mice (Patterson et al.,

2003). In these experiments chemokine transcripts for CXCL10 and CCL5 were elevated in CNS tissues isolated from MV infected immunodeficient NSE-CD46/ Rag-2 knock-out (ko) mice, as well as from MV-infected primary hippocampal neuron cultures established from embryonic NSE-CD46 mice (Patterson et al., 2003). Neutralization of CXCL10 and CCL5 with polyclonal neutralizing antisera resulted in significant reductions in CD4 + and CD8 + T cell infiltration into the CNS following MV infection of NSE-CD46 mice (Patterson et al., 2003). Therefore, these findings highlight the functional relevance of chemokines in the neuropathogenesis of MV infection and reveal potential targets for therapy in SSPE patients.

2.5 Cytomegalovirus

Cytomegalovirus (CMV) infection of humans can result in pathological manifestations within the CNS that include microglial nodule formation and ventriculoencephalitis (Arribas et al., 1995). Glial cells are sensitive to CMV infection and are capable of secreting chemokines. For example, CMV infection of astrocytes did not result in the production of antiviral cytokines but does generate chemokines CCL8 and CCL2 (Cheeran et al., 2001). However, CMV infection of microglial cells promoted production of anti-viral cytokines TNF-α and IL-6, in addition to chemokines, such as CXCL10 (Cheeran et al., 2003). It is possible that CMV-infected astrocytes attract microglial cells to sites of infection to aid in host defense via secretion of anti-viral cytokines. In support of this possibility are data demonstrating an ~60% reduction in viral gene expression in CMV-infected astrocytes co-cultured with microglia (Cheeran et al., 2001). CXCL10 expressed from CMV-infected microglia may aid in host defense by attracting T cells into the CNS (Cheeran et al., 2004). Infection of immunodeficient mice with murine CMV (MCMV) resulted in unrestricted viral replication within the brain and ultimately death (Cheeran et al., 2004). In addition, there were elevated levels of CCL2 and CXCL10 within the brains of MCMV-infected immunodeficient animals, which likely reflect the increase in viral burden. Transfer of splenocytes obtained from MCMV-primed animals into infected immunodeficient mice resulted in protection from lethal disease. These findings indicate that localized expression of chemokines is not sufficient to control viral replication in the absence of an adaptive immune response (Cheeran et al., 2004). Further, these findings suggest that expression of T cell chemoattractant chemokines, such as CXCL10, may exert a protective effect by attracting T cells to sites of viral replication (Cheeran et al., 2004).

2.6 Human Immunodeficiency Virus (HIV)

Human immunodeficiency virus-1 (HIV-1) induces severe damage to the immune system, as it is tropic for both CD4 + T cells and for cells of the monocyte/

macrophage lineage. Ultimately, significant depletion of effector memory CD4 + T cells leads to acquired immunodeficiency syndrome (AIDS) in which susceptibility to opportunistic infection is increased due to a limited adaptive immune system (Brenchley et al., 2006; Grossman et al., 2006). Infection with HIV can also result in damage to the CNS resulting in dementia. HIV-associated dementia (HAD) affects approximately 20% of individuals with advanced HIV disease, and is characterized clinically by numerous neurological and psychiatric symptoms with cognitive and motor impairments being the most prevalent (Albright et al., 2003; Janssen et al., 1991; Marder et al., 1996; Navia et al., 1986). HIV-1 encephalitis (HIVE), the histological correlate of HAD, occurs in the majority of HAD cases and the pathological hallmarks include blood–brain barrier (BBB) damage, productive viral infection, inflammatory infiltrates consisting of lymphocytes and macrophages, astrogliosis, microgliosis and neuronal loss (Budka 1991; Cartier et al., 2005).

Chemokines and chemokine receptors are implicated in both disease pathogenesis and protection during HAD, however, the precise mechanisms governing the damage are presently under investigation. Several chemokine receptors are directly involved in the infection, serving as CD4 coreceptors for viral entry into target cells (Cartier et al., 2005). CCR5 and CXCR4 are predominantly utilized, but additional studies support CCR2a (splice variant), CCR3, CCR8, CXCR6 and CX3CR1 as minor coreceptors for HIV-1 (Cartier et al., 2005).

The initial entry of HIV into the brain is thought to occur via migration of infected monocytes and CD4 + T lymphocytes across the BBB (Cartier et al., 2005; Speth et al., 2005). Although the molecular mechanisms are currently unknown, chemokines are thought to be central regulators of this process (Cartier et al., 2005; Speth et al., 2005). For example, recent studies demonstrate that the presence of CCL2 within the brain promotes migration of HIV-infected leukocytes into the CNS (Eugenin et al., 2006). CXC3CL1 and CXCL12 have also been implicated in the transmigration of lymphocytes and monocytes across the BBB (Nottet 1999). Once HIV enters the brain, virus-host interactions and viral protein release lead to altered chemokine expression that can result in activation or proliferation of microglia and astrocytes or activation of neuronal chemokine receptors resulting in neuronal dysfunction or death (Speth et al., 2005). High CCL2 concentrations in the CSF of SIV-infected macaques correlate with a significantly higher expression of macrophage/microglia and astrocyte activation markers suggesting involvement of CCL2 in microgliosis and astrocytosis (Zink et al., 2001). As an inducer of astrocyte proliferation, enhanced expression of CXCL12 during HIV encephalitis likely contributes to astrocytosis as well (Bonavia et al., 2003). Factors released by activated microglia and astrocytes may indirectly or directly contribute to neuronal damage. Additionally viral proteins such as gp120 can act as chemokine agonists and activate signaling pathways in astrocytes and neurons resulting in neurodegeneration (Kaul et al., 2005). Taken together, these studies support the notion that chemokines and chemokine receptors are serving, in part, as mediators of disease leading to the pathogenesis associated with HAD.

The duplicitous nature of chemokines and chemokine receptors is highlighted by studies demonstrating the protective abilities of these molecules during HAD. Among the protective functions that chemokines and chemokine receptors play are competition for HIV binding, modulation of astrocyte and microglia activation and protection from HIV-induced neurotoxicity (Speth et al., 2005). Corasaniti et al., (2001a, b) showed that CXCL12 elicits protection against gp120-induced neuronal apoptosis in rats and suggests that this protection is conferred through a shift in the competition of gp120 and CXCL12 binding to CXCR4 in favor of CXCL12. Regarding protection from neurotoxicity, studies support a role for CCL2 and CX3CL1 in protecting neurons from Tat-induced apoptosis (Eugenin et al., 2003; Tong et al., 2000). Diminished microglial production of proinflammatory cytokines and reactive oxygen species following CX3CL1 treatment has been observed, yet the mechanism is currently under investigation (Mizuno et al., 2003).

2.7   Coronavirus

Coronaviruses are enveloped positive-strand RNA viruses with a genome ranging from 30–34 kb in size (Masters 2006). Coronaviruses infect numerous vertebrate hosts including humans, chickens, pigs, and mice causing a wide variety of disorders involving a number of different organ systems; however, there are specific tropisms for the CNS, lungs, gastrointestinal tract and liver (Holmes and Lai 1996; McIntosh 1996; Perlman et al., 1999). Receptor use among the varied coronaviruses is restricted to several well-defined proteins. Human coronavirus infections result in acute enteritis as well as 15% of common colds indistinguishable from those caused by other viruses (Holmes and Lai 1996; McIntosh 1996; Perlman et al., 1999). More recently, a human coronavirus (CoV) has been indicated to be the etiologic agent for Severe Acute Respiratory Syndrome (SARS). SARS is a potentially lethal disease and is recognized as a health threat internationally (Holmes 2003). Although normally considered an upper respiratory tract pathogen, SARS CoV was recently isolated from brain tissue of a SARS patient with significant neurologic symptoms and neuropathology associated with neuronal necrosis and glial hyperplasia (Xu et al., 2005). Analysis of blood samples revealed a dramatic increase in CXCL9 and CXCL10 protein levels (Xu et al., 2005). Moreover, immunostaining showed discrete patterns of CXCL9 expression by glial cells that was associated with infiltrating T cells and macrophages. These findings demonstrate that SARS CoV is capable of infecting the CNS and is associated with neurologic disease characterized by immune cell infiltration in which chemokine signaling may be important.

Supporting this possibility are studies utilizing a murine coronavirus, mouse hepatitis virus (MHV), to characterize the functional contributions of chemokine signaling in leukocyte trafficking and accumulation within the CNS (reviewed in Glass et al., 2002; Lane et al., 2006). Instillation of MHV into the

CNS of susceptible mice results in an acute encephalomyelitis characterized by viral infection and replication in either neurons and/or glial cells (Lane et al., 2006). Mice that survive the acute disease often develop a chronic demyelinating disease with virus persisting in white matter tracts and infiltrating T cells and macrophages contribute to myelin damage. MHV infection results in an orchestrated expression of chemokines including CCL2, CCL3, CCL4, CCL5, CXCL9, and CXCL10 (Lane et al., 1998). Analysis of the temporal expression profiles of chemokine transcripts revealed that at all stages of disease examined e.g. acute and chronic, CXCL10 was the prominent chemokine detected.

Inhibition of chemokine signaling through use of genetic knock-out mice or neutralizing antibodies has provided a powerful approach to designate both redundant and non-redundant roles for chemokines in MHV-induced CNS disease. Indeed, MHV infection of CCL3−/− mice results in deficient activation and accumulation of myeloid dendritic cells into draining cervical lymph nodes that results in muted activation of virus-specific T cells (Trifilo et al., 2003; Trifilo and Lane 2004). For example, virus-specific CD8 + T cells are unable to undergo egress from lymphatic tissue and migrate into the CNS as a result of impaired expression of tissue-specific homing receptors e.g. chemokine receptors CXCR3 and CCR5 (Trifilo et al., 2003). In addition, virus-specific CD4 + T cells derived from MHV-infected CCL3−/− mice produce increased levels of IL-10 and diminished IFN-γ when stimulated with viral antigen (Trifilo and Lane 2004). Therefore, these data indicate that early expression of chemokines such as CCL3 serve an important role in linking innate and adaptive immune responses following MHV infection of the CNS.

Blocking either CXCL9 or CXCL10 during acute disease resulted in increased mortality that correlated with reduced infiltration of CXCR3 + T cells into the CNS and increased viral titers, demonstrating an important role for these chemokines in host defense by attracting T cells into the CNS (Dufour et al., 2002; Liu et al., 2000, 2001a). However, blocking CXCL10 during chronic disease was beneficial as animals recovered locomotor activity that was associated with reduced T cell infiltration into the CNS and extensive remyelination (Liu et al., 2001b). Therefore, CXCL10 is either protective or contributes to disease depending on the stage of disease. Further support for the importance of CXCL10 in modulating disease is derived from studies demonstrating that antibody-targeting of CXCR3 improves clinical outcome during chronic disease by reducing T cell accumulation within the CNS (Stiles et al., 2006b). Importantly, the influence of blocking CXCL10 signaling appears to be at the level of trafficking rather than muting specific T cell effector functions e.g. cytokine secretion or proliferation (Stiles et al., 2006a). In addition to CXCL10, the chemokine CCL5 also promotes protection during acute disease as well as amplifies disease severity during chronic disease (Glass et al., 2004, 2001). CCL5 is capable of recognizing the receptors CCR1 and CCR5, which are surface receptors present on activated macrophages and T cells. MHV infection of CCR5−/− did not significantly impact host defense, but demyelination was reduced and this correlated with paucity in macrophage accumulation within the

CNS (Glass et al., 2001). These findings suggest that CCR5 signaling promotes macrophage trafficking but does not significantly impact T cell migration. However, MHV infection of mice deficient in CCR1 signaling (CCR1−/− mice) did not affect viral clearance from the brain, although there was a reduction in CD8 + T cell accumulation (Hickey et al., 2007). Further, MHV-infected CCR1−/− mice ultimately succumbed to fatal disease during the chronic stage of infection in the absence of any significant change in the composition of the cellular infiltrate (Hickey et al., 2007). Therefore, these data demonstrate that neither CCR1 nor CCR5 signaling alone is necessary for optimal host defense during acute disease, but both receptors influence the disease pathogenesis during chronic disease.

2.8   Theiler's Murine Encephalomyelitis Virus (TMEV)

Theiler's Murine Encephalomyelitis Virus (TMEV) is a positive-strand RNA picornavirus that does not infect humans, but does induce an acute encephalomyelitis and chronic demyelinating disease following instillation into the CNS of mice (Oleszak et al., 2004). Importantly, TMEV infection serves as an excellent model for the human demyelinating disease MS due to similarities in neuropathology. TMEV infection results in activation of various chemokine genes within the CNS including CCL2, CCL3, CCL4, CCL5, CXCL9, and CXCL0 that precedes and accompanies leukocyte infiltration into the CNS and is associated with onset of clinical disease (Hoffman et al., 1999; Ransohoff 2002). In vitro studies have revealed that TMEV infection results in a dramatic increase in the synthesis of chemokine mRNA transcripts (Palma and Kim 2001; Palma and Kim 2004; So et al., 2006). Chemokine gene activation was largely independent of type I IFN signaling but completely dependent upon NFκB and IRF/ISRE pathways (Palma and Kim 2004; So et al., 2006). More recently, induction of chemokine gene expression was found to occur in a TLR3-dependent manner (Palma and Kim 2004; So et al., 2006). Functional studies have ruled out CXCL10 as important in either host defense or disease following TMEV infection, indicating that alternative signaling pathways promote these separate events (Tsunoda et al., 2004). Moreover, antibody targeting of either CXCL9 or CCL5 results in increased viral antigen expression within the CNS and increased spinal cord pathology, suggesting that these chemokines serve to restrict viral gene expression (Ure et al., 2005). Therefore, the TMEV model of viral-induced neurologic disease is distinct from others with regards to chemokines participating in leukocyte infiltration into the CNS and/or disease progression. Indeed, CNS-derived chemokine gene expression in TMEV-infected mice during chronic disease appears to be dictated primarily by viral persistence and is independent of genetic factors related to susceptibility, severity of neuropathology, and the presence or absence of regulatory T cells (Ransohoff 2002).

3 Bacterial Infections of the CNS

Several different types of bacteria infect the CNS and cause severe and often fatal diseases, thus highlighting the importance of studying the host immune response following infection (Table 12.2). Unlike viruses, the majority of these bacteria do not require a host cell for survival and replication within the CNS. Rather, many bacterial infections of the CNS stem from the colonization of the CSF. In general, these pathogens gain access to the CNS following septicemia, which facilitates passage of the bacterium and immune cells across the BBB, culminating in diseases such as meningitis, encephalitis and brain abscess. Notably, bacterial meningitis accounts for one of the top ten infectious causes of death throughout the world (Nau and Bruck 2002; Scheld et al., 2002). The high mortality rate associated with these infections and the prevalent neurologic damage incurred by survivors is the result of pathological changes within the CNS due to (1) direct damage elicited by the pathogen itself and (2) the strong host inflammatory response within the CNS during the bacterial infection. For instance, gram-positive bacteria induce a robust neuroinflammatory response via immunoreaction to the peptidoglycan-techoic acid that constitutes the bacterial cell wall. Similarly, the lipopolysaccharide (LPS) contained within the outer membrane of gram-negative bacteria is highly antigenic and contributes to bacterial-mediated CNS pathogenesis. The resulting host inflammatory response contributes to CNS damage at the site of infection due to the toxic effects of immune mediators including cytokines, chemokines, oxidative agents and proteolytic enzymes (Koedel et al., 2002; Pfister and Scheld 1997). Similar to the viral infections discussed in the first section of this chapter, recruitment and activation of leukocytes is a hallmark of acute inflammation in response to bacteria and it is evident that chemokines also play a vital role in this process. Therefore, understanding the role of chemokine/chemokine receptor expression within the CNS of bacteria-infected hosts may uncover potential targets for therapeutic intervention.

Despite the dramatic clinical impact of the neurotropic bacterial pathogens on morbidity and mortality, work dedicated to understanding the mechanisms governing immune cell infiltration following infection are limited compared to viral-based

Table 12.2 Bacteria-Induced CNS Disease

Pathogen	Disease
Bacteria	
Gram-positive	
Streptococcus pneumoniae	Meningitis, meningoencephalitis, brain abscess
Listeria monocytogenes	Meningitis, meningoencephalitis, brain abscess
Staphylococcus aureus	Brain abscess
Gram-negative	
Neisseria meningitidis	Meningitis, meningoencephalitis
Haemophilus influenza	Meningitis
Spirochetes	
Borrelia burgdorferi	Meningitis, encephalitis

studies. Nonetheless, several important studies are beginning to shed light on the subject. From this work it is apparent that bacterial infections within the CNS tend to result in similar chemokine profiles compared to virus (Kielian 2004b; Lahrtz et al., 1998). Indeed, in many instances the level and duration of chemokine expression in response to bacterial infection surpasses what is often observed following viral infections. Presumably, the presence of higher levels of antigens, including cell wall products and/or various secreted virulence factors likely contribute to this feature of the bacteria-induced inflammation. Targeted studies have revealed the presence of elevated levels of CXCL8, CXCL1, CCL2, CCL3, CCL4 and CCL5 in the cerebrospinal fluid (CSF) of patients with bacterial meningitis compared to controls (Lahrtz et al., 1998; Spanaus et al., 1997; Sprenger et al., 1996). More recently, global screening for a broad range of chemokines in the CSF of similar patients has also demonstrated significantly enhanced levels of CXCL5, CXCL7, CXCL10, CCL8 and CCL20 (Kastenbauer et al., 2005). Importantly, CXCL8 expression correlates well with neutrophil infiltration, which are the predominant leukocytes present during bacteria-mediated CNS infections (Halstensen et al., 1993; Lahrtz et al., 1998; Mastroianni et al., 1998; Spanaus et al., 1997; Sprenger et al., 1996). However, the frequent presence of chemokines, such as, CCL2, CCL3 and CCL4, in the context of comparatively low monocyte infiltration during CNS infections caused by bacteria suggests that these chemokines play a broader role in the host anti-bacterial response beyond leukocyte trafficking, and emphasizes the importance of future research dedicated to this field (Inaba et al., 1997; Lahrtz et al., 1998; Mastroianni et al., 1998; Spanaus et al., 1997; Sprenger et al., 1996). The following section of this chapter is intended to highlight the common bacteria associated with CNS disease that have been studied in the context of chemokine regulation and signaling.

3.1 **Streptococcus pneumoniae**

Streptococcus pneumoniae (*S. pneumoniae*) is a gram-positive pathogenic bacterium that is the primary causative agent of otitis media and bacterial pneumoniae (Bridy-Pappas et al., 2005). Importantly, it is also the primary cause of bacterial meningitis in adults, accounting for nearly 50% of all reported cases (Bridy-Pappas et al., 2005; Robinson et al., 2001). *S. pneumoniae* commonly resides within the nasopharynx of healthy individuals, but in many cases, colonization in other areas of the host results in septicemia, subsequent breaching of the BBB and infection of the meninges. Pneumococcal meningitis has a nearly 30% mortality rate and in the majority of cases a high frequency of neurologic defects in survivors (Bohr et al., 1984; Bridy-Pappas et al., 2005; Edwards et al., 1985; Pfister et al., 1993). The high rate of fatal cases is speculated to be a result of an initial suboptimal host immune response against *S. pneumoniae*. The poor immune response can be attributed in part to the polysaccharide capsule surrounding *S. pneumoniae* that has been shown to facilitate immune evasion, thus giving it the advantage of reaching high bacterial titers, which ultimately contributes to an unfavorable inflammatory environment (Musher 1992).

Indeed, chemokine expression is a hallmark of *S. pneumoniae* infection and several chemokines, such as CXCL1, CXCL8, CCL2, CCL3, CCL4 and CCL5 are expressed in the CSF of patients with meningitis (Lahrtz et al., 1998; Spanaus et al., 1997; Sprenger et al., 1996). A similar chemokine protein profile was seen in resident mouse microglia stimulated with the highly immunogenic cell wall components from *S. pneumoniae* (Hausler et al., 2002). In a mouse model of pneumococcal meningitis that mimics the natural route of infection in humans, CXCL2 was highly expressed following intranasal infection (Zwijnenburg et al., 2001). These and other studies have focused on determining the function and regulation of chemokines in the context of pneumococcal meningitis. For instance, CSF from pneumococcal meningitis patients was chemotactic for neutrophils and mononuclear leukocytes (Lahrtz et al., 1998; Spanaus et al., 1997; Sprenger et al., 1996). Given that neutrophils are the predominant immune cell infiltrate during acute disease, other groups have taken a closer look at the function of CXCL8, and its mouse homolog, CXCL2. In a rabbit model of LPS induced meningitis, CXCL8 blockade resulted in reduced leukocyte recruitment (Dumont et al., 2000). Likewise, intravenous treatment with anti-CXCL8 antibody in a rabbit model of *S. pneumoniae*-induced meningitis impaired leukocyte accumulation in the CSF (Ostergaard et al., 2000). CXCL2 expression in mice appears to be dependent on toll-like receptor (TLR) signaling. Mice deficient for MyD88, an integral signal transduction molecule involved in TLR signaling, exhibited a reduced inflammatory host response, including significant reduction in CXCL2 expression, which resulted in higher bacterial titers in the CNS and increased aggravation of disease compared to wild-type mice (Koedel et al., 2004). While these results are interesting with regards to CXCL8, it is apparent that more work needs to be focused on understanding the role of the other chemokines expressed during pneumococcal meningitis. Recent work has demonstrated an even broader array of chemokine protein expression including CXCL5, CXCL7, CXCL10, CCL8 and CCL20 in CSF of meningitis patients (Kastenbauer et al., 2005); and at least one study implicates the cytokine IFN-γ in differentially modulating chemokine production in resident CNS macrophages in response to components of the *S. pneumoniae* cell wall (Hausler et al., 2002). In summary, these studies suggest that inflammation in the CNS following *S. pneumoniae* infection is dynamically regulated by a distinct chemokine profile and highlights the importance of microglia as key mediators of this active process.

3.2   Neisseria meningitidis

Neisseria meningitidis (*N. meningitidis*) is known best as a primary pathogen of meningitis that has the potential to cause epidemic outbreaks (Manchanda et al., 2006). *N. meningitidis* is gram-negative bacterium that elicits a robust host immune response through its LPS-rich cell wall. In fact, the severity of clinical presentation appears to correlate with the level of LPS within the CSF (Moller et al., 2005). Like *S. pneumoniae*, *N. meningitidis* resides within the human nasopharynx and is transmitted by respiratory spread. Its antiphagocytic capsule also gives *N. meningitidis* the advantage of evading immune clearance, which can result in high titers of the pathogen in the blood that ultimately facilitates breaching of the BBB. Upon infection of the CNS, meningeal cells appear to be a prominent source of chemokine production during meningococcal infection (Fowler et al., 2006). Binding of *N. meningitidis* to these cells results in high levels of CCL2, CCL5 and CXCL8 expression (Fowler et al., 2006), BBB damage and access to the CNS. Targeted studies, have demonstrated that the chemokines, CCL2, CCL3, CCL4, CCL5, CXCL1 and CXCL8 are expressed in the CSF from patients with meningococcal meningitis (Moller et al., 2005; Spanaus et al., 1997). Importantly, CSF from these patients is also chemotactic for inflammatory immune cells (Spanaus et al., 1997), supporting the functional role of chemokines for positional migration of immune cells in the context of bacterial meningitis. Additional evidence lies in the observation that the CSF from similar patients infected with *N. meningitidis* also contained significant levels of CCL2, CCL3, CCL5 and CXCL8, which had an inverse correlation with the relative bacterial load (Moller et al., 2005), and in vivo and ex vivo studies demonstrated that LPS was the major mediator of chemokine secretion (Moller et al., 2005). Enhanced CCL2, CCL3 and CXCL8 protein production was associated with increasing concentrations of LPS; however, the plasma CCL5 levels were inversely related (Moller et al., 2005), suggesting that CCL5 does not play a major role in leukocyte recruitment to the site of CNS infection. Meningeal cells appear to be a prominent source of chemokine production during a meningococcal infection, as CCL2, CCL5 and CXCL8 are each highly expressed following in vitro binding studies with *N. meningitidis* (Fowler et al., 2006). This robust host neuroinflammatory response elicited at the site of CNS infection may indeed have consequences beyond direct damage to CNS tissue. Leukopenia is a common feature in patients with systemic meningococcal infections often resulting in fatal septic shock (Flaegstad et al., 1995; Kornelisse et al., 1997). It is interesting to speculate that strong expression of chemokines at the site of infection may effectively deplete the levels of circulating leukocytes, ultimately contributing to high bacterial titers and the resulting fatal disease outcome.

3.3 **Haemophilus influenza**

Pathogenic strains of the gram-negative bacterium *Haemophilus influenza* (*H. influenza*) are defined by the presence of a capsule and are grouped into six types (A-F) that cause a variety of diseases affecting the ear, upper respiratory tract and CNS (Murray et al., 2005). *H. influenza* type B (HiB) was the number one cause of acute meningitis in infant and young children in the United States until widespread use of the HiB vaccine reduced reported cases to less than 10% since 1990

(Schuchat et al., 1997). Similar to *S. pneumoniae* and *N. meningitidis*, the CSF from patients with Hib contains significant amounts of CCL2, CCL3, CCL4, CCL5, CXCL1 and CXCL8 (Spanaus et al., 1997). In addition, similar findings were observed in a rat model of Hib, as significant levels of CCL2, CCL3, CCL5 and CXCL2 were expressed within 24–48 h post infection (Diab et al., 1999). Importantly, expression of these chemokines correlated well with the recruitment of infiltrating neutrophils and macrophages into the meninges, and also with disease severity (Diab et al., 1999). CXCL2 and CCL3 expression was detected in neutrophils and macrophages in the subarachnoid space that separates the meningeal cells from the lateral ventricles. CCL2 expression localized to infiltrating neutrophils and macrophages, and to some extent, astrocytes, whereas CCL5 expression occurred predominantly in astrocytes and resident microglia. The functional roles of CXCL2, CCL2 and CCL3 in immune cell trafficking during HiB infection were also determined by using neutralizing antibody studies (Diab et al., 1999). Treatment with anti-CCL2 significantly reduced macrophage infiltration into the subarachnoid space, while anti-CXCL2 treatment impaired neutrophil trafficking. It is important to note that this study was the first to show that anti-CCL3 also abrogated neutrophil recruitment, despite the observation that this chemokine is not capable of neutrophil recruitment in vitro. However, other studies are starting to show a role of CCL3 in neutrophil trafficking in vivo (Ajuebor et al. 2004; Ramos et al. 2005; Standiford et al. 1995). Interestingly, selective chemokine blockade also modulated the expression of the other chemokines; e.g. anti-CCL3 administration resulted in downregulation of CCL5, but simultaneous upregulation of CCL2 and CXCL2 (Diab et al. 1999). Therefore, it is possible that blocking CCL3 expression may have an indirect effect on neutrophil recruitment by modulating the expression of other chemokines, again highlighting the complex interplay between the regulatory events controlling chemokine regulation and immune infiltration during infection.

3.4 **Listeria monocytogenes**

Listeria monocytogenes (*L. monocytogenes*) is a gram-positive bacterium that is an important human pathogen, causing diseases ranging from mild gastroenteritis to fatal septicemia, encephalitis and up to 15% of the reported cases of meningitis in adults (Calder 1997). Transmission of *L. monocytogenes* and the pathogenesis of listeriosis are facilitated by several key features (Murray et al. 2005). For instance, it is a hardy bacterium that survives in soil, high heat and is able to replicate at low temperatures, thus facilitating transmission as a food-borne pathogen. Furthermore, it is a facultative intracellular bacterium that evades immune surveillance by surviving within macrophages and endothelial cells and can move from cell-to-cell via actin-based motility without exposure to other immune factors, such as antibodies and complement. In fact, Listeria can infect a variety of cell types, including neurons, astrocytes and meningeal cells. Chemokine signaling is considered important for controlling intracellular replication of *Listeria* by recruiting activated monocytes across the BBB and into the subarachnoid space (Frei et al. 1993). In an experimental mouse model, Seebach, et al. (1995) showed that the immune cell infiltration to the meninges is temporally regulated following intracerebral infection with *L. monocytogenes*. Within 24 h p.i., roughly 80% of the infiltrating cells were neutrophils, with monocytes making up approximately 50% of the inflammatory cells after 72 h p.i. (Seebach et al., 1995). The immune cell influx correlated well with time-dependent chemokine expression, as CCL3 and CCL4 were expressed predominantly by neutrophils within the meninges early in the infection (12 h p.i.), and CXCL2, CCL3 and CCL4 were expressed by both neutrophils and monocytes by 24–48 h p.i. Furthermore, CCL3, CCL4 and CXCL2 were present in CSF from infected mice and in vitro antibody treatment with anti-CXCL2 or anti-CCL3 showed that these chemokines were partly responsible for neutrophil and monocyte recruitment, respectively. In light of the in vitro results it is interesting that the temporal expression of CCL3/CCL4, followed by CXCL2/CCL3/CCL4 is associated with a switch from a neutrophil-rich to a monocyte dominant immune cell environment. As we have seen, the in vivo inflammatory environment and in vitro model are often not correlated and can represent two widely distinct situations. The in vivo findings suggest that CCL3 is capable of either directly or indirectly recruiting neutrophils to areas inflammation, including the sites of bacterial-induced meningitis, once again emphasizing the importance of the dynamic regulatory networks that likely modulate chemokine and chemokine receptor function and signaling.

3.5   Borrelia burgdorferi

Borrelia burgdorferi (*B. burgdoferi*) is an obligate intracellular, spirochete bacterium, known best as the causative agent of Lyme's disease (Murray et al., 2005). It is most commonly transmitted to humans by an infected tick. Lyme borreliosis can also affect the CNS, generally manifesting as meningitis. In fact, other spirochetes have long been associated with CNS infection, including the causative agents of syphilis (*Treponema pallidum*) and leptospirosis (*Leptospira*) (Pachner 1986). In the case of borreliosis, host immune evasion is mediated through the ability of *B. burgdorferi* to constantly modify its surface structure by modulating lipoprotein expression. However, the surface structure of borrelia can also induce a strong and often damaging inflammatory response (Morrison et al., 1997), which is mediated in part by chemokine expression. As we have seen, inflammatory changes in the CSF mediate breaching of the BBB and infiltration of immune cells within the CNS, causing subsequent pathogenesis. Indeed, the CSF of patients with neuroborelliosis contained significant amounts of CCL3, CCL4 and CCL8 compared to control patients (Grygorczuk et al., 2003). Furthermore, human monocytes infected with *Borrelia* strongly expressed and secreted CCL2, CCL5, CCL8, CXCL1 (Sprenger et al., 1997). It appears that activated CD8 + T cells expressing the chemokine receptor,

CCR5, may be directed to the CNS via these or other unidentified chemokines during the early phase of neuroborreliosis (Jacobsen et al., 2003). Interestingly, CCR5 signaling on CD8 + T cells during *Borrelia* infection is also speculated to contribute to chronic autoimmune driven disease (Hemmer et al., 1999).

3.6   Staphylococcus aureus

Staphylococcus aureus is a gram-positive bacterium that is well known for causing food poisoning, skin infections and acute endocarditis (Murray et al., 2005). It resides within the human nasal mucosa and/or the skin and is transmitted by skin-to-skin contact and by sneezing. *S. aureus* is a common hospital-acquired infection and methicillin-resistant strains of *S. aureus* (MRSA) complicate effective drug treatment strategies. For example, *S. aureus* represents a significant nosocomial infection and is a prominent cause of brain abscesses, a disease that affects roughly 1 out of every 10,000 hospital-admitted patients in the U.S. (Kielian 2004a; Townsend and Scheld 1998). In addition, *S. aureus* is able to survive intracellularly within neutrophils and endothelial cells further interfering with the success of host-immune clearance (Gresham et al., 2000). Chemokine signaling in response to CNS infection with *S. aureus* has been studied in a murine experimental brain abscess model. Using this model, Keilan, et al. (2001) demonstrated that CCL1, CCL2, CCL3, CCL4, CXCL2 were expressed within 6 h post bacterial exposure. Microglia and astrocytes appear to play a prominent role in immune cell recruitment in this model system as the primary reservoirs for chemokine production and secretion. CCL1, CCL2, CCL3, CCL4, CXCL2 were expressed on primary cultures of murine microglia and astrocytes infected with *S. aureus* (Kielian et al., 2001), and microglia or astrocytes stimulated with heat inactivated *S. aureus* or gram-positive peptidoglycan expressed CCL1, CCL2, CCL3, CCL4, CCL5, CXCL10 (Kielian et al., 2002), and CCL2, CCL4 and CXCL2 (Esen et al., 2004). The functional role of chemokine signaling was further examined using CXCR2 ko mice, which had impaired neutrophil trafficking and increased bacterial burden within the CNS (Kielian et al., 2001), indicating that CXCR2 ligands provide the main chemotactic signal driving neutrophil influx into the CNS. In addition, CXCL2 is chronically expressed in the experimental brain abscess model and correlated with continued infiltration of neutrophils in the presence of low bacterial titers within the CNS (Baldwin and Kielian 2004). It appears that CXCL2 expression in microglia is dependent on MyD88 signaling, suggesting that *S. aureus*-mediated microglia activation and chemokine production are dependent on TLR recognition, in part by TLR2 (Esen and Kielian 2006). These studies indicate that CXCL2:CXCR2 signaling is the dominant cue for neutrophil recruitment to sites of *S. aureus* infection, which in themselves may be the primary mediators of CNS damage. To gain a better understanding of the mechanisms that induce tissue injury within the CNS it is of interest in the future to determine the functional importance of the other chemokines expressed during the course of *S. aureus* infection.

4　Perspectives

Microbial infection of the CNS often results in a tightly regulated inflammatory immune response in which chemokine signaling helps to control leukocyte entry and positional migration to sites of infection. Although functional redundancy is associated with many chemokine ligands as a result of receptor promiscuity, chemokine expression profiles often significantly shape the immune response to infection of the CNS based on various criteria including pathogen, route of infection, and cellular tropism. Careful consideration of chemokine and chemokine receptor expression during the course of disease could help in the derivation of treatments for patients infected with CNS invading pathogens with the ultimate goal of limiting inflammation/pathology without muting specific anti-microbial effector responses. Indeed, numerous treatments have been developed to control inflammation within the context of infection and autoimmune diseases by targeting specific proinflammatory molecules. With this in mind, various approaches are currently being developed and/or are in various stages of clinical trials to disrupt chemokine ligand interactions with specific signaling receptors with the hopes of improving disease outcome (Charo and Ransohoff 2006). Given the relatively rapid pace at which our understanding of the biology of chemokines and chemokine receptors has progressed over the past decade, it is likely that successful clinical approaches will be developed to mute specific chemokine signaling pathways and improve clinical outcome in response to microbial infection of the CNS.

References

Ajuebor MN, Kunkel SL, Hogaboam CM. 2004. The role of CCL3/macrophage inflammatory protein-1alpha in experimental colitis. *Eur J Pharmacol* 497(3):343–349.

Albright AV, Soldan SS, Gonzalez-Scarano F. 2003. Pathogenesis of human immunodeficiency virus-induced neurological disease. *J Neurovirol* 9(2):222–227.

Anlar B, Soylemezoglu F, Aysun S, Kose G, Belen D, Yalaz K. 2001. Tissue inflammatory response in subacute sclerosing panencephalitis (SSPE). *J Child Neurol* 16(12):895–900.

Aravalli RN, Hu S, Rowen TN, Palmquist JM, Lokensgard JR. 2005. Cutting edge: TLR2-mediated proinflammatory cytokine and chemokine production by microglial cells in response to herpes simplex virus. *J Immunol* 175(7):4189–4193.

Arribas JR, Clifford DB, Fichtenbaum CJ, Commins DL, Powderly WG, Storch GA. 1995. Level of cytomegalovirus (CMV) DNA in cerebrospinal fluid of subjects with AIDS and CMV infection of the central nervous system. *J Infect Dis* 172(2):527–531.

Asensio VC, Campbell IL. 1997. Chemokine gene expression in the brains of mice with lymphocytic choriomeningitis. *J Virol* 71(10):7832–7840.

Asensio VC, Kincaid C, Campbell IL. 1999. Chemokines and the inflammatory response to viral infection in the central nervous system with a focus on lymphocytic choriomeningitis virus. *J Neurovirol* 5(1):65–75.

Baldwin AC, Kielian T. 2004. Persistent immune activation associated with a mouse model of *Staphylococcus aureus*-induced experimental brain abscess. *J Neuroimmunol* 151(1–2):24–32.

Bohr V, Paulson OB, Rasmussen N. 1984. Pneumococcal meningitis. Late neurologic sequelae and features of prognostic impact. *Arch Neurol* 41(10):1045–1049.

Bonavia R, Bajetto A, Barbero S, Pirani P, Florio T, Schettini G. 2003. Chemokines and their receptors in the CNS: expression of CXCL12/SDF-1 and CXCR4 and their role in astrocyte proliferation. *Toxicol Lett* 139(2–3):181–189.

Brenchley JM, Price DA, Douek DC. 2006. HIV disease: fallout from a mucosal catastrophe? *Nat Immunol* 7(3):235–239.

Bridy-Pappas AE, Margolis MB, Center KJ, Isaacman DJ. 2005. *Streptococcus pneumoniae*: description of the pathogen, disease epidemiology, treatment, and prevention. *Pharmacotherapy* 25(9):1193–1212.

Buchmeier MJ, Zajac AJ. 1999. Lymphocytic choriomeningitis virus. In *Persistent Viral Infections*, eds. R Ahmed, J Chen, pp. 575–605. London: John Wiley and Sons, Ltd.

Buchmeier MJ, Welsh RM, Dutko FJ, Oldstone MB. 1980. The virology and immunobiology of lymphocytic choriomeningitis virus infection. *Adv Immunol* 30:275–331.

Budka H. 1991. Neuropathology of human immunodeficiency virus infection. *Brain Pathol* 1(3):163–175.

Calder JA. 1997. Listeria meningitis in adults. *Lancet* 350(9074):307–308.

Carr DJ, Tomanek L. 2006. Herpes simplex virus and the chemokines that mediate the inflammation. *Curr Top Microbiol Immunol* 303:47–65.

Carr DJ, Chodosh J, Ash J, Lane TE. 2003. Effect of anti-CXCL10 monoclonal antibody on herpes simplex virus type 1 keratitis and retinal infection. *J Virol* 77(18):10037–10046.

Carr DJ, Ash J, Lane TE, Kuziel WA. 2006. Abnormal immune response of CCR5-deficient mice to ocular infection with herpes simplex virus type 1. *J Gen Virol* 87(Pt 3):489–499.

Cartier L, Hartley O, Dubois-Dauphin M, Krause KH. 2005. Chemokine receptors in the central nervous system: role in brain inflammation and neurodegenerative diseases. *Brain Res Brain Res Rev* 48(1):16–42.

Charo IF, Ransohoff RM. 2006. The many roles of chemokines and chemokine receptors in inflammation. *N Engl J Med* 354(6):610–621.

Cheeran MC, Hu S, Yager SL, Gekker G, Peterson PK, Lokensgard JR. 2001. Cytomegalovirus induces cytokine and chemokine production differentially in microglia and astrocytes: antiviral implications. *J Neurovirol* 7(2):135–147.

Cheeran MC, Hu S, Sheng WS, Peterson PK, Lokensgard JR. 2003. CXCL10 production from cytomegalovirus-stimulated microglia is regulated by both human and viral interleukin-10. *J Virol* 77(8):4502–4515.

Cheeran MC, Gekker G, Hu S, Min X, Cox D, Lokensgard JR. 2004. Intracerebral infection with murine cytomegalovirus induces CXCL10 and is restricted by adoptive transfer of splenocytes. *J Neurovirol* 10(3):152–162.

Chen CJ, Liao SL, Kuo MD, Wang YM. 2000. Astrocytic alteration induced by Japanese encephalitis virus infection. *Neuroreport* 11(9):1933–1937.

Chen CJ, Chen JH, Chen SY, Liao SL, Raung SL. 2004. Upregulation of RANTES gene expression in neuroglia by Japanese encephalitis virus infection. *J Virol* 78(22):12107–12119.

Christensen JE, Nansen A, Moos T, Lu B, Gerard C, Christensen JP, Thomsen AR. 2004. Efficient T-cell surveillance of the CNS requires expression of the CXC chemokine receptor 3. *J Neurosci* 24(20):4849–4858.

Christensen JE, de Lemos C, Moos T, Christensen JP, Thomsen AR. 2006. CXCL10 is the key ligand for CXCR3 on CD8 + effector T cells involved in immune surveillance of the lymphocytic choriomeningitis virus-infected central nervous system. *J Immunol* 176(7):4235–4243.

Corasaniti MT, Bilotta A, Strongoli MC, Navarra M, Bagetta G, Di Renzo G. 2001a. HIV-1 coat protein gp120 stimulates interleukin-1beta secretion from human neuroblastoma cells: evidence for a role in the mechanism of cell death. *Br J Pharmacol* 134(6):1344–1350.

Corasaniti MT, Piccirilli S, Paoletti A, Nistico R, Stringaro A, Malorni W, Finazzi-Agro A, Bagetta G. 2001b. Evidence that the HIV-1 coat protein gp120 causes neuronal apoptosis in the neocortex of rat via a mechanism involving CXCR4 chemokine receptor. *Neurosci Lett* 312(2):67–70.

de Lemos C, Christensen JE, Nansen A, Moos T, Lu B, Gerard C, Christensen JP, Thomsen AR. 2005. Opposing effects of CXCR3 and CCR5 deficiency on CD8 + T cell-mediated inflammation in the central nervous system of virus-infected mice. *J Immunol* 175(3):1767–1775.

Diab A, Abdalla H, Li HL, Shi FD, Zhu J, Hojberg B, Lindquist L, Wretlind B, Bakhiet M, Link H. 1999. Neutralization of macrophage inflammatory protein 2 (MIP-2) and MIP-1alpha attenuates neutrophil recruitment in the central nervous system during experimental bacterial meningitis. *Infect Immun* 67(5):2590–2601.

Dufour JH, Dziejman M, Liu MT, Leung JH, Lane TE, Luster AD. 2002. IFN-gamma-inducible protein 10 (IP-10; CXCL10)-deficient mice reveal a role for IP-10 in effector T cell generation and trafficking. *J Immunol* 168(7):3195–3204.

Dumont RA, Car BD, Voitenok NN, Junker U, Moser B, Zak O, O'Reilly T. 2000. Systemic neutralization of interleukin-8 markedly reduces neutrophilic pleocytosis during experimental lipopolysaccharide-induced meningitis in rabbits. *Infect Immun* 68(10):5756–5763.

Edwards MS, Rench MA, Haffar AA, Murphy MA, Desmond MM, Baker CJ. 1985. Long-term sequelae of group B streptococcal meningitis in infants. *J Pediatr* 106(5):717–722.

Emonet S, Lemasson JJ, Gonzalez JP, de Lamballerie X, Charrel RN. 2006. Phylogeny and evolution of old world arenaviruses. *Virology* 350(2):251–257.

Esen N, Kielian T. 2006. Central role for MyD88 in the responses of microglia to pathogen-associated molecular patterns. *J Immunol* 176(11):6802–6811.

Esen N, Tanga FY, DeLeo JA, Kielian T. 2004. Toll-like receptor 2 (TLR2) mediates astrocyte activation in response to the Gram-positive bacterium *Staphylococcus aureus*. *J Neurochem* 88(3):746–758.

Eugenin EA, D'Aversa TG, Lopez L, Calderon TM, Berman JW. 2003. MCP-1 (CCL2) protects human neurons and astrocytes from NMDA or HIV-tat-induced apoptosis. *J Neurochem* 85(5):1299–1311.

Eugenin EA, Osiecki K, Lopez L, Goldstein H, Calderon TM, Berman JW. 2006. CCL2/monocyte chemoattractant protein-1 mediates enhanced transmigration of human immunodeficiency virus (HIV)-infected leukocytes across the blood–brain barrier: a potential mechanism of HIV-CNS invasion and NeuroAIDS. *J Neurosci* 26(4):1098–1106.

Flaegstad T, Kaaresen PI, Stokland T, Gutteberg T. 1995. Factors associated with fatal outcome in childhood meningococcal disease. *Acta Paediatr* 84(10):1137–1142.

Fowler MI, Ho Wang Yin KY, Humphries HE, Heckels JE, Christodoulides M. 2006. The inflammatory response of human meningeal cells following challenge with *Neisseria lactamica*: comparison with *Neisseria meningitidis*. *Infect Immun* 74(11):6467–6478.

Frei K, Nadal D, Pfister HW, Fontana A. 1993. Listeria meningitis: identification of a cerebrospinal fluid inhibitor of macrophage listericidal function as interleukin 10. *J Exp Med* 178(4):1255–1261.

Glass WG, Liu MT, Kuziel WA, Lane TE. 2001. Reduced macrophage infiltration and demyelination in mice lacking the chemokine receptor CCR5 following infection with a neurotropic coronavirus. *Virology* 288(1):8–17.

Glass WG, Chen BP, Liu MT, Lane TE. 2002. Mouse hepatitis virus infection of the central nervous system: chemokine-mediated regulation of host defense and disease. *Viral Immunol* 15(2):261–272.

Glass WG, Hickey MJ, Hardison JL, Liu MT, Manning JE, Lane TE. 2004. Antibody targeting of the CC chemokine ligand 5 results in diminished leukocyte infiltration into the central nervous system and reduced neurologic disease in a viral model of multiple sclerosis. *J Immunol* 172(7):4018–4025.

Glass WG, Lim JK, Cholera R, Pletnev AG, Gao JL, Murphy PM. 2005. Chemokine receptor CCR5 promotes leukocyte trafficking to the brain and survival in West Nile virus infection. *J Exp Med* 202(8):1087–1098.

Glass WG, McDermott DH, Lim JK, Lekhong S, Yu SF, Frank WA, Pape J, Cheshier RC, Murphy PM. 2006. CCR5 deficiency increases risk of symptomatic West Nile virus infection. *J Exp Med* 203(1):35–40.

Gresham HD, Lowrance JH, Caver TE, Wilson BS, Cheung AL, Lindberg FP. 2000. Survival of *Staphylococcus aureus* inside neutrophils contributes to infection. *J Immunol* 164(7):3713–3722.

Griffin DD, Bellini WJ. 1996. Measles virus. In *Fields Virology*. 3rd edition. Vol. 1, eds. BN Fields, DM Knipe, PM Howley, pp. 1267–1312. Philadelphia, PA: Lippincott-Raven Publishers.

Grossman Z, Meier-Schellersheim M, Paul WE, Picker LJ. 2006. Pathogenesis of HIV infection: what the virus spares is as important as what it destroys. *Nat Med* 12(3):289–295.

Grygorczuk S, Pancewicz S, Kondrusik M, Swierzbinska R, Zajkowska J, Hermanowska-Szpakowicz T. 2003. [Serum and cerebrospinal fluid concentration of inflammatory proteins MIP-1-alpha and MIP-1-beta and of interleukin 8 in the course of borreliosis]. *Neurol Neurochir Pol* 37(1):73–87.

Halstensen A, Ceska M, Brandtzaeg P, Redl H, Naess A, Waage A. 1993. Interleukin-8 in serum and cerebrospinal fluid from patients with meningococcal disease. *J Infect Dis* 167(2):471–475.

Hausler KG, Prinz M, Nolte C, Weber JR, Schumann RR, Kettenmann H, Hanisch UK. 2002. Interferon-gamma differentially modulates the release of cytokines and chemokines in lipopolysaccharide- and pneumococcal cell wall-stimulated mouse microglia and macrophages. *Eur J Neurosci* 16(11):2113–2122.

Hemmer B, Gran B, Zhao Y, Marques A, Pascal J, Tzou A, Kondo T, Cortese I, Bielekova B, Straus SE and others. 1999. Identification of candidate T-cell epitopes and molecular mimics in chronic Lyme disease. *Nat Med* 5(12):1375–1382.

Hickey M.J., Held K.S., Baum E., Gao J.L., Murphy P.M., Lau T.E. 2007. CCR5 deficiency increases susceptibility to fatal corona virus infection of the central nervous system. *Viral Immunology*, In press.

Hoffman LM, Fife BT, Begolka WS, Miller SD, Karpus WJ. 1999. Central nervous system chemokine expression during Theiler's virus-induced demyelinating disease. *J Neurovirol* 5(6):635–42.

Hofman FM, Hinton DR, Baemayr J, Weil M, Merrill JE. 1991. Lymphokines and immunoregulatory molecules in subacute sclerosing panencephalitis. *Clin Immunol Immunopathol* 58(3):331–342.

Holmes KV. 2003. SARS-associated coronavirus. *N Engl J Med* 348(20):1948–1951.

Holmes K, Lai M. 1996. Coronaviridae and their replication. In *Fields Virology*. 3rd edition. Vol. 1, eds. BN Fields, DM Knipe, PM Howley, pp. 1075–1094. Philadelphia, PA: Lippincott-Raven Publishers.

Inaba Y, Ishiguro A, Shimbo T. 1997. The production of macrophage inflammatory protein-1alpha in the cerebrospinal fluid at the initial stage of meningitis in children. *Pediatr Res* 42(6):788–793.

Jacobsen M, Zhou D, Cepok S, Nessler S, Happel M, Stei S, Wilske B, Sommer N, Hemmer B. 2003. Clonal accumulation of activated CD8 + T cells in the central nervous system during early phase of neuroborreliosis. *J Infect Dis* 187(6):963–973.

Jahrling PB, Peters CJ. 1992. Lymphocytic choriomeningitis virus. A neglected pathogen of man. *Arch Pathol Lab Med* 116(5):486–488.

Janssen R, Cornblath D, Epstein L, Foa R, McArthur J, Price R, Asbury A, Beckett A, Benson D, Bridge T. 1991. Nomenclature and research case definitions for neurologic manifestations of human immunodeficiency virus-type 1 (HIV-1) infection. Report of a Working Group of the American Academy of Neurology AIDS Task Force. *Neurology* 41(6):778–785.

Kastenbauer S, Angele B, Sporer B, Pfister HW, Koedel U. 2005. Patterns of protein expression in infectious meningitis: a cerebrospinal fluid protein array analysis. *J Neuroimmunol* 164(1–2):134–139.

Katz M. 1995. Clinical spectrum of measles. *Curr Top Microbiol Immunol* 191:1–12.

Kaul M, Zheng J, Okamoto S, Gendelman HE, Lipton SA. 2005. HIV-1 infection and AIDS: consequences for the central nervous system. *Cell Death Differ* 12 Suppl 1:878–892.

Kielian T. 2004a. Immunopathogenesis of brain abscess. *J Neuroinflammation* 1(1):16.

Kielian T. 2004b. Microglia and chemokines in infectious diseases of the nervous system: views and reviews. *Front Biosci* 9:732–750.

Kielian T, Barry B, Hickey WF. 2001. CXC chemokine receptor-2 ligands are required for neutrophil-mediated host defense in experimental brain abscesses. *J Immunol* 166(7):4634–4643.

Kielian T, Mayes P, Kielian M. 2002. Characterization of microglial responses to *Staphylococcus aureus*: effects on cytokine, costimulatory molecule, and Toll-like receptor expression. *J Neuroimmunol* 130(1–2):86–99.

Klein RS, Lin E, Zhang B, Luster AD, Tollett J, Samuel MA, Engle M, Diamond MS. 2005. Neuronal CXCL10 directs CD8 + T-cell recruitment and control of West Nile virus encephalitis. *J Virol* 79(17):11457–11466.

Koedel U, Rupprecht T, Angele B, Heesemann J, Wagner H, Pfister HW, Kirschning CJ. 2004. MyD88 is required for mounting a robust host immune response to *Streptococcus pneumoniae* in the CNS. *Brain* 127(Pt 6):1437–1445.

Koedel U, Scheld WM, Pfister HW. 2002. Pathogenesis and pathophysiology of pneumococcal meningitis. *Lancet Infect Dis* 2(12):721–736.

Kornelisse RF, Hazelzet JA, Hop WC, Spanjaard L, Suur MH, van der Voort E, de Groot R. 1997. Meningococcal septic shock in children: clinical and laboratory features, outcome, and development of a prognostic score. *Clin Infect Dis* 25(3):640–646.

Lahrtz F, Piali L, Spanaus KS, Seebach J, Fontana A. 1998. Chemokines and chemotaxis of leukocytes in infectious meningitis. *J Neuroimmunol* 85(1):33–43.

Lane TE, Asensio VC, Yu N, Paoletti AD, Campbell IL, Buchmeier MJ. 1998. Dynamic regulation of alpha- and beta-chemokine expression in the central nervous system during mouse hepatitis virus-induced demyelinating disease. *J Immunol* 160(2):970–978.

Lane TE, Hardison JL, Walsh KB. 2006. Functional diversity of chemokines and chemokine receptors in response to viral infection of the central nervous system. *Curr Top Microbiol Immunol* 303:1–27.

Le Y, Zhou Y, Iribarren P, Wang J. 2004. Chemokines and chemokine receptors: their manifold roles in homeostasis and disease. *Cell Mol Immunol* 1(2):95–104.

Lim JK, Glass WG, McDermott DH, Murphy PM. 2006. CCR5: no longer a "good for nothing" gene–chemokine control of West Nile virus infection. *Trends Immunol* 27(7):308–312.

Liu MT, Chen BP, Oertel P, Buchmeier MJ, Armstrong D, Hamilton TA, Lane TE. 2000. The T cell chemoattractant IFN-inducible protein 10 is essential in host defense against viral-induced neurologic disease. *J Immunol* 165(5):2327–2330.

Liu MT, Chen BP, Oertel P, Buchmeier MJ, Hamilton TA, Armstrong DA, Lane TE. 2001a. The CXC chemokines IP-10 and Mig are essential in host defense following infection with a neurotropic coronavirus. *Adv Exp Med Biol* 494:323–327.

Liu MT, Keirstead HS, Lane TE. 2001b. Neutralization of the chemokine CXCL10 reduces inflammatory cell invasion and demyelination and improves neurological function in a viral model of multiple sclerosis. *J Immunol* 167(7):4091–4097.

Looker KJ, Garnett GP. 2005. A systematic review of the epidemiology and interaction of herpes simplex virus types 1 and 2. *Sex Transm Infect* 81(2):103–107.

Luster AD. 1998. Chemokines – chemotactic cytokines that mediate inflammation. *N Engl J Med* 338(7):436–445.

Maertzdorf J, Osterhaus AD, Verjans GM. 2002. IL-17 expression in human herpetic stromal keratitis: modulatory effects on chemokine production by corneal fibroblasts. *J Immunol* 169(10):5897–903.

Manchanda V, Gupta S, Bhalla P. 2006. Meningococcal disease: history, epidemiology, pathogenesis, clinical manifestations, diagnosis, antimicrobial susceptibility and prevention. *Indian J Med Microbiol* 24(1):7–19.

Manchester M, Eto DS, Oldstone MB. 1999. Characterization of the inflammatory response during acute measles encephalitis in NSE-CD46 transgenic mice. *J Neuroimmunol* 96(2):207–217.

Marder K, Albert S, Dooneief G, Stern Y, Ramachandran G, Epstein L. 1996. Clinical confirmation of the American Academy of Neurology algorithm for HIV-1-associated cognitive/motor disorder. The Dana Consortium on Therapy for HIV Dementia and Related Cognitive Disorders. *Neurology* 47(5):1247–1253.

Masters PS. 2006. The molecular biology of coronaviruses. *Adv Virus Res* 66:193–292.

Mastroianni CM, Lancella L, Mengoni F, Lichtner M, Santopadre P, D'Agostino C, Ticca F, Vullo V. 1998. Chemokine profiles in the cerebrospinal fluid (CSF) during the course of pyogenic and tuberculous meningitis. *Clin Exp Immunol* 114(2):210–214.

McIntosh K. 1996. Coronaviruses. In *Fields Virology*. 3rd edition. Vol. 1, eds. BN Fields, DM Knipe, PM Howley, pp. 401–430. Philadelphia, PA: Lippincott-Raven Publishers.

McQuaid S, Cosby SL. 2002. An immunohistochemical study of the distribution of the measles virus receptors, CD46 and SLAM, in normal human tissues and subacute sclerosing panencephalitis. *Lab Invest* 82(4):403–9.

Mizuno T, Kawanokuchi J, Numata K, Suzumura A. 2003. Production and neuroprotective functions of fractalkine in the central nervous system. *Brain Res* 979(1–2):65–70.

Moller AS, Bjerre A, Brusletto B, Joo GB, Brandtzaeg P, Kierulf P. 2005. Chemokine patterns in meningococcal disease. *J Infect Dis* 191(5):768–775.

Morrison TB, Weis JH, Weis JJ. 1997. Borrelia burgdorferi outer surface protein A (OspA) activates and primes human neutrophils. *J Immunol* 158(10):4838–4845.

Murray PR, Pfaller MA, Rosenthal KS. 2005. *Medical Microbiology*. Elsevier Health Sciences, pp. 1–962. Amsterdam, The Netherlands.

Musher DM. 1992. Infections caused by *Streptococcus pneumoniae*: clinical spectrum, pathogenesis, immunity, and treatment. *Clin Infect Dis* 14(4):801–807.

Nagano I, Nakamura S, Yoshioka M, Onodera J, Kogure K, Itoyama Y. 1994. Expression of cytokines in brain lesions in subacute sclerosing panencephalitis. *Neurology* 44(4):710–715.

Nansen A, Marker O, Bartholdy C, Thomsen AR. 2000. CCR2 + and CCR5 + CD8 + T cells increase during viral infection and migrate to sites of infection. *Eur J Immunol* 30(7):1797–1806.

Nansen A, Christensen JP, Andreasen SO, Bartholdy C, Christensen JE, Thomsen AR. 2002. The role of CC chemokine receptor 5 in antiviral immunity. *Blood* 99(4):1237–1245.

Nau R, Bruck W. 2002. Neuronal injury in bacterial meningitis: mechanisms and implications for therapy. *Trends Neurosci* 25(1):38–45.

Navia BA, Cho ES, Petito CK, Price RW. 1986. The AIDS dementia complex: II. Neuropathology. *Ann Neurol* 19(6):525–535.

Noe KH, Cenciarelli C, Moyer SA, Rota PA, Shin ML. 1999. Requirements for measles virus induction of RANTES chemokine in human astrocytoma-derived U373 cells. *J Virol* 73(4):3117–3124.

Nottet HS. 1999. Interactions between macrophages and brain microvascular endothelial cells: role in pathogenesis of HIV-1 infection and blood–brain barrier function. *J Neurovirol* 5(6):659–669.

Oleszak EL, Chang JR, Friedman H, Katsetos CD, Platsoucas CD. 2004. Theiler's virus infection: a model for multiple sclerosis. *Clin Microbiol Rev* 17(1):174–207.

Ostergaard C, Yieng-Kow RV, Larsen CG, Mukaida N, Matsushima K, Benfield T, Frimodt-Moller N, Espersen F, Kharazmi A, Lundgren JD. 2000. Treatment with a monoclonal antibody to IL-8 attenuates the pleocytosis in experimental pneumococcal meningitis in rabbits when given intravenously, but not intracisternally. *Clin Exp Immunol* 122(2):207–211.

Pachner AR. 1986. Spirochetal diseases of the CNS. *Neurol Clin* 4(1):207–222.

Palma JP, Kim BS. 2001. Induction of selected chemokines in glial cells infected with Theiler's virus. *J Neuroimmunol* 117(1–2):166–170.

Palma JP, Kim BS. 2004. The scope and activation mechanisms of chemokine gene expression in primary astrocytes following infection with Theiler's virus. *J Neuroimmunol* 149(1–2):121–129.

Patterson CE, Daley JK, Echols LA, Lane TE, Rall GF. 2003. Measles virus infection induces chemokine synthesis by neurons. *J Immunol* 171(6):3102–3109.

Perlman SR, Lane TE, Buchmeier MJ. 1999. Coronaviruses: Hepatitis, peritonitis, and central nervous system disease. In *Effects of Microbes on the Immune System*, eds. MW Cunningham, RS Fujinami, pp. 331–348. Philadelphia, PA: Lippincott Williams & Wilkins.

Pfister HW, Feiden W, Einhaupl KM. 1993. Spectrum of complications during bacterial meningitis in adults. Results of a prospective clinical study. *Arch Neurol* 50(6):575–581.

Pfister HW, Scheld WM. 1997. Brain injury in bacterial meningitis: therapeutic implications. *Curr Opin Neurol* 10(3):254–259.

Rall GF, Manchester M, Daniels LR, Callahan EM, Belman AR, Oldstone MB. 1997. A transgenic mouse model for measles virus infection of the brain. *Proc Natl Acad Sci U S A* 94(9):4659–4663.

Ramos CD, Canetti C, Souto JT, Silva JS, Hogaboam CM, Ferreira SH, Cunha FQ. 2005. MIP-1alpha[CCL3] acting on the CCR1 receptor mediates neutrophil migration in immune inflammation via sequential release of TNF-alpha and LTB4. *J Leukoc Biol* 78(1):167–177.

Ransohoff RM. 2002. The chemokine system in neuroinflammation: an update. *J Infect Dis* 186 Suppl 2:S152–S156.

Robinson KA, Baughman W, Rothrock G, Barrett NL, Pass M, Lexau C, Damaske B, Stefonek K, Barnes B, Patterson J and others. 2001. Epidemiology of invasive *Streptococcus pneumoniae* infections in the United States, 1995–1998: Opportunities for prevention in the conjugate vaccine era. *JAMA* 285(13):1729–1735.

Saruhan-Direskeneli G, Gurses C, Demirbilek V, Yentur SP, Yilmaz G, Onal E, Yapici Z, Yalcinkaya C, Cokar O, Akman-Demir G and others. 2005. Elevated interleukin-12 and CXCL10 in subacute sclerosing panencephalitis. *Cytokine* 32(2):104–110.

Scheld WM, Koedel U, Nathan B, Pfister HW. 2002. Pathophysiology of bacterial meningitis: mechanism(s) of neuronal injury. *J Infect Dis* 186 Suppl 2:S225–S233.

Schnorr JJ, Xanthakos S, Keikavoussi P, Kampgen E, ter Meulen V, Schneider-Schaulies S. 1997. Induction of maturation of human blood dendritic cell precursors by measles virus is associated with immunosuppression. *Proc Natl Acad Sci U S A* 94(10):5326–5331.

Schuchat A, Robinson K, Wenger JD, Harrison LH, Farley M, Reingold AL, Lefkowitz L, Perkins BA. 1997. Bacterial meningitis in the United States in 1995. Active Surveillance Team. *N Engl J Med* 337(14):970–976.

Seebach J, Bartholdi D, Frei K, Spanaus KS, Ferrero E, Widmer U, Isenmann S, Strieter RM, Schwab M, Pfister H and others. 1995. Experimental Listeria meningoencephalitis. Macrophage inflammatory protein-1 alpha and -2 are produced intrathecally and mediate chemotactic activity in cerebrospinal fluid of infected mice. *J Immunol* 155(9):4367–4375.

Sejvar JJ, Marfin AA. 2006. Manifestations of West Nile neuroinvasive disease. *Rev Med Virol* 16(4):209–224.

So EY, Kang MH, Kim BS. 2006. Induction of chemokine and cytokine genes in astrocytes following infection with Theiler's murine encephalomyelitis virus is mediated by the Toll-like receptor 3. *Glia* 53(8):858–867.

Solomon T, Winter PM. 2004. Neurovirulence and host factors in flavivirus encephalitis – evidence from clinical epidemiology. *Arch Virol* Suppl(18):161–170.

Spanaus KS, Nadal D, Pfister HW, Seebach J, Widmer U, Frei K, Gloor S, Fontana A. 1997. C-X-C and C-C chemokines are expressed in the cerebrospinal fluid in bacterial meningitis and mediate chemotactic activity on peripheral blood-derived polymorphonuclear and mononuclear cells in vitro. *J Immunol* 158(4):1956–1964.

Speth C, Dierich MP, Sopper S. 2005. HIV-infection of the central nervous system: the tightrope walk of innate immunity. *Mol Immunol* 42(2):213–228.

Sprenger H, Rosler A, Tonn P, Braune HJ, Huffmann G, Gemsa D. 1996. Chemokines in the cerebrospinal fluid of patients with meningitis. *Clin Immunol Immunopathol* 80(2):155–161.

Sprenger H, Krause A, Kaufmann A, Priem S, Fabian D, Burmester GR, Gemsa D, Rittig MG. 1997. *Borrelia burgdorferi* induces chemokines in human monocytes. *Infect Immun* 65(11):4384–4388.

Standiford TJ, Kunkel SL, Lukacs NW, Greenberger MJ, Danforth JM, Kunkel RG, Strieter RM. 1995. Macrophage inflammatory protein-1 alpha mediates lung leukocyte recruitment, lung capillary leak, and early mortality in murine endotoxemia. *J Immunol* 155(3):1515–1524.

Stiles LN, Hardison JL, Schuamburg CS, Whitman LM, Lane TE. 2006a. T cell anti-viral effector function is not dependent on CXCL10 following murine coronavirus infection. *J Immunol* 177:8372–8380.

Stiles LN, Hosking MP, Edwards RA, Strieter RM, Lane TE. 2006b. Differential roles for CXCR3 in CD4+ and CD8+ T cell trafficking following viral infection of the CNS. *Eur J Immunol* 36(3):613–622.

Su YH, Yan XT, Oakes JE, Lausch RN. 1996. Protective antibody therapy is associated with reduced chemokine transcripts in herpes simplex virus type 1 corneal infection. *J Virol* 70(2):1277–1281.

Suzuki T, Ogata A, Tashiro K, Nagashima K, Tamura M, Yasui K, Nishihira J. 2000. Japanese encephalitis virus up-regulates expression of macrophage migration inhibitory factor (MIF) mRNA in the mouse brain. *Biochim Biophys Acta* 1517(1):100–106.

ter Meulen V, Stephenson JR, Kreth HW. 1983. *Comprehensive Virology*. New York, NY: Plenum Press, pp. 105–159.

Tishon A, Lewicki H, Andaya A, McGavern D, Martin L, Oldstone MB. 2006. CD4 T cell control primary measles virus infection of the CNS: regulation is dependent on combined activity with either CD8 T cells or with B cells: CD4, CD8 or B cells alone are ineffective. *Virology* 347(1):234–245.

Tong N, Perry SW, Zhang Q, James HJ, Guo H, Brooks A, Bal H, Kinnear SA, Fine S, Epstein LG and others. 2000. Neuronal fractalkine expression in HIV-1 encephalitis: roles for macrophage recruitment and neuroprotection in the central nervous system. *J Immunol* 164(3):1333–1339.

Townsend GC, Scheld WM. 1998. Infections of the central nervous system. *Adv Intern Med* 43:403–447.

Trifilo MJ, Bergmann CC, Kuziel WA, Lane TE. 2003. CC chemokine ligand 3 (CCL3) regulates CD8(+)-T-cell effector function and migration following viral infection. *J Virol* 77(7):4004–4014.

Trifilo MJ, Lane TE. 2004. The CC chemokine ligand 3 regulates CD11c + CD11b + CD8alpha-dendritic cell maturation and activation following viral infection of the central nervous system: implications for a role in T cell activation. *Virology* 327(1):8–15.

Tsunoda I, Lane TE, Blackett J, Fujinami RS. 2004. Distinct roles for IP-10/CXCL10 in three animal models, Theiler's virus infection, EAE, and MHV infection, for multiple sclerosis: implication of differing roles for IP-10. *Mult Scler* 10(1):26–34.

Ure DR, Lane TE, Liu MT, Rodriguez M. 2005. Neutralization of chemokines RANTES and MIG increases virus antigen expression and spinal cord pathology during Theiler's virus infection. *Int Immunol* 17(5):569–579.

Wickham S, Lu B, Ash J, Carr DJ. 2005. Chemokine receptor deficiency is associated with increased chemokine expression in the peripheral and central nervous systems and increased resistance to herpetic encephalitis. *J Neuroimmunol* 162(1–2):51–59.

Winter PM, Dung NM, Loan HT, Kneen R, Wills B, Thu le T, House D, White NJ, Farrar JJ, Hart CA and others. 2004. Proinflammatory cytokines and chemokines in humans with Japanese encephalitis. *J Infect Dis* 190(9):1618–1626.

Wuest T, Austin BA, Uematsu S, Thapa M, Akira S, Carr DJ. 2006. Intact TRL 9 and type I interferon signaling pathways are required to augment HSV-1 induced corneal CXCL9 and CXCL10. *J Neuroimmunol* 179(1–2):46–52.

Xiao BG, Mousa A, Kivisakk P, Seiger A, Bakhiet M, Link H. 1998. Induction of beta-family chemokines mRNA in human embryonic astrocytes by inflammatory cytokines and measles virus protein. *J Neurocytol* 27(8):575–580.

Xu J, Zhong S, Liu J, Li L, Li Y, Wu X, Li Z, Deng P, Zhang J, Zhong N and others. 2005. Detection of severe acute respiratory syndrome coronavirus in the brain: potential role of the chemokine mig in pathogenesis. *Clin Infect Dis* 41(8):1089–1096.

Yan XT, Tumpey TM, Kunkel SL, Oakes JE, Lausch RN. 1998. Role of MIP-2 in neutrophil migration and tissue injury in the herpes simplex virus-1-infected cornea. *Invest Ophthalmol Vis Sci* 39(10):1854–1862.

Zink MC, Coleman GD, Mankowski JL, Adams RJ, Tarwater PM, Fox K, Clements JE. 2001. Increased macrophage chemoattractant protein-1 in cerebrospinal fluid precedes and predicts simian immunodeficiency virus encephalitis. *J Infect Dis* 184(8):1015–1021.

Zwijnenburg PJ, van der Poll T, Florquin S, van Deventer SJ, Roord JJ, van Furth AM. 2001. Experimental pneumococcal meningitis in mice: a model of intranasal infection. *J Infect Dis* 183(7):1143–1146.

13
CNS Dendritic Cells in Inflammation and Disease

Samantha L. Bailey and Stephen D. Miller

1 Introduction – MS and EAE

Multiple sclerosis (MS) is a multi-factorial disease associated with chronic autoimmune inflammation of the central nervous system (CNS). In MS, and the relevant animals models experimental autoimmune encephalomyelitis (EAE), and Theiler's murine encephalitis virus-induced demyelinating disease (TMEV-IDD), myelin destruction is mediated by neuroantigen-specific $CD4^+$ T cells (Gonatas et al., 1986; Miller et al., 1997; Wekerle, 1991). MS and EAE share clinical and histopathological similarities. Mononuclear cells (MNCs) accumulate in demyelinated lesions in the white and grey matter of the brain and spinal cord. Infiltrates are composed of $CD4^+$ and $CD8^+$ T cells, B cells, macrophages and dendritic cells (DCs). $CD4^+$ T cells and DCs are critical for the initiation and progression of EAE as $CD4^+$ T cell depletion renders mice resistant to EAE (Jameson et al., 1994; McDevitt et al., 1987; Sedgwick and Mason, 1986; Waldor et al., 1985), and DCs from the CNS uniquely activate naïve myelin-specific T cells by acquiring and processing endogenous myelin peptides (Bailey et al., 2007). The initiating events in MS are unknown, but studies in the EAE and TMEV-IDD animal models indicate that CNS damage is caused by direct and indirect effects of inflammatory cytokines and chemokines (e.g. TNF, IFN-γ, IL-17, CCL2, etc.) (Begolka et al., 1998; Chen et al., 2006; Karpus et al., 1995; Powell et al., 1990), that induce the activation and recruitment of monocyte/macrophages and resident microglia, that cause axon damage and demyelination by bystander mechanisms (Cammer et al., 1978; Rivers and Schwentker, 1935).

EAE can be induced in a variety of mouse strains by immunizing with mouse spinal cord homogenate or myelin proteins and peptides in complete Freund's adjuvant (CFA), or by the transfer of activated $CD4^+$ T cells expanded in the lymph nodes of myelin/CFA immunized donor mice. For initiation of EAE, activated $CD4^+$ T cells must encounter myelin peptides presented by MHC class II-expressing cells in the CNS (Hickey and Kimura, 1988; Tompkins et al., 2002). T cell activation/re-activation in the CNS is quickly followed by the recruitment of other peripheral immune populations, and subsequently clinical symptoms ranging from tail tone loss and hind weakness to hind and fore-limb paralysis.

T.E. Lane et al. (eds.), *Central Nervous System Diseases and Inflammation.*
© Springer 2008

In this review, we discuss the current understanding of the CNS DCs subsets that drive pathogenic CD4+ T cell activation, responsible for progression of relapsing-remitting EAE (R-EAE).

2   CD4+ Th17 Cells are Critical for the Induction and Progression of EAE

Since the description of CD4+ T helper cells producing either IFN-γ (Th1) or producing IL-4, IL-5, IL-10 (Th2) (Mosmann and Coffman, 1989), EAE was thought to be a Th1-mediated disease because the majority of CD4+ T cells in the inflamed CNS produced IFN-γ, and Th-2 cells, which suppress Th-1 function, were found to be protective for EAE (Cua et al., 1995). However, controversy existed in that mice deficient for critical components of the Th1 pathway such as IFN-γ, IFN-γ receptor, IL-12 receptor β2 and IL-12 p35 (Becher et al., 2002; Ferber et al., 1996; Gran et al., 2002; Willenborg et al., 1999; Zhang et al., 2003) developed an exacerbated course of EAE, implying that Th1 cells may have a regulatory role. More recently, IL-23, an IL-12 family member, was found to be a critical cytokine for EAE development and progression. Mice deficient in the IL-23 subunits p40 or p19 are resistant to EAE induction (Becher et al., 2003; Cua et al., 2003), and transfer of adenovirus expressing IL-23 restored EAE susceptibility to p19/p40 deficient mice (Cua et al., 2003). Cua and colleagues went on to perform gene expression analysis on CD4+ T cells that had been activated in the presence of IL-12 or IL-23, and IL-17A and IL-17F were shown to be specifically and highly up-regulated in the presence of IL-23 (Langrish et al., 2005). IL-17 is a T cell derived cytokine that until that time had been associated with allograft rejection (Antonysamy et al., 1999), and the inflammatory diseases rheumatoid arthritis (Nakae et al., 2003), and asthma (Nakae et al., 2002). IL-17 producing T cells have been shown to be a separate Th lineage, distinct from Th1 and Th2 cells, and as a discrete effector T cell subset have been termed Th17 cells (Park et al., 2005). IL-23 supports the expansion of IL-17+ TNF+ T cells from LNs isolated from mice immunized with myelin peptides in CFA, and these Th17 cells are highly pathogenic, inducing severe EAE when transferred into naïve animals (Langrish et al., 2005). Further investigations have supported a critical role for IL-17 in EAE pathogenesis as neutralizing antibodies against IL-17 ameliorated EAE in animals with established disease (Chen et al., 2006; Langrish et al., 2005; Park et al., 2005).

In vitro studies investigating the activation of IL-17 production during primary stimulation of naïve CD4+ T cells, demonstrated the surprising finding that DCs + TLR stimuli + natural occurring T regulatory cells (Tregs) induced IL-17+ cells (Veldhoen et al., 2006). Using neutralizing antibodies, the cytokines TGF-β and IL-6 were identified as critical for inducing Th17 evolution. IL-23 was found to support the expansion and survival of differentiated Th17 cells, whereas IL-2 pushed the Th17 cells to the IFN-γ-producing Th1 phenotype (Veldhoen et al., 2006). Importantly, the Th1 and Th2 cytokines, IFN-γ and IL-4 inhibit the differentiation of Th17 cells (Harrington et al., 2005; Mangan et al., 2006; Park et al., 2005; Veldhoen et al., 2006), providing an explanation why IFN-γ-deficient animals develop severe EAE, i.e. an absence of Th17 regulation. TGF-β alone acts on naive T cells to induce Foxp3 expression and Tregs that suppress immune responses (Bettelli et al., 2006; Chen et al., 2003), but in the presence of the acute phase cytokine IL-6, TGF-β induces differentiation of Th17 cells. T cells from TGF-β1-deficient animals cannot be induced by mitogen to produce IL-17 (Mangan et al., 2006). One of the IL-17-inducing functions attributed to TGF-β is through the up-regulation of IL-23R on activated T cells (Mangan et al., 2006). Immunization of mice with MOG_{35-55} and CFA in a transgenic system where TGF-β1 expression is driven by the IL-2 promoter in $CD4^+$ T cells that are specific for MOG_{35-55}, induces a rapid and severe EAE, associated with CNS infiltration of that is highly enriched in Th17 cells and reduced $Foxp3^+$ Treg cells, compared with non-TGF-β1-transgenic controls (Bettelli et al., 2006). Similarly, transgenic mice with dominant negative version of TGF-βRII driven by the $CD4^+$ promoter to limit TGF-β1 non-responsiveness to T cells only, are resistant to EAE (Veldhoen et al., 2006). Blocking Th17 priming with local TGF-β antibody prevented EAE, whereas TGF-β neutralization after Th17 evolution had no affect, demonstrating that differentiated Th17 cells do not require further TGF-β signaling and the primary Th17 population is sufficient to induce clinical EAE (Veldhoen et al., 2006). Together these studies show that preventing the generation of Th17 cells, or neutralizing IL-17 after EAE induction protects from EAE induction and progression. Restricting the activation of Th17 cells is an attractive therapeutic target (Schreiner et al., 2007), to avoid immune compromising the patient. We have worked to understand the factors that trigger pathogenic T cell activation during progressive R-EAE and TMEV-IDD, and discuss crucial check points of T cell activation, and the APC that drives pathogenic T cell responses in R-EAE below.

3 Epitope Spreading Drives Pathogenesis of Relapsing EAE

Broadening of autoreactive B cell, $CD8^+$ and $CD4^+$ T cell responses have been described in EAE, where reactivity to myelin epitopes distinct to the inducing epitope develop and have a major role mediating clinical relapses (Katz-Levy et al., 2000; Lehmann et al., 1992; McRae et al., 1995; Neville et al., 2002; Vanderlugt et al., 2000; Yu et al., 1996). $CD4^+$ T cell epitope spreading is crucial for chronic demyelination in both the R-EAE and TMEV-IDD models of MS in susceptible SJL mice as antigen-specific tolerance directed against MHC II-restricted, relapse-associated epitopes halts disease progression (Neville et al., 2002; Vanderlugt et al., 2000). Intramolecular epitope spreading was demonstrated in SJL mice with R-EAE induced with $PLP_{139-151}$, where $CD4^+$ T cells reactive to $PLP_{1781-191}$ were found in the spleens and CNS. T cells reactive to the spread epitopes are pathogenic, because splenic T cells from mice primed with $PLP_{139-151}$, re-stimulated in vitro with $PLP_{178-191}$, induce EAE when transferred to naive SJL mice. Moreover, tolerization of mice during clinical remission with $PLP_{178-191}$, but not the inducing $PLP_{139-151}$, prevents clinical relapses (Vanderlugt et al., 2000). Intermolecular epitope spreading was demonstrated by the development of $PLP_{139-151}$ T cell responses following the acute phase of EAE induced by MBP_{84-104}. Therefore, T cell responses in EAE are dynamic and diversify over time, where pathogenicity of the initial T cell reactivity often disappears (Steinman, 1999; Tuohy et al., 1999). The hierarchy of T cell reactivity in epitope spreading correlates with the extent of frequency of the T cell repertoire (Vanderlugt et al., 2000) and the processing and MHC II-binding efficiency of encephalitogenic epitopes by APCs (Anderton et al., 2002).

4   Epitope Spreading Initiates in the CNS

Given the driving pathogenic force of T cell epitope spreading for clinical relapses, we asked where $CD4^+$ T cells specific for relapse associated epitopes were being initially activated during the acute clinical phases of R-EAE and TMEV-IDD (McMahon et al., 2005). Possible anatomic locations for the activation of T cells specific for endogenous myelin epitopes were the spleen or cervical lymph nodes, where myelin debris may accumulate due to disruption of the blood brain barrier during the acute phase of CNS inflammation (de Vos et al., 2002; Hochwald et al., 1988; Ling et al., 2003; Yamada et al., 1991), and the CNS itself where myelin epitopes may be presented by local APCs such as resident microglia, or peripherally-derived macrophages or DCs (Hickey and Kimura, 1988; Katz-Levy et al., 2000; Mack et al., 2003; Serafini et al., 2000). Naïve transgenic (Tg) $Thy1.1^+$ $CD4^+$ T cells specific for $PLP_{139-151}$ (139TCR cells) were used as reporters for epitope spreading in $PLP_{178-191}$ primed R-EAE (McMahon et al., 2005). CFSE-labeled 139TCR cells were transferred into $PLP_{178-191}$ primed mice, and control mice primed with $OVA_{323-339}$ or $PLP_{139-151}$, and after 3–8 days, the spleen, cervical lymph nodes, peripheral lymph nodes and CNS analyzed for proliferating T cells. The reporter 139TCR cells did not infiltrate the CNS or proliferate in any organ of the healthy $OVA_{323-339}$ primed mice, but proliferated strongly in the peripheral lymph nodes, and spleen, cervical lymph nodes and CNS of $PLP_{139-151}$ primed mice. However, during the acute phase and primary relapse of R-EAE induced with $PLP_{178-191}$, 139TCR cells proliferated only in the CNS, not in any of the peripheral lymphoid organs analyzed. Similar results were found when the 139TCR cells were a component of the immune system, using a mixed bone marrow chimera strategy, i.e. in SJL mice engrafted with a 1:1 mix of wild type and TCR Tg bone marrow followed by EAE induction with $PLP_{178-191}$. TCR cells were identified with Thy1.1 staining, and by measuring CD44, CD45RB, CD49d, CD62L, and CD69 were activated in the CNS, and not spleen or lymph nodes. $PLP_{139-151}$-specific Tg T cell activation and proliferation were also evident in the CNS only of mice with TMEV-IDD, 25 and 50 days after primary infection. These data demonstrate the CNS as the primary site of the activation of naïve T cells specific for relapse-associated antigens. Thus, the CNS can apparently function as a neolymphoid organ, and contains APCs which can present myelin fragments capable of activating naïve T cells (McMahon et al., 2005).

Candidate APCs that express MHC II and co-stimulatory molecules in the inflamed CNS during acute R-EAE include resident microglia, and macrophages and DCs that are also recruited to the CNS during EAE, where they could uptake myelin peptides to present to CD4+ T cells. Highly purified microglia, macrophages and myeloid DCs, and non-myeloid DCs isolated from the CNS of mice with peak acute $PLP_{178-191}$ primed were cultured with a fixed number of CFSE labeled naïve 139TCR cells, or a $PLP_{139-151}$-specific CD4+ T cell line at different APC:T cell ratios, in the presence or absence (to measure presentation of endogenous antigen) of exogenous $PLP_{139-151}$. When $PLP_{139-151}$ was added to the cultures, DCs, macrophages and microglia induced the proliferation of naïve 139TCR cells at the high APC:T cell ratio of 1:1, the microglia only affective at 1:1, macrophages at 1:1 and 1:10, and DCs as the most efficient, inducing naive T cell proliferation at 1 DC: 100 T cells. However, only the DCs induced the proliferation of the naïve Tg T cells with PLP acquired in vivo, no proliferation was seen with macrophages and microglia even at the high APC: T cell ratio's. However, macrophages, microglia and DCs activated a T cell line with endogenously acquired peptides, with microglia being the least efficient, and macrophages and DCs being equivalent. Corroborating analysis also showed DCs in the inflamed CNS, with internalized PLP debris, in close association with CD4+ T cells. The data is consistent with the hypothesis that the CNS can act as a neolymphoid organ, supporting the activation of primary CD4+ T cell responses, and that the inflamed CNS contained DCs that phagocytize myelin proteins and present them to autoreactive T cells that then drive clinical disease. Crucial questions arose as to the ontogeny of the DCs, and their characteristics that distinguish them from microglia and macrophages for activating T cells in the CNS.

5 CNS DC Susbsets Differentially Activate Th17 Cells during Relapsing EAE

DCs provide sentinel function of tissues and organs, are rare in secondary lymphoid organs, yet are potent inducers of T cell activation being proficient at presenting antigens to naïve T cells to induce T cell proliferation and effector differentiation at DC:T cell ratios of 1:100 (Ingulli et al., 1997; Steinman, 1991; Zammit et al., 2005). DCs are a heterologous population of cells, with discrete subtypes that are thought to have distinct precursors (Liu, 2001; Shortman and Liu, 2002). Th cell differentiation is determined, in part, by cytokines produced by the DC presenting peptides to a T cell. The critical factors for DC instruction are the type and amount of antigen processed and presented, innate immune signals received, tissue of residence and stage of DC maturity [reviewed in (McMahon et al., 2006)]. However, DC populations have been attributed with inducing particular types of T-effector cell differentiation. CD11b+ myeloid DCs have been associated with promoting Th2 differentiation and recruitment (Maldonado-Lopez et al., 1999; Penna et al.,

2001; Yasumi et al., 2004), while CD8 DCs promote Th1 differentiation which correlates with high production of IL-12p70 (Hochrein et al., 2001; Maldonado-Lopez et al., 1999; Pulendran et al., 2001; Yasumi et al., 2004). Plasmacytoid DCs (pDC) are potent anti-viral circulating cells that produce high levels of type I interferons during viral infection or stimulation with CpG DNA that mimic viral/bacterial DNA sequences (Asselin-Paturel et al., 2001). pDCs have also been associated with the generation of regulatory T cells (Bilsborough et al., 2003) and may function to support peripheral T cell tolerance (Bilsborough et al., 2003; Martin et al., 2002).

The healthy CNS contains few DCs, and these are situated in vessel-rich areas such as the meninges and choroid plexus (Fischer et al., 2000; Rosicarelli et al., 2005; Serafini et al., 2000; Suter et al., 2000). These few DCs may be sufficient to present endogenous myelin peptides to auto-reactive myelin T cells resulting in the initiation of EAE (Greter et al., 2005). DC numbers in the CNS are enhanced by infection and inflammation (Abreu-Silva et al., 2003; Fischer et al., 2000; Fischer and Reichmann, 2001; Matyszak and Perry, 1996; McMahon et al., 2005), but their role for activating T cells during CNS inflammation has been a contentious issue (Greter et al., 2005; McMahon et al., 2005; Suter et al., 2003).

We identified three discrete populations of $CD11c^+$ DCs in the CNS of mice during peak acute R-EAE (Bailey et al., 2007). The most abundant were $CD11b^+$ $F4/80^+$ myeloid DCs (mDC) comprising 10.8% of the mononuclear cells (MNC) isolated from the CNS; $B220^+$ $PDCA-1^+$ plasmacytoid DCs (pDCs) made up 5.4% of CNS cells, and a rare (0.8% of MNCs) population of $CD8^+$ $CD205^{+/-}$ DCs CD8 DCs. The DCs in the inflamed CNS during the acute phase of EAE could arise from two places, the inflamed CNS itself, deriving from resident precursor cells (Fischer and Bielinsky, 1999; Santambrogio et al., 2001), or infiltrate from a peripheral bone marrow precursor population (Newman et al., 2005). To determine ontogeny of the CNS DCs, we employed bone marrow chimeras with non-allogeneic bone marrow donor (C57Bl/6) expressing CD45.2 into a mixed hybrid (SJL × C57Bl/6)F_1 expressing CD45.1/2 double positive cells. In these chimeras, CNS cells, which are radio-resistant retain the host CD45.1/2 phenotype, while bone marrow cells express donor CD45.2. During the acute phase of $PLP_{178-191}$-induced R-EAE 84% of mDCs and 98% of non-myeloid pDCs and CD8 DCs were of bone marrow origin. Similar results were obtained in experiments using two other chimera strategies (Bailey et al., 2007). Thus, the majority of DCs in the inflamed CNS infiltrate from a peripheral population. Using immunohistochemistry, staining of the DC populations in frozen sections of cerebellum and lumbar spinal cords of mice during the acute phase of R-EAE revealed the DC populations accumulate in discrete locations. mDCs took a central position in demyelinated lesions and perivascular cuffs of inflammatory cells. In mixed bone marrow chimera mice, with 50% of the bone marrow from Thy1.1 139TCR Tg mice, mDCs were visualized surrounding 139TCR cells in the spinal cord during the acute phase of $PLP_{178-191}$-induced R-EAE. Importantly, 61% of the 139TCR cells had down-regulated CD45RB and were activated. Therefore, the majority of mDCs in the inflamed CNS transverse the blood brain barrier from the periphery, and are found in the center of demyelinated lesions interacting with pathogenic $CD4^+$ T cells specific for spread epitopes. We next purified CNS mDCs, pDCs, CD8 DCs and macro-phages during the acute phase of $PLP_{178-191}$-induced R-EAE, and cultured the APCs with naive 139TCR cells with or without added $PLP_{139-151}$, to measure presentation of endogenous $PLP_{139-151}$ (Table 13.1). The extent of 139TCR proliferation in the cultures revealed CNS mDCs as highly efficient at presenting endogenous and exogenous peptides to naive T cells, presenting endogenous $PLP_{139-151}$ to induce 44-fold and 16-fold expansion of the naive 139TCR cells at APC/T cell ratio's of 1:5 and 1:50, respectively. CNS CD8 DCs were the next most efficient, inducing sixfold proliferation with endogenous peptide, and pDCs similar with fivefold expansion of the 139TCR cells. Macrophages were the least effective at activating naïve T cells with endogenous peptides, but were similar to pDCs and CD8 when $PLP_{139-151}$ was added to the cultures. Moreover, macrophages were more effective than pDCs at activating a $CD4^+$ $PLP_{139-151}$-specific T cell line with endogenous peptides, indicating it was not a lack of peptide presentation that accounted for the macrophages being the poorest at activating the naive 139TCR cells ex vivo. To further investigate myelin uptake sorted CNS mDCs, pDCs, CD8 DCs and macro-phages were permeabilized and stained for PLP and all were found to have inter-nalized PLP protein to similar degrees.

We analyzed the supernatants of the naive 139TCR cells activated with the CNS APC populations presenting endogenous PLP for a panel of cytokines and chem-okines. Splenic APCs presenting exogenous $PLP_{139-151}$ to the naive 139TCR cells induced 70 pg/ml IL-17 and 834 pg/ml IFN-γ (IL-17/IFN-γ ratio = 0.09), but when activated with an equivalent number of CNS mDCs presenting endogenous PLP, the 139TCR cell cytokine profile changed to 207 pg/ml IL-17 and 155 pg/ml/IFN-γ (IL-17/IFN-γ ratio = 1.3). The other CNS APC populations induced Th1 polarized cytokines from the 139TCR cells: pDCs induced a IL-17/IFN-γ ratio of 0.6, whereas macrophages induced a IL-17/IFN-γ ratio of 0.13. mDCs induced the highest levels of cytokines and chemokines including IL-2, TNF, IL-10, CCL3, CCL2, CCL5 and CXCL10. This ex vivo assay revealed mDCs as being highly efficient at activating naive $CD4^+$ T cells with relapse associated spread myelin epitopes, and inducing a Th17-biased phenotype. This data supports mDCs as the major drivers of epitope spreading in EAE. To confirm our findings in vivo, 139TCR cells were transferred at the onset of the acute phase of EAE induced by

Table 13.1 Relative efficiency of antigen presenting cells from the inflamed CNS to present endogeneous and exogenous peptides to naïve and activated T cells

	Microglia	Macrophages	CD8 DCs	pDCs	mDCs
Endogenous presentation to Naïve T cells	—	+	++	++	+++++
Peptide presentation to Naïve T cells	—	+++	++	++	+++++
Endogenous presentation to Effector T cells	—	+++	ND	++	+++++
Peptide presentation to Effector T cells	+++	+++	ND	++	+++++

$PLP_{178-191}$, and recovered from the CNS in late acute EAE. Intracellular staining for IFN-γ and IL-17 revealed the 139TCR cells to have a Th17 polarized phenotype with a IL-17/IFN-γ ratio of 2.1, whereas, host, acute phase CNS CD4+ T cells were Th1 polarized with a IL-17/IFN-γ ratio of 0.44.

In an attempt to distinguish the properties that make mDCs uniquely able to activate naive T cells as compared with pDCs and macrophages, the cells were analyzed expression for expression of costimulatory molecules required for CD4+ T cell activation, and for B7-H1 and B7-DC which may have stimulatory or inhibitory functions on T cell activation (Greenwald et al., 2005). Macrophages, mDCs and CD8 DCs infiltrating the inflamed CNS during acute R-EAE display a mature phenotype, expressing high levels of MHC class II, CD80, CD86, CD40, B7-H1 and B7-DC, comparable to their expression in peripheral lymphoid organs (Bailey et al., 2007). pDCs, however, display an immature phenotype in the inflamed CNS characterized by low expression of MHC II, CD80, CD86, CD40, B7-H1 and B7-DC, unlike in peripheral lymphoid compartments such as peripheral and cervical lymph nodes and spleen. Thus, the enhanced ability of mDCs to activate naïve and activated CD4+ T cells is not due to differential expression of co-stimulatory molecules. To determine if mDCs uniquely produced cytokines required for differentiation and survival of Th17 cells, DC subsets and macrophages were purified from the CNS during the acute phase of R-EAE and the production of TGF-β1, IL-6 and IL-23 was determined. As measured directly ex vivo by real-time PCR, TGF-β1 was expressed twofold higher by mDCs than macrophages or pDCs. Upon CD40 ligation, mDCs also secreted more IL-6 than macrophages, whereas all of the CNS DC subsets as well as macrophages produced significant levels of IL-23 following CD40-triggering. Thus, the enhanced ability of mDCs for driving Th17 generation correlates with high production of TGF-β and IL-6. TGF-β production by mDCs has been associated with supporting Treg functions and CD4+ T cell tolerance (Ghiringhelli et al., 2005). However, in the inflamed CNS, mDCs clearly function as inducers of Th17 cells. The dichotomy of Treg vs. Th17 differentiation has been the focus of a series of recent high profiles studies [reviewed in (Bettelli et al., 2007)], and our data is the first to implicate the mDCs, as critical inducers of Th17 differentiation.

6 Summary

CD11c+ DCs play a major role in both the *initiation* and *progression* of autoimmune inflammatory disease in the CNS. Since the CNS serves as the primary site where activation of pathogenic Th1/Th17 cells specific for endogenous myelin epitopes (i.e., epitope spreading), which play a critical role in driving progressive autoimmune disease, the current data suggests that the inflamed CNS can function as a neo-lymphoid organ. In support of this our recent unpublished data indicates that expression of genes encoding multiple receptor:ligand pairs involved in lymphoid organogenesis (including LTα1β2/LTβR, CXCL12/CXCR4, CSCL13/CXCR5,

CCL21/CCR7, and CCL19/CCR7) are highly upregulated in the CNS. Further, mDCs are the main drivers of epitope spreading displaying the unique ability to acquire and present endogenous myelin peptides, to cluster specifically with naïve CD4+ T cells in the inflamed CNS and to polarize towards a Th17 phenotype when presenting endogenous myelin peptides. In conclusion, understanding the cues that determine DC signals to T cells will be crucial to understanding the fate of pathological (auto)immune inflammation in different tissues and diseases. Moreover, strategies targeting inhibition of the migration of myeloid DCs to the CNS may be an effective therapy for chronic immune-mediated CNS demyelinating diseases including MS.

Acknowledgments The authors wish to acknowledge the support of grants from the National Institutes of Health, the National Multiple Sclerosis Society, and the Myelin Repair Foundation.

References

Abreu-Silva, A. L., Calabrese, K. S., Tedesco, R. C., Mortara, R. A., and da Costa, S. C. (2003). Central nervous system involvement in experimental infection with Leishmania (Leishmania) amazonensis. *Am. J. Trop. Med. Hyg.* 68:661–665.

Anderton, S. M., Viner, N. J., Matharu, P., Lowrey, P. A., and Wraith, D. C. (2002). Influence of a dominant cryptic epitope on autoimmune T cell tolerance. *Nat. Immunol.* 3:175–181.

Antonysamy, M. A., Fanslow, W. C., Fu, F., Li, W., Qian, S., Troutt, A. B., and Thomson, A. W. (1999). Evidence for a role of IL-17 in organ allograft rejection: IL-17 promotes the functional differentiation of dendritic cell progenitors. *J. Immunol.* 162:577–584.

Asselin-Paturel, C., Boonstra, A., Dalod, M., Durand, I., Yessaad, N., zutter-Dambuyant, C., Vicari, A., O'Garra, A., Biron, C., Briere, F., and Trinchieri, G. (2001). Mouse type I IFN-producing cells are immature APCs with plasmacytoid morphology. *Nat. Immunol.* 2:1144–1150.

Bailey, S. L., Schreiner, B., McMahon, E. J., and Miller, S. D. (2007). CNS myeloid DCs presenting endogenous myelin peptides 'preferentially' polarize CD4(+) T(H)-17 cells in relapsing EAE. *Nat. Immunol.* 8:172–180.

Becher, B., Durell, B. G., and Noelle, R. J. (2002). Experimental autoimmune encephalitis and inflammation in the absence of interleukin-12. *J. Clin. Invest.* 110:493–497.

Becher, B., Durell, B. G., and Noelle, R. J. (2003). IL-23 produced by CNS-resident cells controls T cell encephalitogenicity during the effector phase of experimental autoimmune encephalomyelitis. *J. Clin. Invest.* 112:1186–1191.

Begolka, W. S., Vanderlugt, C. L., Rahbe, S. M., and Miller, S. D. (1998). Differential expression of inflammatory cytokines parallels progression of central nervous system pathology in two clinically distinct models of multiple sclerosis. *J. Immunol.* 161:4437–4446.

Bettelli, E., Carrier, Y., Gao, W., Korn, T., Strom, T. B., Oukka, M., Weiner, H. L., and Kuchroo, V. K. (2006). Reciprocal developmental pathways for the generation of pathogenic effector TH17 and regulatory T cells. *Nature.* 441:235–238.

Bettelli, E., Oukka, M., and Kuchroo, V. K. (2007). T(H)-17 cells in the circle of immunity and autoimmunity. *Nat. Immunol.* 8:345–350.

Bilsborough, J., George, T. C., Norment, A., and Viney, J. L. (2003). Mucosal CD8alpha + DC, with a plasmacytoid phenotype, induce differentiation and support function of T cells with regulatory properties. *Immunology* 108:481–492.

Cammer, W., Bloom, B. R., Norton, W. T., and Gordon, S. (1978). Degradation of basic protein in myelin by neutral proteases secreted by stimulated macrophages: a possible mechanism of inflammatory demyelination. *Proc. Natl. Acad. Sci. U.S.A.* 75:1554–1558.

Chen, W., Jin, W., Hardegen, N., Lei, K. J., Li, L., Marinos, N., McGrady, G., and Wahl, S. M. (2003). Conversion of peripheral CD4 + CD25- naive T cells to CD4 + CD25 + regulatory T cells by TGF-beta induction of transcription factor Foxp3. *J. Exp. Med.* 198:1875–1886.

Chen, Y., Langrish, C. L., McKenzie, B., Joyce-Shaikh, B., Stumhofer, J. S., McClanahan, T., Blumenschein, W., Churakovsa, T., Low, J., Presta, L., et al. (2006). Anti-IL-23 therapy inhibits multiple inflammatory pathways and ameliorates autoimmune encephalomyelitis. *J. Clin. Invest.* 116:1317–1326.

Cua, D. J., Hinton, D. R., and Stohlman, S. A. (1995). Self-antigen-induced Th2 responses in experimental allergic encephalomyelitis (EAE)-resistant mice. Th2-mediated suppression of autoimmune disease. *J. Immunol.* 155:4052–4059.

Cua, D. J., Sherlock, J., Chen, Y., Murphy, C. A., Joyce, B., Seymour, B., Lucian, L., To, W., Kwan, S., Churakova, T., et al. (2003). Interleukin-23 rather than interleukin-12 is the critical cytokine for autoimmune inflammation of the brain. *Nature* 421:744–748.

de Vos, A. F., van, M. M., Brok, H. P., Boven, L. A., Hintzen, R. Q., Van, d. V., Ravid, R., Rensing, S., Boon, L., t Hart, B. A., and Laman, J. D. (2002). Transfer of central nervous system autoantigens and presentation in secondary lymphoid organs. *J. Immunol.* 169:5415–5423.

Ferber, I. A., Brocke, S., Taylor-Edwards, C., Ridgway, W., Dinisco, C., Steinman, L., Dalton, D., and Fathman, C. G. (1996). Mice with a disrupted IFN-gamma gene are susceptible to the induction of experimental autoimmune encephalomyelitis (EAE). *J. Immunol.* 156:5–7.

Fischer, H. G., and Bielinsky, A. K. (1999). Antigen presentation function of brain-derived dendriform cells depends on astrocyte help. *Int. Immunol.* 11:1265–1274.

Fischer, H. G., and Reichmann, G. (2001). Brain dendritic cells and macrophages/microglia in central nervous system inflammation. *J. Immunol.* 166:2717–2726.

Fischer, H. G., Bonifas, U., and Reichmann, G. (2000). Phenotype and functions of brain dendritic cells emerging during chronic infection of mice with *Toxoplasma gondii*. *J. Immunol.* 164:4826–4834.

Ghiringhelli, F., Puig, P. E., Roux, S., Parcellier, A., Schmitt, E., Solary, E., Kroemer, G., Martin, F., Chauffert, B., and Zitvogel, L. (2005). Tumor cells convert immature myeloid dendritic cells into TGF-{beta}-secreting cells inducing CD4 + CD25 + regulatory T cell proliferation. *J. Exp. Med.* 202:919–929.

Gonatas, N. K., Greene, M. I., and Waksman, B. H. (1986). Genetic and molecular aspects of demyelination. *Immunol. Today.* 7:121–126.

Gran, B., Zhang, G. X., Yu, S., Li, J., Chen, X. H., Ventura, E. S., Kamoun, M., and Rostami, A. (2002). IL-12p35-deficient mice are susceptible to experimental autoimmune encephalomyelitis: evidence for redundancy in the IL-12 system in the induction of central nervous system autoimmune demyelination. *J. Immunol.* 169:7104–7110.

Greenwald, R. J., Freeman, G. J., and Sharpe, A. H. (2005). The B7 family revisited. *Annu. Rev. Immunol.* 23:515–548.

Greter, M., Heppner, F. L., Lemos, M. P., Odermatt, B. M., Goebels, N., Laufer, T., Noelle, R. J., and Becher, B. (2005). Dendritic cells permit immune invasion of the CNS in an animal model of multiple sclerosis. *Nat. Med.* 11:328–334.

Harrington, L. E., Hatton, R. D., Mangan, P. R., Turner, H., Murphy, T. L., Murphy, K. M., and Weaver, C. T. (2005). Interleukin 17-producing CD4 + effector T cells develop via a lineage distinct from the T helper type 1 and 2 lineages. *Nat. Immunol.* 6:1123–1132.

Hickey, W. F., and Kimura, H. (1988). Perivascular microglial cells of the CNS are bone marrow-derived and present antigen in vivo. *Science* 239:290–292.

Hochrein, H., Shortman, K., Vremec, D., Scott, B., Hertzog, P., and O'Keeffe, M. (2001). Differential production of IL-12, IFN-alpha, and IFN-gamma by mouse dendritic cell subsets. *J. Immunol.* 166:5448–5455.

Hochwald, G. M., Van, D. A., Robinson, M. E., and Thorbecke, G. J. (1988). Immune response in draining lymph nodes and spleen after intraventricular injection of antigen. *Int. J. Neurosci.* 39:299–306.

Ingulli, E., Mondino, A., Khoruts, A., and Jenkins, M. K. (1997). In vivo detection of dendritic cell antigen presentation to CD4(+) T cells. *J. Exp. Med.* 185:2133–2141.

Jameson, B. A., McDonnell, J. M., Marini, J. C., and Korngold, R. (1994). A rationally designed CD4 analogue inhibits experimental allergic encephalomyelitis. *Nature*. 368:744–746.

Karpus, W. J., Lukacs, N. W., McRae, B. L., Streiter, R. M., Kunkel, S. L., and Miller, S. D. (1995). An important role for the chemokine macrophage inflammatory protein-1 alpha in the pathogenesis of the T cell-mediated autoimmune disease, experimental autoimmune encephalomyelitis. *J. Immunol.* 155:5003–5010.

Katz-Levy, Y., Neville, K. L., Padilla, J., Rahbe, S. M., Begolka, W. S., Girvin, A. M., Olson, J. K., Vanderlugt, C. L., and Miller, S. D. (2000). Temporal development of autoreactive Th1 responses and endogenous antigen presentation of self myelin epitopes by CNS-resident APCs in Theiler's virus-infected mice. *J. Immunol.* 165:5304–5314.

Langrish, C. L., Chen, Y., Blumenschein, W. M., Mattson, J., Basham, B., Sedgwick, J. D., McClanahan, T., Kastelein, R. A., and Cua, D. J. (2005). IL-23 drives a pathogenic T cell population that induces autoimmune inflammation. *J. Exp. Med.* 201:233–240.

Lehmann, P. V., Forsthuber, T., Miller, A., and Sercarz, E. E. (1992). Spreading of T-cell autoimmunity to cryptic determinants of an autoantigen. *Nature* 358:155–157.

Ling, C., Sandor, M., and Fabry, Z. (2003). In situ processing and distribution of intracerebrally injected OVA in the CNS. *J. Neuroimmunol.* 141:90–98.

Liu, Y. J. (2001). Dendritic cell subsets and lineages, and their functions in innate and adaptive immunity. *Cell* 106:259–262.

Mack, C. L., Neville, K. L., and Miller, S. D. (2003). Microglia are activated to become competent antigen presenting and effector cells in the inflammatory environment of the Theiler's virus model of multiple sclerosis. *J. Neuroimmunol.* 144:68–79.

Maldonado-Lopez, R., De, S. T., Michel, P., Godfroid, J., Pajak, B., Heirman, C., Thielemans, K., Leo, O., Urbain, J., and Moser, M. (1999). CD8alpha + and CD8alpha- subclasses of dendritic cells direct the development of distinct T helper cells in vivo. *J. Exp. Med.* 189:587–592.

Mangan, P. R., Harrington, L. E., O'Quinn, D. B., Helms, W. S., Bullard, D. C., Elson, C. O., Hatton, R. D., Wahl, S. M., Schoeb, T. R., and Weaver, C. T. (2006). Transforming growth factor-beta induces development of the T(H)17 lineage. *Nature* 441:231–234.

Martin, P., Del Hoyo, G. M., Anjuere, F., Arias, C. F., Vargas, H. H., Fernandez, L., Parrillas, V., and Ardavin, C. (2002). Characterization of a new subpopulation of mouse CD8alpha + B220 + dendritic cells endowed with type 1 interferon production capacity and tolerogenic potential. *Blood* 100:383–390.

Matyszak, M. K., and Perry, V. H. (1996). The potential role of dendritic cells in immune-mediated inflammatory diseases in the central nervous system. *Neuroscience* 74:599–608.

McDevitt, H. O., Perry, R., and Steinman, L. A. (1987). Monoclonal anti-Ia antibody therapy in animal models of autoimmune disease. *Ciba Found. Symp.* 129:184–193.

McMahon, E. J., Bailey, S. L., Castenada, C. V., Waldner, H., and Miller, S. D. (2005). Epitope spreading initiates in the CNS in two mouse models of multiple sclerosis. *Nat. Med.* 11:335–339.

McMahon, E. J., Bailey, S. L., and Miller, S. D. (2006). CNS dendritic cells: critical participants in CNS inflammation? *Neurochem. Int.* 49:195–203.

McRae, B. L., Vanderlugt, C. L., Dal Canto, M. C., and Miller, S. D. (1995). Functional evidence for epitope spreading in the relapsing pathology of experimental autoimmune encephalomyelitis. *J. Exp. Med.* 182:75–85.

Miller, S. D., Vanderlugt, C. L., Begolka, W. S., Pao, W., Yauch, R. L., Neville, K. L., Katz-Levy, Y., Carrizosa, A., and Kim, B. S. (1997). Persistent infection with Theiler's virus leads to CNS autoimmunity via epitope spreading. *Nat. Med.* 3:1133–1136.

Mosmann, T. R., and Coffman, R. L. (1989). Th1 and Th2 cells: different patterns of lymphokine secretion lead to different functional properties. *Annu. Rev. Immunol.* 7:145–174.

Nakae, S., Komiyama, Y., Nambu, A., Sudo, K., Iwase, M., Homma, I., Sekikawa, K., Asano, M., and Iwakura, Y. (2002). Antigen-specific T cell sensitization is impaired in IL-17-deficient mice, causing suppression of allergic cellular and humoral responses. *Immunity* 17:375–387.

Nakae, S., Nambu, A., Sudo, K., and Iwakura, Y. (2003). Suppression of immune induction of collagen-induced arthritis in IL-17-deficient mice. *J. Immunol.* 171:6173–6177.

Neville, K. L., Padilla, J., and Miller, S. D. (2002). Myelin-specific tolerance attenuates the progression of a virus-induced demyelinating disease: implications for the treatment of MS. *J. Neuroimmunol.* 123:18–29.

Newman, T. A., Galea, I., van Rooijen, N., and Perry, V. H. (2005). Blood-derived dendritic cells in an acute brain injury. *J. Neuroimmunol.* 166:167–172.

Park, H., Li, Z., Yang, X. O., Chang, S. H., Nurieva, R., Wang, Y. H., Wang, Y., Hood, L., Zhu, Z., Tian, Q., and Dong, C. (2005). A distinct lineage of CD4 T cells regulates tissue inflammation by producing interleukin 17. *Nat. Immunol.* 6:1133–1141.

Penna, G., Sozzani, S., and Adorini, L. (2001). Cutting edge: selective usage of chemokine receptors by plasmacytoid dendritic cells. *J. Immunol.* 167:1862–1866.

Powell, M. B., Mitchell, D., Lederman, J., Buckmeier, J., Zamvil, S. S., Graham, M., Ruddle, N. H., and Steinman, L. (1990). Lymphotoxin and tumor necrosis factor-alpha production by myelin basic protein-specific T cell clones correlates with encephalitogenicity. *Int. Immunol.* 2:539–544.

Pulendran, B., Kumar, P., Cutler, C. W., Mohamadzadeh, M., Van, D. T., and Banchereau, J. (2001). Lipopolysaccharides from distinct pathogens induce different classes of immune responses in vivo. *J. Immunol.* 167:5067–5076.

Rivers, T. M., and Schwentker, F. F. (1935). Encephalomyelitis accompanied by myelin destruction experimentally produced in monkeys. *J. Exp. Med.* 61:689–695.

Rosicarelli, B., Serafini, B., Sbriccoli, M., Lu, M., Cardone, F., Pocchiari, M., and Aloisi, F. (2005). Migration of dendritic cells into the brain in a mouse model of prion disease. *J. Neuroimmunol.* 165:114–120.

Santambrogio, L., Belyanskaya, S. L., Fischer, F. R., Cipriani, B., Brosnan, C. F., Ricciardi-Castagnoli, P., Stern, L. J., Strominger, J. L., and Riese, R. (2001). Developmental plasticity of CNS microglia. *Proc. Natl. Acad. Sci. U.S.A.* 98:6295–6300.

Schreiner, B., Bailey, S. L., and Miller, S. D. (2007). T-cell response dynamics in animal models of multiple sclerosis: implications for immunotherapies. *Expert Rev. Clin. Immunol.* 3:57–72.

Sedgwick, J. D., and Mason, D. W. (1986). The mechanism of inhibition of experimental allergic encephalomyelitis in the rat by monoclonal antibody against CD4. *J. Neuroimmunol.* 13:217–232.

Serafini, B., Columba-Cabezas, S., Di, R. F., and Aloisi, F. (2000). Intracerebral recruitment and maturation of dendritic cells in the onset and progression of experimental autoimmune encephalomyelitis. *Am. J. Pathol.* 157:1991–2002.

Shortman, K., and Liu, Y. J. (2002). Mouse and human dendritic cell subtypes. *Nat. Rev. Immunol.* 2:151–161.

Steinman, R. M. (1991). The dendritic cell system and its role in immunogenicity. *Annu. Rev. Immunol.* 9:271–296.

Steinman, L. (1999). Absence of "original antigenic sin" in autoimmunity provides an unforeseen platform for immune therapy. *J. Exp. Med.* 189:1021–1024.

Suter, T., Malipiero, U., Otten, L., Ludewig, B., Muelethaler-Mottet, A., Mach, B., Reith, W., and Fontana, A. (2000). Dendritic cells and differential usage of the MHC class II transactivator promoters in the central nervous system in experimental autoimmune encephalitis. *Eur. J. Immunol.* 30:794–802.

Suter, T., Biollaz, G., Gatto, D., Bernasconi, L., Herren, T., Reith, W., and Fontana, A. (2003). The brain as an immune privileged site: dendritic cells of the central nervous system inhibit T cell activation. *Eur. J. Immunol.* 33:2998–3006.

Tompkins, S. M., Padilla, J., Dal Canto, M. C., Ting, J. P., Van Kaer, L., and Miller, S. D. (2002). De novo central nervous system processing of myelin antigen is required for the initiation of experimental autoimmune encephalomyelitis. *J. Immunol.* 168:4173–4183.

Tuohy, V. K., Yu, M., Yin, L., Kawczak, J. A., and Kinkel, R. P. (1999). Spontaneous regression of primary autoreactivity during chronic progression of experimental autoimmune encephalomyelitis and multiple sclerosis. *J. Exp. Med.* 189:1033–1042.

Vanderlugt, C. L., Eagar, T. N., Neville, K. L., Nikcevich, K. M., Bluestone, J. A., and Miller, S. D. (2000). Pathologic role and temporal appearance of newly emerging autoepitopes in relapsing experimental autoimmune encephalomyelitis. *J. Immunol.* 164:670–678.

Veldhoen, M., Hocking, R. J., Atkins, C. J., Locksley, R. M., and Stockinger, B. (2006). TGFbeta in the context of an inflammatory cytokine milieu supports de novo differentiation of IL-17-producing T cells. *Immunity* 24:179–189.

Waldor, M. K., Sriram, S., Hardy, R., Herzenberg, L. A., Lanier, L., Lim, M., and Steinman, L. (1985). Reversal of experimental allergic encephalomyelitis with monoclonal antibody to a T-cell subset marker. *Science* 227:415–417.

Wekerle, H. (1991). Immunopathogenesis of multiple sclerosis. *Acta Neurol.(Napoli)* 13:197–204.

Willenborg, D. O., Fordham, S. A., Staykova, M. A., Ramshaw, I. A., and Cowden, W. B. (1999). IFN-gamma is critical to the control of murine autoimmune encephalomyelitis and regulates both in the periphery and in the target tissue: a possible role for nitric oxide. *J. Immunol.* 163:5278–5286.

Yamada, S., DePasquale, M., Patlak, C. S., and Cserr, H. F. (1991). Albumin outflow into deep cervical lymph from different regions of rabbit brain. *Am. J. Physiol.* 261:H1197–H1204.

Yasumi, T., Katamura, K., Yoshioka, T., Meguro, T. A., Nishikomori, R., Heike, T., Inobe, M., Kon, S., Uede, T., and Nakahata, T. (2004). Differential requirement for the CD40–CD154 costimulatory pathway during Th cell priming by CD8 alpha + and CD8 alpha- murine dendritic cell subsets. *J. Immunol.* 172:4826–4833.

Yu, M., Johnson, J. M., and Tuohy, V. K. (1996). A predictable sequential determinant spreading cascade invariably accompanies progression of experimental autoimmune encephalomyelitis: a basis for peptide-specific therapy after onset of clinical disease. *J. Exp. Med.* 183:1777–1788.

Zammit, D. J., Cauley, L. S., Pham, Q. M., and Lefrancois, L. (2005). Dendritic cells maximize the memory CD8 T cell response to infection. *Immunity* 22:561–570.

Zhang, G. X., Gran, B., Yu, S., Li, J., Siglienti, I., Chen, X., Kamoun, M., and Rostami, A. (2003). Induction of experimental autoimmune encephalomyelitis in IL-12 receptor-beta 2-deficient mice: IL-12 responsiveness is not required in the pathogenesis of inflammatory demyelination in the central nervous system. *J. Immunol.* 170:2153–2160.

14
MHC Class I Expression and CD8 T Cell Function: Towards the Cell Biology of T-APC Interactions in the Infected Brain

Cornelia Bergmann and Pedro Lowenstein

1 Introduction

Antigen presentation by major histocompatibility complex class I (MHC-I) and class II (MHC-II) molecules is a prerequisite for T cell engagement during the activation as well as the effector phase. The central nervous system (CNS) is unique in that cells resident in the parenchyma, glia and neurons, do not constitutively (or very sparsely at best) express MHC molecules (Aloisi et al., 2000; Sedgwick and Hickey, 1997; Xiao and Link, 1998), making them invisible to T cells. Additional restrictions on T cell surveillance are imposed by the absence of classical lymphatic drainage, the blood–brain barrier (BBB), and the unique anatomy of the brain microvasculature (Bechmann et al., 2007; Galea et al., 2007; Hickey, 2001; Xiao and Link, 1998; Lowenstein, 2002). Infiltrating cells not only have to cross the vascular wall to penetrate into the perivascular space, but more importantly overcome the barrier formed by the glia limitans to access the CNS parenchyma. While the first step is generally not associated with pathology, penetration from the perivascular space of postcapillary venules into the parenchyma is more restricted and once overcome, associated with clinical consequences (Bechmann et al., 2007). In the resting state, perivascular macrophages are maintained by replacement with circulating monocytes. However the glial limitans is not breached, and thus, these cells are considered to be located outside the confines of the BBB. While diffusion of soluble factors and antibodies is restricted by the BBB, especially by the tight, continuous, unfenestrated capillary epithelium, leukocyte infiltration preferentially occurs at distal sites in postcapillary venules (Bechmann et al., 2007). BBB permeability and leukocyte infiltration are thus not necessarily functionally nor physically linked. The barriers separating CNS parenchyma from the circulation explains the rare presence of T cells in the normal CNS parenchyma, despite the ability of peripherally activated and memory T cells to traffic to non lymphoid tissues independent of antigen presentation (Masopust et al., 2004). Nevertheless, T cells activated during an infection or auto-immune response, are able to enter into, and migrate within the brain parenchyma, even in the presence of an intact, non-inflamed BBB (Cabarrocas et al., 2003; Chen et al., 2005; Evans et al., 1996; Hickey, 2001). However, although entry of activated T cells into the CNS is independent of their

T.E. Lane et al. (eds.), *Central Nervous System Diseases and Inflammation.*
© Springer 2008

antigen specificity, only those T cells that recognize antigen are retained. Thus, barriers limiting T cell surveillance of the brain are rapidly overcome following CNS infections and other inflammatory conditions (Griffin, 2003; Ransohoff et al., 2003). Under such conditions, it is also likely that a number of non-activated, bystander T cells, as well as other leukocytes, including B cells are able to penetrate into the brain parenchyma. Mechanisms propagating protective versus pathogenic potential of T cells in varying disease models are complex and require more in depth exploration. This chapter highlights recent advances relating to antigen presentation functions by resident CNS cells and effects exerted by CD8 T cells in vivo with an emphasis on anti-viral functions.

2 Requirement of MHC-I Antigen Recognition for CD8 T Cell Induction and Function

CD8 T cells are present in the CNS during many inflammatory diseases including infections and autoimmune diseases (Dorries, 2001; Neumann et al., 2002). They are primary mediators in clearing viral infections, in anti tumor defense, and transplant rejection. However, they are also associated with a number of pathologies, including demyelinating and degenerative diseases. Physiological and pathological consequences are sometimes inextricably intertwined. Examples of T cell induced pathology are multiple sclerosis (Neumann et al., 2002; Sospedra and Martin, 2005), as well as experimental models of demyelination (Brisebois et al., 2006; Evans et al., 1996; Huseby et al., 2001; Ip et al., 2006), while examples of viral clearing and associated pathology are brain atrophy in Borna virus encephalitis (Bilzer and Stitz, 1994), Rasmussen's encephalitis (Bauer et al., 2002; Bien et al., 2002), and HTLV-1 myelopathy (Nakamura, 2000). Trafficking of CD8 T cells into the CNS is antigen independent and regulated by both chemokines and activation state of the T cells. However, their accumulation and retention is dependent on MHC-I restricted antigen recognition (Cabarrocas et al., 2003; Carson et al., 1999; Chen et al., 2005; Reuter et al., 2005). As constitutive MHC-I expression is restricted to the meninges, choroid plexus, and perivascular spaces, engagement of CD8 T cells in the parenchyma requires induction of MHC-I on parenchymal cells. This is generally thought to be transcriptionally regulated resulting in de novo expression of MHC-I molecules.

2.1 Class I Antigen Processing Pathways

Class I surface expression involves numerous proteolytic, transport, and assembly events, which are prominently influenced by type I and type II IFNs and TNFα (Garbi et al., 2005; Koch and Tampe, 2006; Strehl et al., 2005; Trombetta and Mellman, 2005). In the quiescent, constitutive state, the major pool of peptides are derived from short lived nascent proteins in the cytoplasm, which are ubiquitinated and targeted to a multicatalytic protease complex, designated the 26S proteasome (Strehl et al., 2005). The proteasome complex, located within the cytoplasm, releases peptide precursors with a length and composition suitable for transport into the endoplasmic reticulum (ER). Transport is conducted by members of the ABC transporter family, TAP1 and TAP2, which form a heterodimer. However, 99% of peptides are degraded by aminopeptidases resident in the cytosol prior to transport into the ER (Reits et al., 2003). In the ER, nascent class I heavy chains, $\beta 2$ microglobulin ($\beta 2$ m), and peptides for presentation are assembled in a complex process involving further N-terminal peptide trimming for optimal class I binding and cellular chaperones, including tapasin (Garbi et al., 2005). Once a stable tripartite complex is formed, it is transported through the secretory pathway to the cell surface. The quantity and quality of peptides presented by MHC molecules thus reflects cellular protein turnover, proteolytic specificities of the proteasome and activity of aminopeptidases (Strehl et al., 2005; Yewdell et al., 2003).

The 26S proteasome is composed of a catalytic 20S core complex and two 19S regulator complexes (Strehl et al., 2005). The building blocks for the 20S proteasome are four rings, each formed by seven α or β subunits, respectively. Proteolytic activities and specificities are conferred by three β subunits localized to the two inner rings. The 19S regulator serves as a gate controlling entry into the proteolytic chamber. A second regulator is the IFNγ inducible PA28 complex, which also forms a ring composed of PA28α and PA28β subunits. Antigen processing and presentation is significantly increased in the presence of IFNγ, which enhances transcription of multiple genes associated with the processing machinery. Importantly, IFNγ induces a distinct subset of 20S proteasomal β immunosubunits, which can replace their $\beta 1$, $\beta 2$ and $\beta 5$ homologs during assembly to form immunoproteasomes. Altered cleavage site preferences of IFNγ-induced immunoproteasomes result in enhanced formation of peptides carrying an optimal C-terminus compatible with class I binding, and an extended N-terminus that facilitates TAP transport. Independent of immunoproteasome β subunits, P28 can also modulate presentation of epitopes, albeit in a more restricted, epitope dependent manner. Additional IFNγ upregulated modulators of the antigen processing machinery are POMP, a facilitator of proteasome maturation, the transporters TAP1 and TAP2, and a subset of cytosolic and ER-specific aminopeptidases. The relative expression of all these components thus determines the efficiency and epitope diversity presented by MHC-I.

Not surprisingly, different tissues and cell types display a distinct array of constitutive and immunoproteasomes in the quiescent state, which is altered during inflammation. Lymphoid organs, e.g. thymus, lymph nodes and spleen, in normal adult mice contain high levels of immunoproteasomes and reduced levels of constitutive subunits, while constitutive subunits predominate in the brain, with little expression of immunosubunits (Stohwasser et al., 1997). Although beneficial for presentation of microbial epitopes, immunoproteasomes appear strictly controlled. Their half life is ~5-fold shorter than that of constitutive proteasomes, and they rarely completely replace constitutive proteasomes (Heink et al., 2005). Nevertheless, at sites of infection subunit exchange can be very prominent, leading to an almost complete exchange as shown during hepatotropic lymphocytic choriomeningitis virus (LCMV) and bacterial *Listeria monocytogenes* infection (Khan et al., 2001). Replacement of constitutive by immunoproteasomes serves to focus immune reactivity towards anti-infectious CD8 T cell responses.

While the role of IFNγ in enhancing class I Ag processing and presentation efficacy is well established, other factors regulating the cell-type specific and/or processing kinetics are less well studied. For example, IFNα/β does not significantly impact immunoproteasomes in liver during murine LCMV infection, but does stimulate generation of immunoproteasomes in hepatocytes in vitro and in vivo following hepatitis C virus (HCV) infection (Shin et al., 2006). TNFα is also an inducer of immunoproteasomes in vitro and may contribute to IFNγ independent induction of immunoproteasomes early during infection (Khan et al., 2001).

In addition to the endogenous class I antigen processing pathway used by infected cells to present microbial epitopes, antigen can also be processed and presented following cellular uptake from exogenous sources, a process termed cross-presentation (Rock and Shen, 2005; Trombetta and Mellman, 2005). The mechanisms can involve both TAP-dependent, or endocytic loading of class I molecules. The capture and subsequent presentation of exogenous antigen may be an effective pathway for dendritic cells to prime naïve CD8 T cells if they are not directly infected (Chen et al., 2004). Peptides can also access the cytoplasm of adjacent cells through gap junctions formed by connexins (Neijssen et al., 2005). These intercellular channels are used for transport of nutrients and second messengers, but also allow diffusion of 4–10 amino acid peptides. Neighboring cells connected by gap junctions may thus provide an alternative peptide source for dendritic cells or activated monocytes to present peptide during the priming or the effector phase.

2.2   *MHC Class I Regulation on Parenchymal CNS Cells*

The paucity of class I surface expression on quiescent CNS cells in adult rodents, non-human primates and humans, suggests strict control of class I heavy chains and components of the antigen processing machinery compared to nucleated cells in other non lymphoid tissues and the vasculature. However, despite extensive characterization of the molecules involved in the regulation of expression of class I antigen presentation in other cell types, little is known about regulation of MHC-I expression in vivo in mature resident CNS cells, and their consequent capacity to present antigens to CD8 T cells. The mechanisms regulating MHC expression and T cell engagement have largely been studied using glial or neuronal cultures derived from neonatal rodents or immortalized cell lines (Aloisi et al., 2000; Dong and Benveniste, 2001; Sedgwick and Hickey 1997). Even when the cells studied are obtained from normal animals, placing glial cells in culture already acts as a stimulus for their activation even in the absence of any further stimuli, and this is likely to determine their reactivity with immune cells. Furthermore, the majority of these data pertain to MHC class II presentation based on the association of demyelinating autoimmune disease with CD4 T cells (Bailey et al., 2006; Dong and Benveniste, 2001). Nevertheless, there is ample evidence for a role of CD8 T cells in demyelinating and neurodegenerative disease, as well as microbial infection (Neumann et al., 2002; Dorries, 2001). In vitro studies have revealed that resident CNS cells including microglia, astrocytes, oligodendrocytes and neurons are susceptible to CD8 T cell mediated killing (Neumann et al., 2002; Sedgwick and Hickey 1997; Xiao and Link, 1998). CD8 T cells are thus considered prime suspects in propagating CNS pathology. However, in vitro killing assays necessitate the use of cells derived from neonatal brains. CNS cells in their natural environment are more stringently regulated in their ability to express MHC and this is age dependent. Furthermore, regulation of MHC-I antigen presentation as well as susceptibility to CD8 T cell function is distinct among parenchymal CNS cell types in vitro and in vivo. Neurons are the most restricted in their capacity to express MHC-I (Neumann et al., 1995, 1997a, 1997b; Sedgwick and Hickey, 1997). Furthermore, even overexpression of MHC-I in the CNS did not trigger detectable killing in vivo, although cell death of cultured neonatal neurons in vitro could be demonstrated (Rall et al., 1995). Thus, the conditions under which cell killing in vitro has been detected, does not allow to draw conclusions on the potential immune-killing in the adult brain.

The capacities of glia and neurons to present antigens and engage CD8 T cells in vivo are not only dictated by MHC expression, but also by processing components and activating and inhibitory interactions. The nature of the insult, cell types infected, innate and bystander responses thus all contribute in shaping a protective or detrimental response. On one hand, the success of adaptive T cell responses in clearing intracellular pathogens is governed by processing and recognition of antigen in the context of MHC molecules and subsequent release of antimicrobial factors, including IFNγ, TNFα, perforin and granzymes (Harty et al., 2000). On the other side, excessive activation and dysregulation of CD8 T cell function contributes to pathology (Armstrong and Lampson, 1997; Barmak et al., 2003; Brisebois et al, 2006; Neumann et al., 2002; Stitz et al., 2002). Despite their capacity to act as targets for effector CD8 T cells in vitro, neuronal and glia susceptibility to "direct, not indirect destruction" by CD8 T cells in vivo remains to be determined beyond reasonable doubt. Concomitantly, the contribution of antigen processing components in antigen presentation and expression of MHC-I in vivo in different types of brain cells needs to be determined. A set of complementary experimental approaches (e.g. immunohistochemistry, flow cytometry, molecular) will be required, since single experimental approaches can be inconclusive.

2.2.1   Regulation of Antigen Processing in Glia

Among resident cells of the CNS parenchyma, microglia are implicated as the most potent efferent APC to brain infiltrating, activated CD8 T cells, with regard to MHC and costimulatory molecule expression. Quiescent mouse microglia in vivo do not or only sparsely express surface MHC-I and do not express MHC-II (Sedgwick and Hickey, 1997). In vitro and in vivo activation of microglia by inflammatory cytokines upregulates expression of MHC-I and MHC-II molecules, as well as costimulatory molecules (Aloisi et al., 2000). During viral infection in vivo, surface MHC-I upregulation precedes MHC-II upregulation and does not require IFNγ (Bergmann et al., 2003).

Regulation of MHC-I at the level of processing genes following inflammation in vivo has not been explored. However, analysis of proteasomes from primary mouse microglia cultures revealed that expression of constitutive proteasomes resembles that of fibroblasts or lymphoblastoid cells (Stohwasser et al., 2000). The immunoproteasome subunits iβ1/LMP2 and iβ5/LMP7, but not iβ2/MECL1, were also detectable at low levels, contrasting with their absence from control cell lines. IFNγ as well as LPS treatment both induced immunoproteasome subunits concomitant with a decrease in incorporated constitutive subunits, although at distinct levels (Stohwasser et al., 2000). While LPS also induced TNFα, IL-6 and KC, the individual roles of these cytokines/chemokines in modulating proteasomes were not determined in this study. Analysis of the expression signature of either unstimulated or IFNγ stimulated primary rat microglia revealed expression of TAP1, class I heavy chains, constitutive proteasome subunits and the protease activator PA28a subunit in unstimulated cultures. Transcript levels of these genes were further upregulated by IFNγ, predominantly TAP 1 (Moran et al., 2004). De novo transcribed genes included the immunoproteasome subunits iβ1/LMP2 and iβ5/LMP7, the proteasome activator PA28 β chain and 26S components, heterologous MHC-I heavy chains and the MHC-II invariant chain.

Similar to microglia, expression of MHC-I and -II is undetectable in astrocytes from naïve animals. Whereas MHC-I is upregulated during inflammation, detection of MHC-II has been more controversial (Sedgwick and Hickey, 1997). In vitro, neonatal astrocytes spontaneously express MHC-I in culture and are readily induced to express MHC-II upon IFNγ treatment or virus infection (Massa et al., 1993; Sedgwick and Hickey, 1997). Similar to microglia, microarray analysis of primary murine astrocyte cultures revealed that a large proportion of genes upregulated by IFNγ were immune response genes (Halonen et al., 2006). Specifically, genes involved in MHC-I and MHC-II antigen processing, e.g. TAP1, proteasome components, class I heavy chains and β2 m, MHC-II molecules, invariant chains, and costimulatory molecules were most strongly induced.

The regulation of antigen processing and MHC-I expression on oligodendrocytes is also important in elucidating their role as targets of antimicrobial or autoimmune CD8 T cell cytolysis during demyelinating diseases, but also in glial graft rejection during remyelination therapies. Nevertheless, studies on MHC regulation and processing capabilities in this glia subset are limited. Oligodendrocytes in naïve adult mice do not express detectable surface MHC-I. However, IFNγ stimulates MHC-I expression dramatically in vitro and in vivo (Popko and Baerwald, 1999; Sedgwick and Hickey, 1997). Interestingly, transgenic expression of class I heavy chains beyond a certain threshold leads to their accumulation in the ER and severe defects in myelination (Baerwald et al., 2000; Popko and Baerwald, 1999). This may be due to low basal levels of proteins involved in antigen processing, thus impeding assembly and transport of MHC molecules to the cell membrane. The concerted regulation of antigen processing genes may thus be particularly important in oligodendrocytes to allow continued deployment of MHC-I at the cell membrane, while securing a functional secretory pathway for the myelination process. However, regulation of components of the antigen processing machinery have only been sparsely studied in oligodendrocytes.

Analysis of Schwann cells from sciatic nerve revealed no MHC surface expression (Tsuyuki et al., 1998). Although mRNA encoding class I heavy chains was faintly detected, mRNA for $\beta 2\,m$, TAP-1, and LMP2 was undetectable (Tsuyuki et al., 1998). Similar to neonatal derived microglia and astrocytes, IFNγ predominantly induced transcripts associated with genes of the MHC-I antigen processing machinery. Consistent with these data, IFNγ also induced strong MHC-I surface expression. It is interesting that no induction of MHC expression was mediated by TNFα in Schwann cells. Unlike MHC-I expression, MHC-II remained undetectable both at the level of surface expression as well as mRNA after IFNγ treatment. Consistent with the IFNγ inducible MHC-I expression, mouse Schwann cells are susceptible to CD8 T cell mediated cytolysis in vitro (Steinhoff et al., 1990). Of note, CD8 T cells primed to mycobacterial antigen displayed crossreactivity to host determinants displayed by Schwann cells, suggesting a possible role for molecular mimicry. Supporting MHC-I upregulation under conditions of IFNγ supplementation, MHC-I is also upregulated on oligodendrocytes and Schwann cells during virus induced inflammation (Pereira et al., 1994; Ramakrishna et al., 2006; Redwine et al., 2001). However, similar to microglia and astrocytes, the relative roles of IFNα/β and IFNγ in MHC-I regulation in vivo have not been elucidated. Thus, rather than focusing exclusively on MHC expression, future studies may have to address the concerted expression of components of the antigen presenting machinery, as well as their kinetics of induction, since the quality, quantity and longevity of presented antigens is likely to determine microbial control and disease outcome.

MHC-I expression in vivo is also observed on oligodendrocytes in adult mice mildly overexpressing the myelin component proteolipid protein (PLP) (Ip et al., 2006). These mice develop a late onset progressive demyelination associated with axonopathic changes and infiltrating CD8 T cells and CD11b macrophage like cells. Whereas MHC-I expression was undetectable in wt mice, PLP mutants exhibited substantial increase in MHC-I expression in white matter. Closer analysis co-localized MHC expressing structures with myelin markers but not neuronal markers. Importantly, as the vast majority of T cells were CD44$^+$ CD62L$^-$ CD69$^+$ effector CD8 T cells, localization of ~25–30% of T cells in close proximity to MHC–I suggested direct interaction of CD8 T cells with targets (Ip et al., 2006). Whether CD69 expression on T cells reflects MHC-I driven activation or provides a signature for CD8 T cells that have entered non lymphoid parenchymal organs is unclear (Bergmann et al., 1999; Hawke et al., 1998). Furthermore, the mechanisms underlying MHC-I surface expression and a role of T cell receptor (TCR) mediated CD8 contact and/or stress responses also remain to be elucidated in this model.

Relevant to oligodendrocyte transplant mediated remyelination, both the haplotype combination and cellular graft composition determined graft survival and efficacy of remyelination (Tepavcevic and Blakemore, 2005, 2006), suggesting a role for MHC mediated rejection. Using a rat model of toxin induced demyelination, grafts enriched for oligodendrocyte precursor cells (OPCs) persisted longer in allogeneic recipients than unenriched glial cultures. Persisting remyelination was attributed to undetectable MHC-I expression on OPC cultures under basal conditions. MHC-I was only induced on OPC following IFNγ treatment, whereas MHC-II was never detected on OPC or oligodendrocytes. By contrast, MHC-I expression was constitutive on astrocytes and microglia in the implant cultures, reducing graft survival and remyelinating potential.

Another intriguing role for MHC-I expression in brain synaptic function has been proposed by a number of authors (Boulanger and Shatz, 2004; Corriveau et al., 1998; Goddard et al., 2007; Huh et al., 2000; Oliveira et al., 2004). MHC-I expression has been found in neurons of the visual, and olfactory system, and in dorsal root ganglion neurons. Results from β2m deficient mice have shown that these animals have alterations in the development of visual connectivity, synaptic remodeling in response to insults in dorsal root neurons, and hippocampal neuron synaptic function. In addition, MHC peptides have been shown to modulate the function of olfactory system neurons (Leinders-Zufall et al., 2004). Taken together, these results suggest the hypothesis that MHC-I expression in the CNS may regulate wider aspects of brain physiology and behavior. If so, MHC-I expression may provide a further link between immune reactivity and brain function through structural and synaptic effects of MHC-I expression. Although the precise mechanisms on how this may occur remains to be explored, how MHC-I expression in normal brain physiology relates to the function of MHC-I during inflammatory and immune brain diseases opens up further exciting avenues for future research.

3 Priming of Adaptive Responses and CD8 Effector Function in Response to Viral CNS Infection

3.1 Location Determines Brain Immune Reactivity

There are two fundamentally different immune compartments in the brain. The brain ventricles, meninges and choroid plexi contain all cellular, vascular and lymphatic components necessary for immune function that are also associated with most other organs. This includes the dendritic cells (DC), the major cell type capable of inducing primary T cell responses, which can be found within the meninges, choroid plexus and cerebrospinal fluid (CSF) under non inflammatory conditions (McMenamin, 1999; McMenamin et al., 2003; Pashenkov et al., 2003). These anatomical sites are strategic for capture of foreign or self antigens, which can trigger their migration to deep cervical lymph nodes (CLN), the primary lymph nodes draining the brain and CSF (Cserr and Knopf, 1992). However, the brain parenchyma itself is devoid of DC in its naive state, lacks classical lymphatic drainage, and its endothelial cells form a tight diffusion barrier between the vascular and brain compartment (Bechmann et al., 2007). In addition to these structural differences a number of molecular mechanisms intervene in the brain parenchyma to display a very peculiar type of 'dampened' immune reactivity.

Administration of antigens into either CNS immune-compartment demonstrated profound differences in immunogenicities. Injection of a particulate antigen or infectious agent (e.g. live influenza virus, BCG, non-replicative adenoviral vectors) exclusively and selectively into the brain parenchyma only causes innate inflammatory responses, but fails to stimulate systemic adaptive immune responses (Cartmell et al., 1999; Lowenstein, 2002; Matyszak and Perry, 1996; Stevenson et al., 1997; Thomas et al., 2001). By contrast, injection of the same type of antigen into the ventricular system, nonetheless, induces both an innate inflammatory and a systemic adaptive immune response (Matyszak, 1998; Matyszak and Perry, 1996; Stevenson et al., 1997). However, injection of a soluble diffusible antigen (e.g. OVA) into either compartment, does induce a systemic B cell response (Knopf et al., 1998).

This differential immune-reactivity is thought to reside, at least partly, in the distribution of DC. DC localize predominantly to lymphoid tissue, where they take up antigen and mature to potent antigen presenting cells (APC). Alternatively they acquire antigen at inflamed sites and traffic back to lymphoid tissue to activate T cells (Caux et al., 2000). Recent data has also suggested that monocytes recruited to sites of acute inflammation can acquire the phenotype of DC and present antigen to primed T cells thus propagating T cell responses (Leon et al., 2007). DC uptake of foreign antigen in the ventricular system is likely to trigger migration of DCs to CLN. Alternatively, antigens can drain directly into deep CLN. An explanation for differential priming of lymphocytes in the distinct CNS compartments may reside in the inability of particulate antigen to drain from the brain parenchyma, either through a cellular or diffusible route. Thus, particulate antigens injected into the brain parenchyma cause inflammation, but are never transported to the lymph nodes to prime a systemic immune response. Soluble antigen can diffuse from the brain to the ventricles, and thus, eventually reach the lymph nodes, where it will stimulate a systemic immune response.

Irrespective of the antigen transport and delivery mode, naïve T cells are primed in the CLN, expand and traffic to the site of insult, where they exert effector function upon antigen re-encounter. Thus, although DCs can enter the CNS parenchyma during inflammation to sustain T cell function, initial T cell activation preceding disease onset likely occurs in the CLN. The role of DCs during chronic inflammation is discussed in more detail in Chap. 13.

3.2 Cell Type Dependent CD8 Effector Functions During CNS Infection

3.2.1 Immune Control of Neuronal Infections

T cells are critical anti-viral effectors in controlling replication of various RNA and DNA viruses infecting the CNS and PNS (Dorries, 2001; Griffin, 2003; Divito et al., 2006). However, evidence in the past decade revealed that anti viral mechanisms depend on viral tropism for neuronal as well as glia subsets. As in other tissues, viral clearance can be achieved via perforin or Fas ligand dependent lytic pathways resulting in cell death, or through the release of 'curative' antiviral cytokines which clear virus, while sparing the infected cells (Guidotti and Chisari, 2001). Although CNS recruitment of CD8 T cells is MHC-I independent, local expression of perforin mediated killing, as well as secretion of perforin, IFNγ, TNFα and CCL5 (Rantes), are dependent on MHC-I/T cell receptor (TCR) interactions (Slifka et al., 1999; Slifka and Whitton, 2000). Interruption of cell-cell interactions by physical separation leads to rapid down-regulation of IFNγ, and TNFα, but not perforin at the transcriptional level (Slifka et al., 1999). These data suggest that expression of T cell effector function also requires continuous TCR engagement in vivo. Successful control of CNS pathogens may thus depend on prolonged MHC dependent target cell contact. Whereas perforin/granzyme can only act directly on the infected target cells through physical contact with T cells, soluble mediators can potentially act distally on adjacent cells, irrespective of their ability to physically engage T cells directly (Guidotti and Chisari, 2001). At least in theory, the curative pathway is thus especially beneficial for the host in preserving non regenerative cells vital for host function such as neurons and the cells which synthesize and maintain myelin.

Both cytolytic and noncytolytic mechanisms are employed by T cells to control virus infected CNS cells (Dorries, 2001; Griffin, 2003). Nevertheless, noncytolytic mechanisms would be favored to avoid extensive cell destruction. This is most evident in non-lytic neuronal infections, which establish viral persistence without killing the host cell, e.g. LCMV or herpes simplex virus (HSV-1). Neurons persistently infected by LCMV can be cured by adoptive transfer of LCMV specific CD8 T cells without suffering detectable cell loss (Oldstone et al., 1986; Tishon et al., 1993). By contrast, infection of cells in the leptomeninges during the acute infection leads to severe neurological disease following CD8 T cell transfer (Mucke and Oldstone, 1992). In the LCMV study the acute phase infection was associated with class I expression, whereas persistence was not.

Class I surface expression has also been detected on neurons infected with Sindbis virus (Kimura and Griffin, 2000), and during acute, but not persistent HSV infection (Pereira and Simmons, 1999). CD8 T cells contribute to reduction of Sindbis viral RNA from neurons, although humoral immunity is credited with primarily clearing Sindbis virus from neurons (Kimura and Griffin, 2000). In antibody deficient mice Sindbis virus infection of neurons is also primarily controlled by T cell derived IFNγ mediated noncytolytic mechanisms (Binder and Griffin, 2001). It is interesting that motor neurons were more sensitive to antiviral IFNγ mediated effects compared to cortical neurons, indicating regional differences in IFNγ susceptibility. Measles virus CNS disease is also controlled by non cytolytic, IFNγ mediated mechanisms without apparent neuronal loss (Patterson et al., 2002). CD8 T cells are also critical in controlling HSV-1 replication in a non-cytolytic manner in trigeminal ganglia sensory neurons and establishing latency (Divito et al., 2006). Latent infection is associated with ongoing CD8 T cell stimulation (van Lint et al., 2005) and reactivation from latency is inhibited in an MHC-restricted and antigen specific manner without notable neuronal loss or pathology (Khanna et al., 2003; Theil et al., 2003). In this case CD8 T cells require both perforin and IFNγ to block reactivation in TG cultures ex vivo. Nevertheless, the lytic pathway is tempered by expression of the inhibitory receptor CD94-NKG2a on CD8 T cells and its major ligand Qa-1b on a subset of neurons (Suvas et al., 2006). In addition to distinct regulation of CD8 T cell function by latently infected neuronal populations, protection in neurons is mediated by viral anti-apoptotic activity and IFNγ mediated blockade of HSV genes playing a role in reactivation (Divito et al., 2006).

Contrasting the preferential non cytolytic clearance in these infections, perforin is essential to clear a virulent West Nile virus from neurons in an MHC-I dependent manner (Shrestha et al., 2006a). IFNγ plays a role in limiting viral spread in the periphery and dampening CNS infection (Shrestha et al., 2006b). The fact that clearance of a less virulent strain of WNV is perforin independent supports the concept that damage may predispose infected neurons to an enhanced capacity for MHC-I surface expression and thus susceptibility to CD8 T cell mediated lysis (Neumann et al., 1995, 1997a).

A role for both perforin and IFNγ is evident during Theilers's murine encephalomyelitis virus (TMEV) infection, an experimental model of virus induced demyelinating disease (Drescher et al., 1997). Infection by TMEV, a picornavirus family member, is characterized by two types of diseases. In 'resistant' C57BL/6 mice the viral DA strain primarily infects neurons. These mice develop acute encephalitis and clear virus within 2 weeks with no pathological sequelae. In 'susceptible' SJL mice acute virus infection in gray matter is controlled, but persists in white matter of the spinal cord mainly in macrophages, but also oligodendrocytes and astrocytes (Drescher et al., 1997). MHC- I restricted CD8 T cells are the effectors providing protection from chronic disease in resistant mice. Perforin deficiency in resistant mouse strains is associated with mortality within 20 days, decreased viral clearance with virus spread to motor neurons in the spinal cords, and tissue destruction (Rossi et al., 1998). Lytic mechanisms targeting neurons thus play a dominant role in controlling early neuronal infection in the gray matter. Nevertheless, IFNγ plays an equally critical role in protection from chronic TMEV demyelinating disease, as evidenced by the failure to clear virus, extensive demyelination, and severe neurological symptoms in the absence of IFNγ signaling in mice normally resistant to chronic infection (Drescher et al., 1997; Pullen et al., 1994; Fiette et al., 1995; Rodriguez et al., 2003). Although a direct comparison is tentative due to administration of different virus doses in different laboratories, survival rates were significantly enhanced in IFNγ deficient compared to perforin deficient mice (Rodriguez et al., 1995; Rossi et al., 1998). The protective IFNγ effects clearly reside in the ability of CNS resident cells to respond to IFNγ (Murray et al., 2002). Results from adoptive CD8 T cell transfers into Rag$^{-/-}$ mice were consistent with bone marrow chimeras harboring somatic cells intact for IFNγR reconstituted with bone marrow cells deficient in IFNγ signaling and vice versa. Although virus antigen positive cells were similar in Rag$^{-/-}$ recipients of IFNγ deficient CD4 or CD8 T cells, demyelination was significantly enhanced in the CD4 T cell recipient group (Murray et al., 2002). How prevention of IFNγ mediated downregulation of T cells or distinct T cell functions contribute to disease remains unresolved.

Establishment of TMEV persistence in H-2D^b MHC-I deficient mice, in addition to the detection of MHC-I molecules on infected neurons in TMEV infected brains of resistant mice supports MHC-I dependent viral clearance mechanisms (Azoulay-Cayla et al., 2000; Fiette et al., 1993; Njenga et al., 1997a, b). Surprisingly, in contrast to MHC-I sufficient susceptible mice, class I deficiency results in the absence of neurological deficits despite extensive demyelination, implicating a detrimental role of MHC-I in development of neurological deficits (Rivera-Quinones et al., 1998).

CD8 T cells also contribute to TMEV clearance and control of acute disease in susceptible SJL mice (Begolka et al., 2001). The inability of CD8 T cells to effectively control acute infection in β2 m deficient SJL mice not only results in an earlier onset of acute disease, but also a more rapid onset of chronic demyelinating disease, and epitope spreading associated with the autoimmune nature of the chronic disease (Begolka et al., 2001). In both resistant and susceptible strains of mice, virus specific CD8 T cells comprise over 50% of CD8 T cells within the CNS (Johnson et al., 1999; Kang et al., 2002). The inability to prevent persistent infection and disease in susceptible mice, despite similar magnitudes of CD8 responses has been speculated to reside in less efficient presentation of K^s vs D^b restricted epitopes in the respective hosts (Kang et al., 2002).

Borna disease virus (BDV) is another highly neurotropic virus causing persistent CNS infection associated with severe tissue destruction (Planz et al., 1993; Stitz et al., 2002). Similar to LCMV, BDV is a noncytolyic virus in which immune pathology is mediated by CD8 T cells. Neurons and astrocytes are primary targets for infection. MHC-I is expressed in BDV infected rat brains (Stitz et al., 1991) and on infected neurons propagated in vitro from infected rats (Planz et al., 1993). CD8 T cells isolated from infected rat brains during acute infection have vigorous cytolytic activity ex vivo (Planz et al., 1993); however, similar to other CNS infections, the CD8 T cells loose cytolytic activity and ability to suppress infectious virus during persistence (Sobbe et al., 1997). Therefore, viral persistence appears to be related to the inability of CD8 T cells to exert anti-viral function rather than the lack of MHC-I target structures on infected cells. Following transfer of T cells isolated from brains of infected rats into immunosuppressed BDV infected recipients, degenerative disease corresponded to CD8 T cell entry into the parenchyma and detection of CD8 and perforin RNA (Sobbe et al., 1997). By contrast, CLN and splenic T cells only induced neurological symptoms if re-stimulated in vitro with BDV antigen. Irrespective of the implication of CD8 T cells in BDV-induced brain atrophy in rats, similar pathogenesis in BDV infected perforin deficient and wt mice indicates perforin plays no role in viral control nor development of neurological disease in the mouse model (Hausmann et al., 2001). Nevertheless, adoptive transfer of activated CD4 T cells prior to infection mediated protection and virus clearance. Protection coincided with earlier and enhanced parenchymal infiltration of CD8 T cells (Noske et al., 1998). An antiviral and protective role for CD8 T cell secreted IFNγ was also demonstrated in mice vaccinated with vectors expressing the nucleocapsid protein to establish CD8 memory T cells (Hausmann et al., 2005). IFNγ not only eliminated virus from neurons, but also limited pathological damage mediated by immune responses in unvaccinated animals.

These data highlight the diverse responses of T cells to primarily neuronotropic infections and consequences of ineffective early control resulting in viral persistence and pathology. In conclusion, susceptibility of neurons to perforin mediated lysis appears to be dependent on virulence, neuronal subpopulations affected, and the presence of IFNγ. Neuronal damage mediated by virus replication itself and high IFNγ make infected neurons more susceptible to CD8 mediated cytolysis rather than curative mechanisms of clearance. This balance is beneficial to the host to block rapidly spreading virulent viruses at the sacrifice of few neurons. Nevertheless, rapidly spreading viruses such as BDV may not be susceptible to CD8 effector function after a threshold of viral spread has been exceeded. The curative process during LCMV persistence is rather slow (Oldstone et al., 1986), suggesting a paucity in IFNγ production in vivo, or low sensitivity to the effects of IFNγ. Irrespective of the mechanisms, the signals triggering IFNγ secretion by CD8 T cells in the absence of detectable class I expression in persistently infected neurons remains an enigma. Possible explanations may reside in class I independent IFNγ secretion observed in unconventional CD8 T cells (Braaten et al., 2006) or dysregulated CD8 T cells as observed in anti-tumor responses (De Geer et al., 2006; Maccalli et al., 2003). Alternatively, viral epitopes may be presented by cell types other than the primary targets of infection via crosspriming (Rock and Shen, 2005; Neijssen et al., 2005; also see Chapter 13). This mechanisms has been demonstrated for CNS recruitment of tumor-specific CD8 T cells (Calzascia et al., 2003). An examination of this problem at the cellular level will allow to determine whether IFNγ is being secreted by T cells in direct contact with infected neurons or by T cells contacting other brain cells. The recent demonstration of the existence of immunological synapses between CD8 T cells and infected astrocytes during the clearance of virally infected astrocytes provides a direct cellular approach to the questions discussed above (Barcia et al., 2006, 2007). From a teleological point of view it would make sense that the immune system clears brain infections by a non-cytolytic mechanism. It will be important to examine this hypothesis more thoroughly utilizing detailed combined anatomical and molecular approaches. If virus is truly cleared by non-cytolytic mechanisms, stimulating anti-viral brain immune responses would be acceptable as therapeutic intervention; however, if the immune system were to indeed kill infected brain cells, a therapeutic approach would have to take potential direct brain toxicity into account. Also, it is possible that the mechanisms of clearance may differ for RNA vs. DNA viruses, depending on viral virulence or potential for latent infection, level of T cell activation, antigen presentation capacity of infected CNS cells, as well as infected cell type and/or neuroanatomical location. While strong evidence exists in favor of non-cytolytic clearing mechanisms, we believe that the potential existence of brain cytotoxicity ought to be reexamined in detail.

3.2.2 Immune Control of Glial Infections

Cell type dependent and pathogen specific mechanisms of CD8 T cell effector function are also evident in infections primarily targeting glial cells in the acute phase of infection. Glial tropic coronaviruses, represented by neurotropic mouse hepatitis virus strains, are controlled by CD8 T cells during the acute phase (Bergmann et al., 2003, 2004; Lin et al., 1997); nevertheless virus persists in the form of RNA for the life of the mouse, indicating escape from T cell effector function (Bergmann et al., 2006). Is it not known how the RNA genomes persist. Persisting infectious virus cannot be recovered from CNS explants or from immunosuppressed mice perfused to remove neutralizing antibody prior to explant. Nevertheless, persistence in a replication competent form is indicated by the necessity of intrathecal humoral immune responses to maintain infectious virus below detection thresholds (Bergmann et al., 2006; Ramakrishna et al., 2002, 2003; Tschen et al., 2006). The cell types harboring persisting virus are astrocytes and oligodendrocytes (Perlman and Ries, 1987; Ramakrishna et al., 2003; Gonzalez et al., 2006). Virus-specific CD8 T cells are enriched in the CNS comprising at least up to 50% of total CD8 T cells (Bergmann et al., 1999), similar to acute TMEV infection. The vast majority express a $CD44^{hi}$, $CD62L^{-/lo}$, $CD11a^{hi}$, and $CD49d^+$ (VLA-4) activation/memory phenotype. Consistent with their activated effector phenotype, virus-specific CD8 T cells from the acutely inflamed CNS exert cytolytic effector function and produce IFN-γ and to a lesser degree TNFα upon short term antigen stimulation ex vivo (Bergmann et al., 1999; Ramakrishna et al., 2004, 2006). Expression of granzyme B and high levels of CD43 are consistent with expression of effector function in vivo. Studies in mice genetically deficient for CD8 effector molecules demonstrate that antiviral activity is mediated by both a perforin-dependent and an IFNγ dependent pathway (Lin et al., 1997; Parra et al., 1999). However, IFNγ played a more prominent protective role compared to perforin, as demonstrated by diminished virus control and enhanced mortality in the absence of IFNγ function. Mice deficient for both functions were even more susceptible to viral-induced disease and were unable to control infection, despite increased expansion and recruitment of CD8 T cells to the CNS (Bergmann et al., 2003). Immunohistochemical analysis revealed that perforin sufficed to control replication in astrocytes and microglia, while oligodendrocytes appeared insensitive to perforin mediated cytolysis in the absence of IFN-γ (Parra et al., 1999). By contrast, mice deficient in perforin controlled replication in oligodendrocytes, but could not eliminate virus from astrocytes and microglia (Lin et al., 1997). Adoptive transfer of virus specific memory CD8 [+] T cells deficient in either cytolytic activity or IFN-γ secretion into infected immuno-deficient hosts confirmed distinct susceptibilities of glial cell types to T cell antiviral functions in vivo (Bergmann et al., 2003, 2004). Even though detection of class I expression on astrocytes during MHV infection has been inconsistent (Redwine et al., 2001; Hamo et al, 2007), perforin-mediated control of virus replication in astrocytes and microglia supports direct MHC-I/TCR interactions in these cells.

Mechanisms of T cell mediated control of intracerebral murine cytomegalovirus (MCMV) replication are controversial. Adoptive transfer studies using splenocytes from MCMV primed donors revealed that CD8 T cell mediated perforin, but not IFNγ function, is critical in reducing virus from adult, immune-compromised SCID mice (Cheeran et al., 2005). In this experimental model MCMV replicates productively in astrocytes, but other CNS cell types have not been ruled out. By contrast, reconstitution of SCID mice with CD4 T cells, but not CD8 T cells from immunized mice reduced virus and provided protection from disease, without preventing establishment of latency in another study (Reuter et al., 2005). These distinct outcomes have not been reconciled, but may reside in distinct virus strains utilized. Brains from fetal and perinatal mice are most susceptible to MCMV with the predominant infected cells localized to the subventricular zone (Tsutsui et al., 2005). This region primarily harbors undifferentiated neural stem/progenitor cells, giving rise to neuronal and glial cells during brain development. Prominent susceptibility of neural stem cells to MCMV infection may thus lead to differential expression of viral genes in the developing brain, depending on differentiation into immature glial cells or neurons. Both glial cells and neurons may be sources of latent infection.

So far, interactions between immune cells and target brain cells have been studied at the population level. Thus, there is relatively little information on the in vivo cell biology of T cell interactions with infected brain cells studied at the single cell level (Barcia et al., 2006; McGavern et al., 2002). Over the last 10 years immunological synapses have been characterized as the cellular substrate of intercellular communication in the immune system. Immunological synapses that form at the junction between T cells and antigen presenting cells consist of a rearrangement of membrane proteins (i.e., intercellular adhesion molecules [e.g. ICAM-1], TCR), intracellular TCR downstream signaling tyrosine kinases, as well as cytoskeletal structures, and intracellular organelles of the secretory pathway of the T cells (Cemerski et al., 2007; Davis et al., 2007; Dustin and Cooper, 2000; Friedl et al., 2005; Grakoui et al., 1999; Huppa and Davis, 2003; Huse et al., 2006; Lee et al., 2002; Monks et al., 1998). Although various types of arrangements of T cell proteins have been found at these intercellular junctions, a canonical structure, known as the mature immunological synapse has been described as consisting of the following arrangement: a peripheral supramolecular activation cluster (pSMAC), consists of a ring of adhesion molecules that anchor the membrane of the T cell to the membrane of the APC, while a central-SMAC (cSMAC), consists of a higher concentration of TCR and signaling molecules. Immunological synapses have been described for both CD4 and CD8-T cells, and NK cells in contact with various types of APCs, e.g. dendritic cells, B cells, or target cells.

Evidence for multi-cellular CD8 T cell engagement has been provided in the LC MV-infected CNS (McGavern et al., 2002), although target cells were not identified. Furthermore, it was recently shown that anti-adenoviral CD8 T cells infiltrating the brain form classical mature immunological synapses with class I expressing astrocytes (Barcia et al., 2006, 2007). These immunological synapses were characterized through the formation of the classical supramolecular activation clusters (SMACS), which constitute the hallmark of immunological synapses (Fig. 14.1). In this model a non-replicating adenoviral vector was used to predominantly target astrocytes in the rat brain, resulting in a fixed number of astrocytes harboring viral genomes. This virus is replication-defective and thus unable to directly kill infected cells. As the parenchymal CNS infection itself does not induce significant inflammation nor a systemic anti-adenoviral immune response, systemic anti-adenoviral immunization was induced with a different Ad vector injected systemically. Systemic

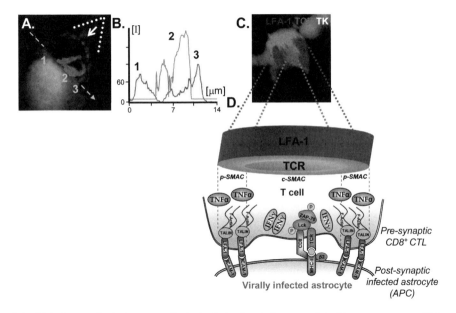

Fig. 14.1 Immunological synapses in the brain in vivo during an antiviral immune response. This figure shows an immunological synapse formed between CD8 T cells and an adenovirally infected astrocyte. (**a**) shows the synapse as seen under the confocal microscope; the infected cell is detected through its expression of a marker gene expressed from the viral genome (i.e. HSV1-TK; white), and the CD8 + T cell is detected through its expression of LFA-1 (red) and TCR (green). (**b**) shows the relative quantification of fluorescence across the yellow arrow in (**a**); 1,2, 3 are the areas indicated in the fluorescence intensity graph in (**b**). Notice the typical distribution of increased intensity of LFA-1 at the p-SMAC (in red), and the peak of TCR intensity at the c-SMAC. (**c**) shows the view from the 3-D reconstruction of the image stack illustrated in (**a**). The view shown in (**c**) is viewed from the white triangle and white arrow in (**a**). (**d**) illustrates schematically the distribution of molecules at the p-SMAC and c-SMAC, and also illustrates the intracellular re-distribution and potential secretion of T cell effector molecules either towards the immunological synapse (i.e. IFN-γ), or outside the immunological synapse (i.e. TNF-α). (**d**) is based on results from Huse et al. (2006), and Lowenstein et al. (2007) (*See also color plates*).

anti-adenoviral immunization triggered a systemic anti-adenoviral immune response, which led to brain inflammation. Brain inflammation consisted mainly in an infiltration of the brain parenchyma of CD8 T cells and macrophages, and a perivascular infiltration of CD4 T cells.

The systemic anti-adenoviral immune response resulted in a significant reduction in the number of brain astrocytes that express adenoviral proteins, and a concomitant reduction in the number of viral genome copy numbers present in the CNS. Loss of infected cells was dependent on both CD4 and CD8 T cells. The presence of CD8 T cells within the brain parenchyma suggested the operation of direct cytolytic mechanisms in the elimination of infected cells. Although no direct evidence for apoptotic astrocytes was obtained, macrophages containing remains of infected astrocytes were found throughout the area of the brain that had been cleared of infected cells. This suggested that the formation of immunological synapses may represent the microanatomical substrate underlying CD8 T cell effector functions in the CNS, and mediate the anti-viral clearing of CD8 T cells. The importance of these studies is the demonstration that immunological synapse do form indeed in vivo in the brain during the clearing of virally infected astrocytes by the adaptive immune response. Their in vivo description in the context of an anti-viral immune response highlights their physiological role as the structure underlying neuroimmune interactions in vivo. Also, the existence of immunological synapses in the brain during the clearing of virally infected brain cells opens up the examination of neuroimmune interactions at the single cell level.

3.2.3   Immune Clearing of Infected Oligodendrocytes

The apparent resistance of oligodendrocytes to perforin mediated clearance mechanisms in coronavirus-brain infections is not resolved. Nevertheless, glial tropic coronavirus infection of mice with an IFN-γ signaling defect selectively in oligodendroglia directly confirmed the importance of IFN-γ signaling in this cell type for controlling oligodendroglial viral clearance (Gonzalez et al., 2005, 2006). Virus clearance was delayed and viral antigen almost exclusively localized to oligodendrocytes resulting in at least tenfold higher antigen load compared to wt mice in the waning period of acute infection. These data implied either differences in antigen presentation by glial cell subsets and/or inherent resistance of oligodendrocytes to contact dependent CD8 T cell function. Although previously controversial, MHC-I upregulation on oligodendrocytes has recently been demonstrated in two independent models of neurotropic MHV infection (Redwine et al, 2001; Ramakrishna et al., 2006) and in mice expressing IFNγ as a transgene in the CNS (Horwitz et al., 1999). However, the kinetics of class I upregulation during infections may be delayed compared to other glia. Both IFNα/β and IFNγ can favor MHC-I antigen processing by inducing expression of proteasomal subunits and peptide transporters TAP1 and TAP2. Microglia upregulate surface MHC-I as early as 3–4 days following glial tropic MHV infection, prior to detection of significant amounts of IFNγ (Bergmann et al., 2003; Zuo et al., 2006). Expression on microglia is indeed reduced and more transient in the absence of T cell produced IFNγ (Bergmann et al., 2003). Although IFNγ independent MHC upregulation can be mediated on cultured glial cells by type I IFNs, TNFα or IL-6 in vitro, their contribution in vivo is unknown. Similar to neurons, oligodendrocytes may have more stringent regulation of class I presentation to sustain their vital function in maintaining myelin. Preliminary analysis indeed suggests that class I is not upregulated on oligodendrocytes in IFNγ deficient mice (Bergmann, unpublished). Furthermore, surface expression is associated with vigorous de novo induction of mRNA species encoding antigen processing genes (Malone et al., 2006). More detailed studies are required to assess the possibility of impaired antigen processing, distinct thresholds of oligodendrocytes to initiate and sustain antigen presentation, or regulation by inhibitory receptors.

An antigen specific model of chronic CD8 T cell inflammation is provided by transgenic mice constitutively expressing a viral protein in oligodendrocytes (Evans et al., 1996). Transgenic mice expressing the LCMV protein nucleoprotein or glycoprotein under the MBP promoter exhibit no clinical or pathological abnormalities. Peripheral LCMV infection is cleared similar to wt mice, with no apparent involvement of CNS infection. However, although transient accumulation of CD8 T cells in meninges and ventricular linings was similar in both infected mice, CD8 T cells accumulated to significantly higher levels in the CNS of transgenic mice compared to wt mice, in which CD8 T cells waned by 3 weeks p.i. CD8 T cells persisted at constant levels throughout 1 year p.i. and was associated with MHC-I expression. A second viral challenge resulted in enhanced CD8 and CD4 T cell inflammation, and upregulation of proinflammatory cytokines including IFNγ, leading to demyelination and enhanced clinical disease. These data implied chronic CD8 T cell inflammation was triggered by class I mediated oligodendrocyte CD8 T cell interactions.

3.3 Long Term CD8 T Cell Survival in the CNS

Following successful T cell mediated control of acute viral replication, T cells decline but significant numbers are nevertheless maintained in the CNS for many months or life span of infected mice. This is observed both following primary responses or recall responses (Hawke et al., 1998; Marten et al., 2000b; van der Most et al., 2003). After infectious virus is eliminated in the glial tropic coronavirus model, both CD4 and CD8 T cells persist concomitantly with detectable, persisting viral mRNA, but undetectable viral protein or infectious virus titers (Bergmann et al., 1999, 2006; Marten et al., 2000b). Despite the drop in total T cell numbers, the percentages of virus specific T cells within the CD8 T cell compartment within the CNS remain remarkably stable throughout infection (Bergmann et al., 1999; Marten et al., 2000b; Ramakrishna et al., 2004). This suggests that survival/retention signals do not discriminate between virus specific CD8 T cell populations and populations of unknown specificities potentially recruited as bystander cells.

There is no evidence that T cells exert antiviral function during coronavirus persistence. Cytolytic activity by primary virus-specific CD8 T cells is readily detected ex vivo during acute virus replication, but barely, if at all detectable by 2 weeks p.i., after clearance of infectious virus (Bergmann et al., 1999; Marten et al., 2000a; Ramakrishna et al., 2004). IFN-γ mRNA levels also decline, as virus is cleared (Parra et al., 1997; Zuo et al., 2006). Loss of cytolytic function is independent of demyelination associated factors (Marten et al., 2000a). Ex vivo cytolysis can also not be recovered during virus recrudescence in B cell deficient mice (Ramakrishna et al., 2002), despite MHC-I expression on microglia and oligodendrocytes (Ramakrishna et al., 2006). However, there is no evident impairment of IFN-γ secretion upon antigen re-exposure in vitro, suggesting CD8 T cells are not anergized (Bergmann et al., 1999; Ramakrishna et al., 2002). Comparison of coronavirus-specific primary versus memory CD8 T cells in response to CNS challenge revealed that CNS derived CD8 T cells from immunized mice exhibit enhanced cytolysis at a single cell level as well as increased IFN-γ and granzyme B production, compared to naïve mice after challenge (Ramakrishna et al., 2004). Importantly, reactivated memory CD8 T cells retained cytolytic function coincident with increased granzyme B levels for prolonged periods compared to primary CD8 T cells. Retention of cytolytic activity in reactivated memory cells persisting in the CNS was first demonstrated in a model of neurotropic influenza virus challenge (Hawke et al., 1998). The lethality of this influenza virus infection in naïve mice prevented longitudinal analysis of primary CD8 T cell function. Loss of virus-specific cytolytic function thus reflects distinct differentiation states of primary compared to memory CD8 T cells rather than an intrinsic property of the inflamed CNS environment. Importantly, despite retention of a cytolytic phenotype, there was no evidence for ongoing pathology in either challenge model.

Enhanced effector function by reactivated memory CD8 T cells directly translated to more effective virus control upon challenge compared to naïve mice in both the coronavirus and neurotropic influenza virus models. Nevertheless, despite a nearly 3 week phase of apparent clearance during which infectious virus remained undetectable, CD8 T cell persistence at higher numbers and a pre-armed state was insufficient to prevent coronavirus reactivation in the absence of antiviral antibody (Ramakrishna et al., 2006). Interestingly, MHC expression was significantly downregulated during the period of apparent virus clearance. At the time of recrudescence MHC expression on microglia and oligodendrocytes was low compared to the acute response, suggesting too few if any target MHC molecules presenting viral epitope to re-initiate IFNγ mediated class I upregulation and trigger antiviral function. Despite being fully armed to exert anti-viral activity upon antigen exposure in vitro, memory T cells in the CNS in vivo may thus not be responsive in an environment in which too few cells are persistently infected to engage antiviral T cell function. These observations highlight the functional controversies encountered upon the comparative analysis of CNS CD8 T cell function in vitro and in vivo. Furthermore, they emphasize the complex regulation of MHC-I/TCR interactions and a potential feedback loop involving IFNs in sustaining or downregulating MHC-I. As viral clearance progresses fewer MHC- I microbial antigen complexes are recognized during viral control, IFNγ levels drop, resulting in decreased transcription of genes associated with MHC-I antigen processing and presentation.

The mechanisms of prolonged T cell persistence in the CNS in the absence of overt chronic inflammation and CNS disease have not been extensively explored. Comparison of the fate of T cells following infection with a persisting and non persisting glial tropic coronavirus variant suggested a role for persisting viral RNA in sustaining T cell retention (Marten et al., 2000b). Active maintenance is also supported by selection of CD8 T cell populations with limited T cell receptor specificities during the persistent phase of coronavirus infection compared to the acute phase (Marten et al., 1999). By contrast, following challenge of T cell immune mice with a neurovirulent influenza virus infection, residual virus was undetectable, demonstrating that retention of CD8 T cells in the CNS was independent of persisting viral antigen as measured by PCR analysis for viral RNA (Hawke et al., 1998). MHC expression within the CNS was not analyzed in either study. It is possible that immunological synapses provide an anatomical substrate to sustain long term T cell survival in the CNS even after MHC expression drops. The longevity of immunological synapses in vivo during the clearing of viral infections is currently being examined.

The contribution of local homeostatic proliferation or ongoing recruitment to T cell maintenance in the CNS remains unclear. Memory cells traffic poorly into the quiescent CNS (Masopust et al., 2004) and activated T cells are only retained within the CNS upon cognate antigen recognition (Hickey, 2001; Chen et al., 2005). In the influenza virus model, memory CD8 T cells retained in the CNS after viral clearance had very slow turnover compared to peripheral memory CD8 T cells (Hawke et al., 1998). The memory T cell survival factor IL-15, which regulates homeostasis of CD8 memory cells in lymphoid organs (Masopust and Ahmed, 2004), does also not appear to be required for prolonged CD8 T cell survival in the coronavirus persistently infected CNS (Zuo and Bergmann, unpublished), supporting low turnover.

Persisting infection is associated with prolonged detection of CXCR3 ligands, potentially mediating ongoing lymphocyte recruitment (Tschen et al., 2006; also see Chap. 12). However, blockade of CXCR3 signaling early during persistence selectively reduces CD4 but not CD8 T cells (Stiles et al., 2006). These observations suggest that neither peripheral chemokine-mediated recruitment nor local division contribute significantly to CNS CD8 T cell maintenance. These findings implied that a small subpopulation of T cells responding to infection are recruited into a long lived, persisting pool with enhanced survival in the CNS. It is unknown whether T cell retention within the CNS is associated with enhanced expression of anti-apoptotic factors and/or survival factors in the CNS environment. Higher propensity for apoptosis by T cells in lymphoid compared to non lymphoid tissues (Wang et al., 2003) suggests that CD8 T cells surviving a selection process during the decline phase may have a prolonged lifespan in the CNS. Thus although T cells are the primary cells undergoing apoptosis during acute antiviral responses in the CNS (Gonzalez et al., 2006), those that do not become susceptible to antigen induced cell death may survive long term in the CNS.

4 Concluding Remarks

The communication between CD8 T cells and resident cells of the CNS appears highly sophisticated, yet our understanding is still very limited. Major limitations reside in the unique interactions and crosstalk between CNS cells amongst themselves, their fully differentiated state in mature individuals, and their anatomical location. This network amongst a very large number of different cells types makes it difficult to extrapolate results obtained from in vitro studies using primary cultures from neonatal tissue to events in vivo. Although glial cells and neurons are capable of expressing MHC-I in vitro and thus acting as targets for in vitro stimulated CD8 T cells, the mode of CD8 T cell function in vivo has largely been deducted from indirect evidence, such as viral clearance, loss of detection of viral genomes, demyelination or tissue atrophy. How T cells eliminate viral infections from the CNS remains contested. It is likely to depend crucially on the individual virus, its capacity to remain latent or persistent in CNS cells, the exact nature of the infected cell type, the anatomical region infected, and the level of T cell activation. Additional characteristics, such as sex and age are also likely to be crucial determinants of the outcome of T cell mediated clearing of viral infections. While teleologically, non-cytolytic clearing of non-dividing infected brain cells may be preferable to killing of postmitotic neurons, the alternative needs to be thoroughly examined. Direct contact or even proximity of class I expressing cells and CD8 T cells in situ has only been demonstrated in isolated reports. Similarly, the notion of curative rather than cytolytic virus clearance should be regarded critically, as demonstration of apoptotic or dead CNS resident cells in situ, especially if the numbers are sparse are technically challenging. The technical challenge of determining cell death in vivo is substantial. Immune-mediated killing cell assays rely on removing T cells from the target organ and exposing them to artificial target cells in vitro. While such studies have demonstrated the capacity of T cells isolated from infected brains to kill pre-selected target cells, this comes short of demonstrating that these cells indeed kill brain cells in vivo. In vivo killing assays using CFSE labeled target cells have been developed, yet their application to CNS tissue has not yet been achieved. While such assays come closer to demonstrating that T cells can kill within the animal, whether T cells can directly kill CNS target cells in situ, remains to be explored further. It is likely that novel methods will need to be developed to examine directly the capacity of T cells to do so.

The involvement of activating and inhibitory receptors on T cells and expression of their ligands in the CNS is not well understood. Similarly, regulation of anti-apoptotic factors requires further investigation. A better understanding of the relative kinetics of MHC presentation, T cell accumulation, and regulation of effector function at the target site during inflammatory reactions may lead to novel preventive and therapeutic intervention in enhancing, microbial clearance, while minimizing pathological damage. Finally, it is likely that a combination of novel microanatomical imaging techniques will encourage new analyses of T cell activity in the CNS. We believe that a novel examination of the interactions of T cells with target cells at the individual cellular level in vivo, will allow new perspectives on T cell function in the CNS. Until now, even quantitative assays on T cell function in the CNS have relied on studies of the whole population of brain infiltrating T cells. Many times, due to unavoidable technical constraints, functional analysis has relied on the ex vivo functional analysis of brain infiltrating T cells. Equally, the examination of target brain cell function has relied on the analysis of large populations of cells, for which their past direct interactions with effector cells remained unknowable. New morpho-functional approaches, such as the study of immunological synaptic function in vivo utilizing confocal microscopy, or direct in vivo analysis of T cell – CNS cell interactions using two photon microscopical approaches will usher in further understandings of T cell function in the brain in vivo, either during the beneficial clearing of viral infections, immunopathology in response to brain infection, or autoimmune attack.

Acknowledgements We thank Stephen Stohlman for reading this review and providing helpful comments. The authors of this article were funded by NIH grants PO1 NS18146, RO1 AI47249 and NMSS grant RG 3808 (C. C. B) and NIH/NINDS grants R01 NS44556, 1 RO1 NS42893, U54 NS045309, R21 NS047298, and R01 NS054193, and The Bram and Elaine Goldsmith Chair In Gene Therapeutics (P.R.L.). PRL also thanks the generous support the BOG-GTRI receives from the Board of Governors at Cedars Sinai Medical Center and Kurt Kroeger for his assistance in preparation of Fig. 14.1.

References

Aloisi, F., Serafini, B., and Adorini, L. (2000). Glia-T cell dialogue. *J. Neuroimmunol., 107*, 111–117.

Armstrong, W. S and Lampson, L. A. (1997). Direct cell-mediated responses in the nervous system: CTL vs. NK activity, and their dependence upon MHC expression and modulation. In: R. W. Keane and W. F. Hickey (Eds.), *Immunology of the Central Nervous System* (pp. 493–547). Oxford: Oxford University Press.

Azoulay-Cayla, A., Dethlefs, S., Perarnau, B., Larsson-Sciard, E. L., Lemonnier, F. A., Brahic, M., and Bureau, J. F. (2000). H-2D(b–/–) mice are susceptible to persistent infection by Theiler's virus. *J. Virol., 74*, 5470–7476.

Baerwald, K. D., Corbin, J. G., and Popko, B. (2000). Major histocompatibility complex heavy chain accumulation in the endoplasmic reticulum of oligodendrocytes results in myelin abnormalities. *J. Neurosci. Res., 59*, 160–169.

Bailey, S. L., Carpentier, P. A., McMahon, E. J., Begolka, W. S., and Miller, S. D. (2006). Innate and adaptive immune responses of the central nervous system. *Crit. Rev. Immunol., 26*, 149–188.

Barcia, C., Thomas, C. E., Curtin, J. F., King, G. D., Wawrowsky, K., Candolfi, M., Xiong, W. D., Liu, C., Kroeger, K., Boyer, O., Kupiec-Weglinski, J., Klatzmann, D., Castro, M. G., and Lowenstein, P. R. (2006). In vivo mature immunological synapses forming SMACs mediate clearance of virally infected astrocytes from the brain. *J. Exp. Med., 203*, 2095–2107.

Barcia, C., Gerdes, C., Xiong, W.-D., Thomas, C. E., Liu, C., Kroeger, K. Castro. M. G., and Lowenstein, P. R. (2007). Immunological thresholds in neurological gene therapy: highly efficient elimination of transduced cells may be related to the specific formation of immunological synapses between T cells and virus-infected brain cells. *Neuron Glia Biol., 2*:309–322.

Barmak, K., Harhaj, E. W., and Wigdahl, B. (2003). Mediators of central nervous system damage during the progression of human T-cell leukemia type I-associated myelopathy/tropical spastic paraparesis. *J. Neurovirol., 9*, 522–529.

Bauer, J., Bien, C. G., and Lassmann, H. (2002). Rasmussen's encephalitis: a role for autoimmune cytotoxic T lymphocytes. *Curr. Opin. Neurol., 15,* 2197–2200.

Bechmann, I., Galea, I., and Perry, V. H. (2007). What is the blood–brain barrier (not)?. *Trends Immunol., 28,* 5–11.

Begolka, W. S., Haynes, L. M., Olson, J. K., Padilla, J., Neville, K. L., Dal Canto, M., Palma, J., Kim, B. S., and Miller, S. D. (2001). CD8-deficient SJL mice display enhanced susceptibility to Theiler's virus infection and increased demyelinating pathology. *J. Neurovirol., 7,* 409–420.

Bergmann, C. C., Altman, J. D., Hinton, D., and Stohlman, S. A. (1999). Inverted immunodominance and impaired cytolytic function of CD8 + T cells during viral persistence in the central nervous system. *J. Immunol., 163*, 3379–3387.

Bergmann, C. C., Parra, B., Hinton, D. R., Chandran, R., Morrison, M., and Stohlman, S. A. (2003). Perforin-mediated effector function within the central nervous system requires IFN-gamma-mediated MHC up-regulation. *J. Immunol., 170*, 3204–3213.

Bergmann, C. C., Parra, B., Hinton, D. R., Ramakrishna, C., Dowdell, K. C., and Stohlman, S. A. (2004). Perforin and interferon gamma mediated control of coronavirus central nervous system infection by CD8 T cells in the absence of CD4 T cells. *J. Virol., 78*, 1739–1750.

Bergmann, C. C., Lane, T. E., and Stohlman, S. A. (2006). Coronavirus infection of the central nervous system: host-virus stand-off. *Nat. Rev. Microbiol., 4*, 121–132.

Bien, C. G., Bauer, J., Deckwerth, T. L., Wiendl, H., Deckert, M., Wiestler, O. D., Schramm, J., Elger, C. E., and Lassmann, H. (2002). Destruction of neurons by cytotoxic T cells: a new pathogenic mechanism in Rasmussen's encephalitis. *Ann. Neurol., 51*, 311–318.

Bilzer, T. and Stitz, L. (1994). Immune-mediated brain atrophy. CD8 + T cells contribute to tissue destruction during Borna disease. *J. Immunol., 153*, 818–823.

Binder, G. K. and Griffin, D. E. (2001). Interferon-gamma-mediated site-specific clearance of alphavirus from CNS neurons. *Science, 293*, 303–306.

Boulanger, L. M. and Shatz, C. J. (2004). Immune signalling in neural development, synaptic plasticity and disease. *Nat. Rev. Neurosci., 5*, 521–531.

Braaten, D. C., McClellan, J. S., Messaoudi, I., Tibbetts, S. A. McClellan, K. B., Nikolich-Zugich, J., and Virgin, H. W. (2006). Effective control of chronic gamma-herpesvirus infection by unconventional MHC Class Ia-independent CD8 T cells. *PLoS Pathog., 2*, e37.

Brisebois, M., Zehntner, S. P., Estrada, J., Owens, T., and Fournier, S. (2006). A pathogenic role for CD8 + T cells in a spontaneous model of demyelinating disease. *J. Immunol., 177*, 2403–2411.

Cabarrocas, J., Bauer, J., Piaggio, E., Liblau, L., and Lassman, H. (2003). Effective and selective immune surveillance of the brain by MHC class I-restricted cytotoxic T lymphocytes. *Eur. J. Immunol., 33*, 1174–1182.

Calzascia, T., Di Berardino-Besson, W., Wilmotte, R., Masson, F., de Tribolet, N., Dietrich, P. Y., and Walker, P. R. (2003). Cutting edge: cross-presentation as a mechanism for efficient recruitment of tumor-specific CTL to the brain. *J. Immunol., 171*, 2187–2191.

Carson, M. J., Reilly, C. R., Sutcliffe, J. G., and Lo, D. (1999). Disproportionate recruitment of CD8 + T cells into the central nervous system by professional antigen-presenting cells. *Am. J. Pathol., 154*, 481–494.

Cartmell, T., Southgate, T., Rees, G. S., Castro, M. G., Lowenstein, P. R., and Luheshi, G. N. (1999). Interleukin-1 mediates a rapid inflammatory response after injection of adenoviral vectors into the brain. *J. Neurosci., 19*, 1517–1523.

Caux, C., Ait-Yahia, S., Chemin, K., de Bouteiller, O., Dieu-Nosjean, M. C., Homey, B., Massacrier, C., Vanbervliet, B., Zlotnik, A., and Vicari, A. (2000). Dendritic cell biology and regulation of dendritic cell trafficking by chemokines. *Springer Semin. Immunopatho., 22*, 345–369.

Cemerski, S., Das, J., Locasale, J., Arnold, P., Giurisato, E., Markiewicz, M. A., Fremont, D., Allen, P. M., Chakraborty, A. K., and Shaw, A. S. (2007). The stimulatory potency of T cell antigens is influenced by the formation of the immunological synapse. *Immunity, 26*, 345–355.

Cheeran, M. C., Gekker, G., Hu, S., Palmquist, J. M., and Lokensgard, J. R. (2005). T cell-mediated restriction of intracerebral murine cytomegalovirus infection displays dependence upon perforin but not interferon-gamma. *J. Neurovirol., 11*, 274–280.

Chen, W., Masterman, K. A., Basta, S., Haeryfar, S. M., Dimopoulos, N., Knowles, B., Bennink, J. R., and Yewdell, J. W. (2004). Cross-priming of CD8 + T cells by viral and tumor antigens is a robust phenomenon. *Eur. J. Immunol., 34*, 194–199.

Chen, A. M., Khanna, N. Stohlman, S. A., and Bergmann, C. C. (2005). Virus-specific and bystander CD8 T cells recruited during virus induced encephalomyelitis. *J. Virol., 79*, 4700–4708.

Corriveau, R. A., Huh, G. S., and Shatz, C. J. (1998). Regulation of class I MHC gene expression in the developing and mature CNS by neural activity. *Neuron, 21*, 505–520.

Cserr, H. F. and Knopf, P. M. (1992). Cervical lymphatics, the blood–brain barrier and the immunoreactivity of the brain: a new view. *Immunol. Today, 13*, 507–512.

Davis, M. M., Krogsgaard, M., Huse, M., Huppa, J., Lillemeier, B. F., and Li, Q. J. (2007). T cells as a self-referential, sensory organ. *Annu. Rev. Immunol., 25*, 681–95.

De Geer, A., Kiessling, R., Levitsky, V., and Levitskaya, J. (2006). Cytotoxic T lymphocytes induce caspase-dependent and -independent cell death in neuroblastomas in a MHC-nonrestricted fashion. *J. Immunol., 177*, 7540–7550.

Divito, S., Cherpes, T. L., and Hendricks, R. L. (2006). A triple entente: virus, neurons, and CD8 + T cells maintain HSV-1 latency. *Immunol. Res., 36*, 119–126.

Dong, Y. and Benveniste, E. N. (2001). Immune function of astrocytes. *Glia, 36*, 180–190.

Dorries, R. (2001). The role of T-cell-mediated mechanisms in virus infections of the nervous system. *Curr. Top Microbiol. Immunol., 253*, 219–245.

Drescher, K. M., Pease, L. R., and Rodriguez, M. (1997). Antiviral immune responses modulate the nature of central nervous system (CNS) disease in a murine model of multiple sclerosis. *Immunol. Rev., 159*, 177–193.

Dustin, M. L. and Cooper, J. A. (2000). The immunological synapse and the actin cytoskeleton: molecular hardware for T cell signaling. *Nat. Immunol., 1*, 23–29.

Evans, C. F., Horwitz, M. S., Hobbs, M. V., and Oldstone, M. B. (1996). Viral infection of transgenic mice expressing a viral protein in oligodendrocytes leads to chronic central nervous system autoimmune disease. *J. Exp. Med., 184*, 2371–2384.

Fiette, L., Aubert, C., Brahic, M., and Rossi, C. P. (1993). Theiler's virus infection of beta 2-microglobulin-deficient mice. *J. Virol., 67*, 589–592.

Fiette, L., Aubert, C., Muller, U., Huang, S., Aguet, M., Brahic, M., and Bureau, J. F. (1995). Theiler's virus infection of 129Sv mice that lack the interferon alpha/beta or interferon gamma receptors. *J. Exp. Med., 181*, 2069–2076.

Friedl, P., den Boer, A. T., and Gunzer, M. (2005). Tuning immune responses: diversity and adaptation of the immunological synapse. *Nat. Rev. Immunol., 5*, 532–545.

Galea, I., Bechmann, I., and Perry, V. H. (2007). What is immune privilege (not)? *Trends Immunol., 28*, 12–18.

Garbi, N., Tanaka, S., van den Broek, M., Momburg, F., and Hammerling, G. J. (2005). Accessory molecules in the assembly of major histocompatibility complex class I/peptide complexes: how essential are they for CD8(+) T-cell immune responses? *Immunol. Rev., 207*, 77–88.

Guidotti, L. G. and Chisari, F. V. (2001). Noncytolytic control of viral infections by the innate and adaptive immune response. *Annu. Rev. Immunol., 19*, 65–91.

Goddard, C. A., Butts, D. A., and Shatz, C. J. (2007). Regulation of CNS synapses by neuronal MHC class I. *Proc. Natl. Acad. Sci. U.S.A., 104*, 6828–6833.

Gonzalez, J. M., Bergmann, C. C., Fuss, B., Hinton, D. R., Kangas, C., Macklin, W. B., and Stohlman, S. A. (2005). Expression of a dominant negative IFN-gamma receptor on mouse oligodendrocytes. *Glia, 51*, 22–34.

Gonzalez, J. M., Bergmann, C. C., Ramakrishna, C., Hinton, D. R., Atkinson, R., Hoskin, J., Macklin, W. B., and Stohlman, S. A. (2006). Inhibition of interferon-gamma signaling in oligodendroglia delays coronavirus clearance without altering demyelination. *Am. J. Pathol., 168,* 796–804.

Grakoui, A., Bromley, S. K., Sumen, C., Davis, M. M., Shaw, A. S., Allen, P. M., and Dustin. M. L. (1999). The immunological synapse: a molecular machine controlling T cell activation. *Science, 285,* 221–227.

Griffin, D. E. (2003). Immune responses to RNA-virus infections of the CNS. *Nat. Rev. Immunol., 3,* 493–502.

Halonen, S. K., Woods, T., McInnerney, K., and Weiss, L. M. (2006). Microarray analysis of IFN-gamma response genes in astrocytes. *J. Neuroimmunol, 175,* 19–30.

Hamo, L., Stohlman, S. A., Otto-Duessel, M., and Bergmann C. C. (2007). Distinct regulation of MHC molecule expression on astrocytes and microglia during viral encephalomyelitis. *Glia,* 55, 1169–77.

Harty, J. T., Tvinnereim, A. R., and White, D. W. (2000). CD8 + T cell effector mechanisms in resistance to infection. *Annu. Rev. Immunol., 18,* 275–308.

Hausmann, J., Schamel, K., and Staeheli, P. (2001). CD8(+) T lymphocytes mediate Borna disease virus-induced immunopathology independently of perforin. *J. Virol., 75,* 10460–10466.

Hausmann, J., Pagenstecher, A., Baur, K., Richter, K., Rziha, H. J., and Staeheli, P. (2005). CD8 T cells require gamma interferon to clear Borna disease virus from the brain and prevent immune system-mediated neuronal damage. *J. Virol., 79,* 13509–13518.

Hawke, S., Stevenson, P. G., Freeman, S., and Bangham, C. R. (1998). Long-term persistence of activated cytotoxic T lymphocytes after viral infection of the central nervous system. *J. Exp. Med., 187,* 1575–1582.

Heink, S., Ludwig, D., Kloetzel, P. M., and Kruger, E. (2005). IFN-gamma-induced immune adaptation of the proteasome system is an accelerated and transient response. *Proc. Natl. Acad. Sci. U.S.A., 102,* 9241–9246.

Hickey, W. F. (2001). Basic principles of immunological surveillance of the normal central nervous system. *Glia, 36,* 118–128.

Horwitz, M. S., Evans, C. F., Klier, F. G., and Oldstone, M. B. (1999). Detailed in vivo analysis of interferon-gamma induced major histocompatibility complex expression in the central nervous system: astrocytes fail to express major histocompatibility complex class I and II molecules. *Lab Invest., 79,* 235–242.

Huh, G. S., Boulanger, L. M., Du, H., Riquelme, P. A., Brotz, T. M., and Shatz, C. J. (2000). Functional requirement for class I MHC in CNS development and plasticity. *Science, 290,* 2155–2159.

Huppa, J. B. and Davis, M. M. (2003). T-cell-antigen recognition and the immunological synapse. *Nat. Rev. Immunol., 3,* 973–983.

Huse, M., Lillemeier, B. F., Kuhns, M. S.,Chen, D. S., and Davis. M. M. (2006). T cells use two directionally distinct pathways for cytokine secretion. *Nat. Immunol., 7,* 247–255.

Huseby, E. S., Liggitt, D., Brabb, T., Schnabel, B., Ohlen, C., and Goverman, J. (2001). A pathogenic role for myelin-specific CD8(+) T cells in a model for multiple sclerosis. *J. Exp. Med., 194,* 669–676.

Ip, C. W., Kronerk, A., Bendszus, M., Leder, C., Kobsar, I., Fischer, S., Wiendl, H., Nave, K. A., and Martini, R. (2006). Immune cells contribute to myelin degeneration and axonopathic changes in mice overexpressing proteolipid protein in oligodendrocytes. *J. Neurosci., 26,* 8206–8216.

Johnson, A. J., Njenga, M. K., Hansen, M. J., Kuhns, S. T., Chen, L., Rodriguez, M., andPease, L. R. (1999). Prevalent class I-restricted T-cell response to the Theiler's virus epitope Db:VP2121-130 in the absence of endogenous CD4 help, tumor necrosis factor alpha, gamma interferon, perforin, or costimulation through CD28. *J. Virol., 73,* 3702–3708.

Kang, B. S., Lyman, M. A., and Kim, B. S. (2002). The majority of infiltrating CD8 + T cells in the central nervous system of susceptible SJL/J mice infected with Theiler's virus are virus specific and fully functional. *J. Virol., 76,* 6577–6585.

Khan, S., van den Broek, M., Schwarz, K., de Giuli, R., Diener, P. A., and Groettrup, M. (2001). Immunoproteasomes largely replace constitutive proteasomes during an antiviral and antibacterial immune response in the liver. *J. Immunol., 167*, 6859–6868.

Khanna, K. M., Bonneau, R. H., Kinchington, P. R., and Hendricks, R. L. (2003). Herpes simplex virus-specific memory CD8 + T cells are selectively activated and retained in latently infected sensory ganglia. *Immunity, 18*, 593–603.

Kimura, T. and Griffin, D. E. (2000). The role of CD8(+) T cells and major histocompatibility complex class I expression in the central nervous system of mice infected with neurovirulent Sindbis virus. *J. Virol., 74*, 6117–6125.

Knopf, P. M., Harling-Berg, C. J., Cserr, H. F., Basu, D., Sirulnick, E. J., Nolan, S. C., Park, J. T., Keir, G., Thompson, E. J., and Hickey, W. F. (1998). Antigen-dependent intrathecal antibody synthesis in the normal rat brain: tissue entry and local retention of antigen-specific B cells. *J. Immunol., 161*, 692–701.

Koch, J. and Tampe, R. (2006). The macromolecular peptide-loading complex in MHC class I-dependent antigen presentation. *Cell Mol. Life Sci., 63*, 653–662.

Lee, K. H., Holdorf, A. D., Dustin, M. L., Chan, A. C., Allen, P. M., and Shaw, A. S. (2002). T cell receptor signaling precedes immunological synapse formation. *Science, 295*, 1539–1542.

Leinders-Zufall, T., Brennan, P., Widmayer, P., S PC, Maul-Pavicic, A., Jager, M., Li, X. H., Breer, H., Zufall, F., and Boehm, T. (2004). MHC class I peptides as chemosensory signals in the vomeronasal organ. *Science, 306*, 1033–1037.

Leon, B., Lopez-Bravo, M., and Ardavin, C. (2007). Monocyte-derived dendritic cells formed at the infection site control the induction of protective T helper 1 responses against Leishmania. *Immunity, 26*, 519–531.

Lin, M. T., Stohlman, S. A., and Hinton, D. R. (1997). Mouse hepatitis virus is cleared from the central nervous systems of mice lacking perforin-mediated cytolysis. *J. Virol., 71*, 383–391.

Lowenstein, P. R. (2002). Immunology of viral-vector-mediated gene transfer into the brain: an evolutionary and developmental perspective. *Trends Immunol., 23*, 23–30.

Lowenstein, P. R., Barcia, C., Liu, C., Wawrowsky, K., Kroeger, K. M., Castro, M. G. (2007). IFN-γ secretion is polarized at effector immunological synapses in vivo. *J. Immunol., 178*, 87.52

Maccalli, C., Pende, D., Castelli, C., Mingari, M. C., Robbins, P. F., and Parmiani, G. (2003). NKG2D engagement of colorectal cancer-specific T cells strengthens TCR-mediated antigen stimulation and elicits TCR independent anti-tumor activity. *Eur. J. Immunol., 33*, 2033–2043.

Malone, K. E., Ramakrishna, C., Gonzalez, J. M., Stohlman, S. A., and Bergmann, C. C. (2006). Glia expression of MHC during CNS infection by neurotropic coronavirus. *Adv. Exp. Med. Biol., 581*, 543–546.

Marten, N. W., Stohlman, S. A., Atkinson, R. D., Hinton, D. R., Fleming, J. O., and Bergmann, C. C. (2000a). Contributions of CD8 + T cells and viral spread to demyelinating disease. *J. Immunol., 164*, 4080–4088.

Marten, N. W., Stohlman, S. A, and Bergmann, C. C. (2000b). Role of viral persistence in retaining CD8(+) T cells within the central nervous system. *J. Virol., 74*, 7903–7910.

Marten, N. W., Stohlman, S. A., Smith-Begolka, W., Miller, S. D., Dimacali, E., Yao, Q., Stohl, S., Goverman, J., Bergmann, C. C. (1999). Selection of CD8+ T cells with highly focused specificity during viral persistence in the central nervous system. *J. Immunol., 162*, 3905–14.

Masopust, D. and Ahmed, R. (2004). Reflections on CD8 T-cell activation and memory. *Immunol. Res., 29*, 151–60.

Masopust, D., Vezys, V., Usherwood, E. J., Cauley, L. S., Olson, S., Marzo, A. L., Ward, R. L., Woodland, D., and Lefrancois, L. (2004). Activated primary and memory CD8 T cells migrate to nonlymphoid tissues regardless of site of activation or tissue of origin. *J. Immunol., 172*, 4875–4882.

Matyszak, M. K. (1998). Inflammation in the CNS: balance between immunological privilege and immune responses. *Prog. Neurobiol., 56*, 19–35.

Matyszak, M. K. and Perry, V. H. (1996). A comparison of leucocyte responses to heat-killed bacillus Calmette-Guerin in different CNS compartments. *Neuropathol. Appl. Neurobiol., 22*, 44–53.

Massa, P. T., Ozato, K., and McFarlin, D. E. (1993). Cell type-specific regulation of major histocompatibility complex (MHC) class I gene expression in astrocytes, oligodendrocytes, and neurons. *Glia, 8*, 201–207.

McGavern, D. B., Christen, U., and Oldstone, M. B. (2002). Molecular anatomy of antigen-specific CD8(+) T cell engagement and synapse formation in vivo. *Nat. Immunol., 3*, 918–925.

McMenamin, P. G. (1999). Distribution and phenotype of dendritic cells and resident tissue macrophages in the dura mater, leptomeninges, and choroid plexus of the rat brain as demonstrated in wholemount preparations. *J. Comp. Neurol., 405*, 553–562.

McMenamin, P. G., Wealthall, R. J., Deverall, M., Cooper, S. J., and Griffin, B. (2003). Macrophages and dendritic cells in the rat meninges and choroid plexus: three-dimensional localisation by environmental scanning electron microscopy and confocal microscopy. *Cell Tissue Res., 313*, 259–269.

Monks, C. R., Freiberg, B. A. Kupfer, H. Sciaky, N., and Kupfer, A. (1998). Three-dimensional segregation of supramolecular activation clusters in T cells. *Nature, 395*, 82–86.

Moran, L. B., Duke, D. C., Turkheimer, F. E., Banati, R. B., and Graeber, M. B. (2004). Towards a transcriptome definition of microglial cells. *Neurogenetics, 5*, 95–108.

Mucke, L. and Oldstone, M. B. (1992). The expression of major histocompatibility complex (MHC) class I antigens in the brain differs markedly in acute and persistent infections with lymphocytic choriomeningitis virus (LCMV). *J. Neuroimmunol., 36*, 193–198.

Murray, P. D., McGavern, D. B., Pease, L. R., and Rodriguez, M. (2002). Cellular sources and targets of IFN-gamma-mediated protection against viral demyelination and neurological deficits. *Eur. J. Immunol., 32*, 606–615.

Nakamura, T. (2000). Immunopathogenesis of HTLV-I-associated myelopathy/tropical spastic paraparesis. *Ann. Med., 32*, 600–607.

Neijssen, J., Herberts, C., Drijfhout, J. W., Reits, E., Janssen, L., and Neefjes, J. (2005). Cross-presentation by intercellular peptide transfer through gap junctions. *Nature, 434*, 83–88.

Neumann, H., Cavalie, A., Jenne, D. E., and Wekerle, H. (1995). Induction of MHC class I genes in neurons. *Science, 269*, 549–552.

Neumann, H., Schmidt, H., Cavalie, A., Jenne, D., and Wekerle, H. (1997a). Major histocompatibility complex (MHC) class I gene expression in single neurons of the central nervous system: differential regulation by interferon (IFN)-gamma and tumor necrosis factor (TNF)-alpha. *J. Exp. Med., 185*, 305–316.

Neumann, H., Schmidt, H., Wilharm, E., Behrens, L., and Wekerle, H. (1997b). Interferon gamma gene expression in sensory neurons: evidence for autocrine gene regulation. *J. Exp. Med., 186*, 2023–2031.

Neumann, H., Medana, I. M., Bauer, J., and Lassmann, H. (2002). Cytotoxic T lymphocytes in autoimmune and degenerative CNS diseases. *Trends Neurosci., 25*, 313–319.

Njenga, M. K., Asakura, K., Hunter, S. F., Wettstein, P., Pease, L. R., and Rodriguez, M. (1997a). The immune system preferentially clears Theiler's virus from the gray matter of the central nervous system. *J. Virol., 71*, 8592–8601.

Njenga, M. K., Pease, L. R., Wettstein, P., Mak, T., and Rodriguez, M. (1997b). Interferon alpha/beta mediates early virus-induced expression of H-2D and H-2K in the central nervous system. *Lab Invest., 77*, 71–84.

Noske, K., Bilzer, T., Planz, O., and Stitz, L. (1998). Virus-specific CD4 + T cells eliminate Borna disease virus from the brain via induction of cytotoxic CD8 + T cells. *J. Virol., 72*, 4387–4395.

Oldstone, M. B., Blount, P., Southern, P. J., and Lampert, P. W. (1986). Cytoimmunotherapy for persistent virus infection reveals a unique clearance pattern from the central nervous system. *Nature, 321*, 239–243.

Oliveira, A. L., Thams, S., Lidman, O., Piehl, F., Hokfelt, T., Karre, K., Linda, H., and Cullheim, S. (2004). A role for MHC class I molecules in synaptic plasticity and regeneration of neurons after axotomy. *Proc. Natl. Acad. Sci. U.S.A., 101*, 17843–18748.

Pashenkov, M., Teleshova, N., and Link, H. (2003). Inflammation in the central nervous system: the role for dendritic cells. *Brain Pathol., 13*, 23–33.

Parra, B., Hinton, D. R., Lin, M. T., Cua, D. J., and Stohlman, S. A. (1997). Kinetics of cytokine mRNA expression in the central nervous system following lethal and nonlethal coronavirus-induced acute encephalomyelitis. *Virology, 233*, 260–270.

Parra, B., Hinton, D. R., Marten, N. W., Bergmann, C. C., Lin, M. T., Yang, C. S., and Stohlman, S. A. (1999). IFN-gamma is required for viral clearance from central nervous system oligoden-droglia. *J. Immunol., 162*, 1641–1647.

Patterson, C. E., Lawrence, D. M., Echols, L. A., and Rall, G. F. (2002a). Immune-mediated protection from measles virus-induced central nervous system disease is noncytolytic and gamma interferon dependent. *J. Virol., 76*, 4497–4506.

Pereira, R. A. and Simmons, A. (1999). Cell surface expression of H2 antigens on primary sensory neurons in response to acute but not latent herpes simplex virus infection in vivo. *J. Virol., 73*, 6484–6489.

Pereira, R. A., Tscharke, D. C., Simmons, A. (1994). Upregulation of class I major histocompati-bility complex gene expression in primary sensory neurons, satellite cells, and Schwann cells of mice in response to acute but not latent herpes simplex virus infection in vivo. *J. Exp. Med., 180*, 841–850.

Perlman, S. and Ries, D. (1987). The astrocyte is a target cell in mice persistently infected with mouse hepatitis virus, strain JHM. *Microb. Pathog., 3*, 309–314.

Perry, V. H., Bell, M. D., Brown, H. C., and Matyszak, M. K. (1995). Inflammation in the nervous system. *Curr. Opin. Neurobiol., 5*, 636–641.

Planz, O., Bilzer, T., Sobbe, M., and Stitz, L. (1993). Lysis of major histocompatibility complex class I-bearing cells in Borna disease virus-induced degenerative encephalopathy. *J. Exp. Med., 178*, 163–174.

Popko, B. and Baerwald, K. D. (1999). Oligodendroglial response to the immune cytokine inter-feron gamma. *Neurochem Res., 24*, 331–338.

Pullen, L. C., Miller, S. D., Dal Canto, M. C., Van der Meide, P. H., and Kim, B. S. (1994). Alteration in the level of interferon-gamma results in acceleration of Theiler's virus-induced demyelinating disease. *J. Neuroimmunol., 55*, 143–152.

Rall, G. F., Mucke, L., and Oldstone, M. B. (1995). Consequences of cytotoxic T lymphocyte interaction with major histocompatibility complex class I-expressing neurons in vivo. *J. Exp. Med., 182*, 1201–1212.

Ramakrishna, C., Stohlman, S. A., Atkinson, R. D., Shlomchik, M. J., and Bergmann, C. C. (2002). Mechanisms of central nervous system viral persistence: the critical role of antibody and B cells. *J. Immunol., 168*, 1204–1211.

Ramakrishna, C., Bergmann, C. C., Atkinson, R., and Stohlman, S. A. (2003). Control of central nervous system viral persistence by neutralizing antibody. *J. Virol., 77*, 4670–4678.

Ramakrishna, C., Stohlman, S. A., Atkinson, R. A., Hinton, D. R., and Bergmann, C. C. (2004). Differential regulation of primary and secondary CD8 + T cells in the central nervous system. *J. Immunol., 173*, 6265–6273.

Ramakrishna, C., Atkinson, R., Stohlman, S., and Bergmann, C. (2006). Vaccine induced memory CD8 + T cells cannot prevent virus reactivation. *J. Immunol., 176*, 3062–3069.

Ransohoff, R. M., Kivisakk, P., and Kidd, G. (2003). Three or more routes for leukocyte migration into the central nervous system. *Nat. Rev. Immunol., 3*, 569–581.

Redwine, J. M., Buchmeier, M. J., and Evans, C. F., (2001). In vivo expression of major histocom-patibility complex molecules on oligodendrocytes and neurons during viral infection. *Am. J. Pathol., 159*, 1219–1224.

Reits, E., Griekspoor, A., Neijssen, J., Groothuis, T., Jalink, K., van Veelen, P., Janssen, H., Calafat, J., Drijfhout, J. W., and Neefjes, J. (2003). Peptide diffusion, protection, and degradation in nuclear and cytoplasmic compartments before antigen presentation by MHC class I. *Immunity, 18*, 97–108.

Reuter, J. D., Wilson, J. H., Idoko, K. E., and van den Pol, A. N. (2005). CD4 + T-cell reconstitution reduces cytomegalovirus in the immunocompromised brain. *J. Virol., 79*, 9527–9539.

Rivera-Quinones, C., McGavern, D., Schmelzer, J. D., Hunter, S. F., Low, P. A., and Rodriguez, M. (1998). Absence of neurological deficits following extensive demyelination in a class I-deficient murine model of multiple sclerosis. *Nat. Med., 4*, 187–193.

Rock, K. L. and Shen, L. (2005). Cross-presentation: underlying mechanisms and role in immune surveillance. *Immunol. Rev., 207*, 166–183.

Rodriguez, M., Pavelko, K., and Coffman, R. L. (1995). Gamma interferon is critical for resistance to Theiler's virus-induced demyelination. *J. Virol., 69*, 7286–7290.

Rodriguez, M., Zoecklein, L. J., Howe, C. L., Pavelko, K. D., Gamez, J. D., Nakane, S., and Papke, L. M. (2003). Gamma interferon is critical for neuronal viral clearance and protection in a susceptible mouse strain following ear ly intracranial Theiler's murine encephalomyelitis virus infection. *J. Virol., 77*, 12252–12265.

Rossi, C. P., McAllister, A., Tanguy, M., Kagi, D., and Brahic, M. (1998). Theiler's virus infection of perforin-deficient mice. *J. Virol., 72*, 4515–4519.

Sedgwick, J. D. and Hickey, W. (1997). Antigen presentation in the central nervous system. In: R. W. Keane and W. F. Hickey (Eds.), *Immunology of the Central Nervous System* (pp. 364–418). Oxford: Oxford University Press.

Shin, E. C., Seifert, U., Kato, T., Rice, C. M., Feinstone, S. M., Kloetzel, P. M., and Rehermann, B. (2006). Virus-induced type I IFN stimulates generation of immunoproteasomes at the site of infection. *J. Clin. Invest.,116*, 3006–3014.

Shrestha, B., Samuel, M. A., and Diamond, M. S. (2006a). CD8 + T cells require perforin to clear West Nile virus from infected neurons. *J. Virol., 80*, 119–129.

Shrestha, B., Wang, T., Samuel, M. A., Whitby, K., Craft, J., Fikrig, E., and Diamond, M. S. (2006b). Gamma interferon plays a crucial early antiviral role in protection against West Nile virus infection. *J. Virol., 80*, 5338–5348.

Slifka, M. K. and Whitton, J. L. (2000). Antigen-specific regulation of T cell-mediated cytokine production. *Immunity, 12*, 451–457.

Slifka, M. K., Rodriguez, F., and Whitton, J. L. (1999). Rapid on/off cycling of cytokine production by virus-specific CD8 + T cells. *Nature, 401*, 76–79.

Sobbe, M., Bilzer, T., Gommel, S., Noske, K., Planz, O., and Stitz, L. (1997). Induction of degenerative brain lesions after adoptive transfer of brain lymphocytes from Borna disease virus-infected rats: presence of CD8 + T cells and perforin mRNA. *J. Virol., 71*, 2400–2407.

Sospedra, M. and Martin, R. (2005). Immunology of multiple sclerosis. *Annu. Rev. Immunol., 23*, 683–747.

Steinhoff, U., Schoel, B., and Kaufmann, S. H. (1990). Lysis of interferon-gamma activated Schwann cell by cross-reactive CD8 + alpha/beta T cells with specificity for the mycobacterial 65 kd heat shock protein. *Int. Immunol., 2*, 279–284.

Stevenson, P. G., Hawke, S., Sloan, D. J., and Bangham, C. R. (1997). The immunogenicity of intracerebral virus infection depends on anatomical site. *J. Virol., 71*, 145–151.

Stiles, L. N., Hosking, M. P., Edwards, R. A., Strieter, R. M., and Lane, T. E. (2006). Differential roles for CXCR3 in CD4 + and CD8 + T cell trafficking following viral infection of the CNS. *Eur. J. Immunol., 36*, 613–622.

Stitz, L., Planz, O., Bilzer, T., Frei, K., and Fontana, A. (1991). Transforming growth factor-beta modulates T cell-mediated encephalitis caused by Borna disease virus. Pathogenic importance of CD8 + cells and suppression of antibody formation. *J. Immunol., 147*, 3581–3586.

Stitz, L., Bilzer, T., and Planz, O. (2002). The immunopathogenesis of Borna disease virus infection. *Front Biosci., 7*, d541–d555.

Stohwasser, R., Standera, S., Peters, I., Kloetzel, P. M., and Groettrup, M. (1997). Molecular cloning of the mouse proteasome subunits MC14 and MECL-1: reciprocally regulated tissue expression of interferon-gamma-modulated proteasome subunits. *Eur. J. Immunol., 27*, 1182–1187.

Stohwasser, R., Giesebrecht, J., Kraft, R., Muller, E. C., Hausler, K. G., Kettenmann, H., Hanisch, U. K., and Kloetzel, P. M. (2000). Biochemical analysis of proteasomes from mouse microglia: induction of immunoproteasomes by interferon-gamma and lipopolysaccharide. *Glia, 29*, 355–365.

Strehl, B., Seifert, U., Kruger, E., Heink, S., Kuckelkorn, U., and Kloetzel, P. M. (2005). Interferon-gamma, the functional plasticity of the ubiquitin-proteasome system, and MHC class I antigen processing. *Immunol. Rev., 207*, 19–30.

Suvas, S., Azkur, A. K., and Rouse, B. T. (2006). Qa-1b and CD94-NKG2a interaction regulate cytolytic activity of herpes simplex virus-specific memory CD8 + T cells in the latently infected trigeminal ganglia. *J. Immunol., 176*, 1703–1711.

Tepavcevic, V. and Blakemore, W. F. (2005). Glial grafting for demyelinating disease. *Philos. Trans. R. Soc. Lond. B Biol. Sci., 360*, 1775–1795.

Tepavcevic, V. and Blakemore, W. F. (2006). Haplotype matching is not an essential requirement to achieve remyelination of demyelinating CNS lesions. *Glia, 54*, 880–890.

Theil, D., Derfuss, T., Paripovic, I., Herberger, S., Meinl, E., Schueler, O., Strupp, M., Arbusow, V., and Brandt, T. (2003). Latent herpesvirus infection in human trigeminal ganglia causes chronic immune response. *Am. J. Pathol., 163*, 2179–2184.

Thomas, C. E., Birkett, D., Anozie, I., Castro, M. G., and Lowenstein, P. R. (2001). Acute direct adenoviral vector cytotoxicity and chronic, but not acute, inflammatory responses correlate with decreased vector-mediated transgene expression in the brain. *Mol. Ther., 3*, 36–46.

Tishon, A., Eddleston, M., de la Torre, J. C., and Oldstone, M. B. (1993). Cytotoxic T lymphocytes cleanse viral gene products from individually infected neurons and lymphocytes in mice persistently infected with lymphocytic choriomeningitis virus. *Virology, 197*, 463–467.

Trombetta, E. S. and Mellman, I. (2005). Cell biology of antigen processing in vitro and in vivo. *Annu. Rev. Immunol., 23*, 975–1028.

Tschen, S. I., Stohlman, S. A., Ramakrishna, C., Hinton, D. R., Atkinson, R. D., and Bergmann, C. C. (2006). CNS viral infection diverts homing of antibody-secreting cells from lymphoid organs to the CNS. *Eur. J. Immunol., 36*, 603–612.

Tsuyuki, Y., Fujimaki, H., Hikawa, N., Fujita, K., Nagata, T., and Minami, M. (1998). IFN-gamma induces coordinate expression of MHC class I-mediated antigen presentation machinery molecules in adult mouse Schwann cells. *Neuroreport, 9*, 2071–2075.

Tsutsui, Y., Kosugi, I., and Kawasaki, H. (2005). Neuropathogenesis in cytomegalovirus infection: indication of the mechanisms using mouse models. *Rev. Med. Virol., 15*, 327–345.

van der Most, R. G., Murali-Krishna, K., and Ahmed, R. (2003). Prolonged presence of effector-memory CD8 T cells in the central nervous system after dengue virus encephalitis. *Int. Immunol., 15*, 119–125.

van Lint, A. L., Kleinert, L., Clarke, S. R., Stock, A., Heath, W. R., and Carbone, F. R. (2005). Latent infection with herpes simplex virus is associated with ongoing CD8 + T-cell stimulation by parenchymal cells within sensory ganglia. *J. Virol., 79*, 14843–14851.

Wang, X. Z., Stepp, S. E., Brehm, M. A., Chen, H. D., Selin, L. K., and Welsh, R. M. (2003). Virus-specific CD8 T cells in peripheral tissues are more resistant to apoptosis than those in lymphoid organs. *Immunity, 18*, 631–642.

Xiao, B. G. and Link, H. (1998). Immune regulation within the central nervous system. *J. Neurol. Sci., 157*, 1–12.

Yewdell, J. W., Reits, E., Neefjes, J. (2003). Making sense of mass destruction: quantitating MHC class I antigen presentation. *Nat. Rev. Immunol., 3*, 952–61.

Zuo, J., Stohlman, S. A., Hoskin, J. B., Hinton, D. R., Atkinson, R., Bergmann, C. C. (2006). Mouse hepatitis virus pathogenesis in the central nervous system is independent of IL-15 and natural killer cells. *Virology, 350*, 206–15.

Index

Printed in the United States of America.